CHEMICAL AND APPLIED ENGINEERING MATERIALS

Interdisciplinary Research and Methodologies

T0262782

CHEMICAL AND APPLIED ENGINEERING MATERIALS

Interdisciplinary Research and Methodologies

Edited by
Maria Rajkiewicz, DSc

Gennady E. Zaikov, DSc, and A. K. Haghi, PhD
Reviewers and Advisory Board Members

APPLE ACADEMIC PRESS

Apple Academic Press Inc. | Apple Academic Press Inc.
3333 Mistwell Crescent | 9 Spinnaker Way
Oakville, ON L6L 0A2 | Waretown, NJ 08758
Canada | USA

©2015 by Apple Academic Press, Inc.

First issued in paperback 2021

Exclusive worldwide distribution by CRC Press, a member of Taylor & Francis Group
No claim to original U.S. Government works

ISBN 13: 978-1-77463-206-2 (pbk)
ISBN 13: 978-1-77188-074-9 (hbk)

Library of Congress Control Number: 2015937656

Library and Archives Canada Cataloguing in Publication

Chemical and applied engineering materials: interdisciplinary research and methodologies / edited by Maria Rajkiewicz, DSc; Gennady E. Zaikov, DSc, and A.K. Haghi, PhD, Reviewers and Advisory Board Members.

Includes bibliographical references and index.
ISBN 978-1-77188-074-9 (bound)
1. Chemical engineering. 2. Materials. I. Rajkiewicz, Maria, editor

---

TP155.C33 2015   660'.04   C2015-902542-7

Apple Academic Press also publishes its books in a variety of electronic formats. Some content that appears in print may not be available in electronic format. For information about Apple Academic Press products, visit our website at **www.appleacademicpress.com** and the CRC Press website at **www.crcpress.com**

ABOUT THE EDITOR

Maria Rajkiewicz, DSc

Professor Maria Rajkiewicz is Head of the Division of the Institute for Engineering of Polymer Materials and Dyes, in Warsaw, Poland. She is a well-known specialist in the field of synthesis, investigation of properties and applications of low molecular compounds, oligomers, polymers, composites, and nanocomposites. She is a contributor or co-contributor to several monographs and the author of about 100 original papers.

REVIEWERS AND ADVISORY BOARD MEMBERS

Gennady E. Zaikov, DSc

Gennady E. Zaikov, DSc, is Head of the Polymer Division at the N. M. Emanuel Institute of Biochemical Physics, Russian Academy of Sciences, Moscow, Russia, and Professor at Moscow State Academy of Fine Chemical Technology, Russia, as well as Professor at Kazan National Research Technological University, Kazan, Russia. He is also a prolific author, researcher, and lecturer. He has received several awards for his work, including the Russian Federation Scholarship for Outstanding Scientists. He has been a member of many professional organizations and on the editorial boards of many international science journals.

A. K. Haghi, PhD

A. K. Haghi, PhD, holds a BSc in urban and environmental engineering from University of North Carolina (USA); a MSc in mechanical engineering from North Carolina A&T State University (USA); a DEA in applied mechanics, acoustics and materials from Université de Technologie de Compiègne (France); and a PhD in engineering sciences from Université de Franche-Comté (France). He is the author and editor of 65 books as well as 1000 published papers in various journals and conference proceedings. Dr. Haghi has received several grants, consulted for a number of major corporations, and is a frequent speaker to national and international audiences. Since 1983, he served as a professor at several universities. He is currently Editor-in-Chief of the *International Journal of Chemoinformatics and Chemical Engineering* and *Polymers Research Journal* and on the editorial boards of many international journals. He is a member of the Canadian Research and Development Center of Sciences and Cultures (CRDCSC), Montreal, Quebec, Canada.

CONTENTS

LIST OF CONTRIBUTORS

Arezoo Afzali
University of Guilan, Rasht, Iran

S. A. Bogdanova
Department of Plastics Technology, Kazan State Technological University, Kazan, Russia

A. A. Bokarev
Moscow State University of Applied Biotechnology, Moscow, Russia.

A. Yu. Bondar
Basic Research-High Educational Center of Chemical Physics and Mesoscopy, Udmurt Scientific Center, Russian Academy of Sciences, Izhevsk, Udmurt Republic, Russia

M. M. Doustdar
I. H. U, Thermal Engine Research Center, E-mail: mdostdar@ihu.ac.ir

Arash Esmaili
Tarbiat Modares University, Faculty of Technical and Engineering, Mechanical Engineering Section, Energy Conversion Group

A. Goudarzi
I. H. U, Thermal Engine Research Center, E-mail: kpgoudarzi@ihu.ac.ir

A. K. Haghi
University of Guilan, Rasht, Iran

V. A. Ilatovsky
N. N. Semenov Institute of Chemical Physics, Russian Academy of Sciences, 119991 Moscow, Russia

S. G. Karpova
Emanuel Institute of Biochemical Physics, Russian Academy of Sciences

Azamat A. Khashirov
Kabardino-Balkarian State University a. Kh. M. Berbekov, 360004, Nalchik, Russian Federation, Email: new_kompozit@mail.ru

Svetlana Yu. Khashirova
Kabardino-Balkarian State University a. Kh. M. Berbekov, 360004, Nalchik, Russian Federation

V. I. Kodolov
Basic Research – High Educational Centre of Chemical Physics & Mesoscopy, Udmurt Scientific Center, Ural Division, Russian Academy of Sciences, Izhevsk, Russia; M. T. Kalashnikov Izhevsk State Technical University, Izhevsk, Russia

G. G. Komissarov
N. N. Semenov Institute of Chemical Physics, Russian Academy of Sciences, 119991 Moscow , Russia. E-mail: gkomiss@yandex.ru; komiss@chph.ras.ru

G. A. Korablev
Izhevsk State Agricultural Academy, Russia, Izhevsk 426000, E-mail: korablev@udm.net

N. G. Korableva
Izhevsk State Agricultural Academy, Russia, Izhevsk 426000, E-mail: korablev@udm.net

O. A. Kovyazina
Basic Research-High Educational Center of Chemical Physics and Mesoscopy, Udmurt Scientific Center, Russian Academy of Sciences, Izhevsk, Udmurt Republic, Russia

O. A. Legonkova
Moscow State University of Applied Biotechnology, Moscow, Russia

A. M. Lipanov
Institute of Mechanics, Ural Branch of the Russian Academy of Sciences, T. Baramsinoy 34, Izhevsk, Russia E-mail: postmaster@ntm.udm.ru

Shima Maghsoodlou
University of Guilan, Rasht, Iran

Maerefat Mehdi
Tarbiat Modares University, Faculty of Technical and Engineering, Mechanical Engineering Section, Energy Conversion Group

Vadim Z. Mingaleev
Institute of Organic Chemistry, Ufa Scientific Center of Russian Academy of Sciences, Ufa, Bashkortostan, 450054, Russia

V. V. Molokin
Department of Plastics Technology, Kazan State Technological University, Kazan, Russia

A. A. Popov
Emanuel Institute of Biochemical Physics, Russian Academy of Sciences, Moscow, Russia

G. A. Ptitsyn
N. N. Semenov Institute of Chemical Physics, Russian Academy of Sciences, 119991 Moscow, Russia

G. V. Sinko
N. N. Semenov Institute of Chemical Physics, Russian Academy of Sciences, 119991 Moscow, Russia

M. Ghanbarnia Sooteh
I. H. U, Thermal Engine Research Center, E-mail: mohsen.ghanbarnia@yahoo.com

V. V. Trineeva
Basic Research – High Educational Centre of Chemical Physics & Mesoscopy, Udmurt Scientific Center, Ural Division, Russian Academy of Sciences, Izhevsk, Russia; Institute of Mechanics, Russian Academy of Sciences, Izhevsk, Udmurt Republic, Russia

A. V. Vakhrushev
Institute of Mechanics, Ural Branch of the Russian Academy of Sciences, Izhevsk, Russia E-mail: postmaster@ntm.udm.ru

Yu. M. Vasil'chenko
Basic Research – High Educational Center of Chemical Physics and Mesoscopy, Udmurt Scientific Center, Ural Division, Russian Academy of Sciences, Russia; Basic Research – High Educational Center of Chemical Physics and Mesoscopy, Udmurt Scientific Center, Ural Division, Russian Academy of Sciences, Russia

A. E. Zaikin
Department of Plastics Technology, Kazan State Technological University, Kazan, Russia

G. E. Zaikov
N. M. Emanuel Institute of Biochemical Physics of Russian Academy of Sciences, 119991, Moscow, Russian Federation, E-mail: chembio@sky.chph.ras.ru

Vadim P. Zakharov
Bashkir State University, Zaki Validi str. 32, Ufa, 450076 Bashkortostan, Russia, E-mail: zaharovvp@mail.ru

Elena M. Zakharova
Institute of Organic Chemistry, Ufa Scientific Center of Russian Academy of Sciences, Ufa, Bashkortostan, 450054, Russia

Azamat A. Zhansitov
Kabardino-Balkarian State University a. Kh. M. Berbekov, 173 Chernyshevskogo st., 360004, Nalchik, Russian Federation.

LIST OF ABBREVIATIONS

AG	acrylate guanidine
BMEP	brake mean effective pressure
CNTs	carbon nanotubes
DAGA	N,N-Diallylguanidine acetate
DAGTFA	N,N-Diallylguanidine trifluoroacetate
HOMO	highest occupied state
LUMO	lowest free state
MAG	methacrylate guanidine
MAO	methylalumi- noxane
MCC	microcrystalline cellulose
PEDA	phosphorus-boron-nitrogen-containing oligomer
PFT	polymerization- filling technique
PVP	polyvinyl pyrrolidone
SDS	sodium dodecyl- sulfate
SWCNTs	single-walled carbon nanotubes
TBP	tetrabenzoporphyrin
TEG	thermal expanded graphite
TPC	tetrapyrrole compounds
TPP	tetraphenyl porphyrin

LIST OF SYMBOLS

h	Planck constant
I_{ph}	photocurrent in Ma
W	absorbed light power
λ	wavelength m
c	speed of light in m/s.
PR_{SC}	pressure ratio at surge line
PR_{SC}	pressure ratio at surge control line
E_i	orbital energy
W_i	bond energy of an electron
n_i	number of elements of the given orbital
K	maxing or hybridization coefficient
N_0	number of particles
r	radius of rotating bodies
h	film thickness
N	number changing depending on the nanostructure shape
υ	crystallinity degree
τ	duration
k	value corresponding to specific process rate
a	nanoreactor activity
ε_S	surface energy reflecting the energy
ε_V	nanoreactor volume energy
$\varepsilon^0_S d$	multiplication of surface layer energy by its thickness
ε^0_V	energy of nanoreactor volume unit
S	surface of nanoreactor walls
V	nanoreactor volume

PREFACE

This new research book explores and discusses a range of topics on the physical and mechanical properties of chemical engineering materials. Chapters from prominent researchers in the fields of physics, chemistry and engineering science present new research on composite materials, blends, carbon nanotubes, and nanocomposites along with their applications in technology. Discussing the processing, morphology, structure, properties, performance, and applications, the book highlights the diverse and multidisciplinary nature of the field.

In the first chapter a study on highly filled composite materials with regulated physical and mechanical properties based on synthetic polymers and organic and inorganic fillers is presented. For predication of photoelectrochemical properties of selected molecules by their structure, chapter 2 could be used as an advanced review. In chapter 3 performance of turbocharged spark ignition engine equipped with anti-surge valve and bypass flow control mechanism at various working conditions is presented in detail. Chapter 4 describes the dependence of some thermodynamic characteristics upon initial spatial-energy parameters of free atoms. Fire retardant coatings based on perchlorovinyl resin with improved adhesive properties to protect fiberglass plastics are presented in chapter 5. As an multidisciplinary engineering subject, of course, the modification of peculiarities of microcrystalline cellulose and its oxidized form Guanidine-containing monomers and polymers of vinyl and Diallyl series in chapter 6 could be very interesting for the readers. Research progress on carbon nanotube–polymer composites as well as CNT/polymer composites from the chemistry, mechanics and physics aspects are well developed in chapters 7 and 8. In the next two chapters trends in nanochemistry for metal-carbon nanocomposites as well as production technology of carbon-metal containing nanoproducts in nanoreactors of polymeric matrix are described in detail. In chapter 11 a note on Redox processes in polymeric matrix nanoreactors and in chapter 12 conditions for carbon black accumulation at the interface in heterogeneous binary polymer blends are well presented. Performance analysis of multilayer insulations in cryogenic applications is presented in chapter 13. The effect of particle size of microheterogenous catalyst TiCl4−Al(iso- C4H9)3 on the basic patterns of isoprene polymerization is shown in chapter 14. Internal structure and the equilibrium configuration of separate non-interacting nanoparticles by the molecular mechanics and dynamics methods is another multidisciplinary subject that is presented in chapter 15. A very detailed review on membrane filtration technology is selected for chapter 16. In this chapter theory and application presented step-by-step for readers in science and engineering.

CHAPTER 1

INVESTIGATION ON HIGHLY FILLED COMPOSITE MATERIALS WITH REGULATED PHYSICAL AND MECHANICAL PROPERTIES BASED ON SYNTHETIC POLYMERS AND ORGANIC AND INORGANIC FILLERS

O. A. LEGONKOVA[1]*, A. A. POPOV[2], A. A. BOKAREV[1], and S. G. KARPOVA[2]

[1]Moscow State University of Applied Biotechnology

[2]Emanuel Institute of Biochemical Physics, Russian Academy of Sciences

*E-mail: OALegonkovaPB@mail.ru

CONTENTS

1.1 AIMS AND BACKGROUNDS

The aim of the carried out work was to elaborate the highly filled composite materials with regulated physical and mechanical properties based on synthetic polymers and organic and inorganic fillers that can be used in creation of biodegradable in the environment goods of different purposes. Aspects, essential for technology, such as physical and chemical, structural, rheological properties, were investigated. Methods of electron paramagnetic probe, IR-spectrometry were used to prove that inorganic filler plays the role of plasticizer in creation of hybrid composites. Microbiological aspects will be the subject of another article.

1.2 INTRODUCTION

One of the promising trends from the viewpoint of ecology is the development of biodegradable polymer composites. These materials, along with the polymer base more resistant to biodegradation, comprise fillers, which are not only accessible for microbial degraders but are also agro-industrial wastes to be utilized [1]. Search for cheap fillers and development of polymer composites makes it possible not only to reduce the cost of product, but also contributes to the solution of ecological problems.

1.3 EXPERIMENTAL

Third-grade threshed grain wastes (size of particles, 63–240 μm; bulk density, 350 kg/m³; humidity, 4%) were used as organic filler. As inorganic filler, we took a Rastvorin-A water-soluble mineral fertilizer (OST 10-193-96, produced by the Buysk Mineral Fertilizer Plant) of the following composition (in %): $(NH_4)_2SO_4$, 35; $NH_4H_2PO_4$, 6; KNO_3, 32; $MgSO_4 \cdot 7H_2O$, 27.

The physicomechanical properties of specimens within a broad range of ingredient ratios were determined according to GOST 14236-82; the rheological characteristics of filled compositions, by the method of capillary viscosimetry.

In the work, use was made of the method of electron paramagnetic probe, which was stable nitroxyl radical 2,2,6,6-tetramethylpiperidine-1-oxyl. The radical was introduced into films from vapors at $T = 25°C$ up to a concentration of 10^{-3} mol/l. A reference solution of the radical in CCl_4 with a known number of spins was used in the determination of the concentration of radicals in films. The number of spins in a specimen was determined by comparing the areas under the absorption curves of the specimen studied and the reference. The rotational mobility of the probe was characterized by the correlation time τ. The values of τ were assessed from the EPR spectra by the method outlined in [2].

1.4 RESULTS AND DISCUSSIONS

At the introduction of fillers into SEVA irrespective of its grade, we found that the strength and relative elongation decrease with an increase of the content both of organic and inorganic filler in two-component polymer–filler systems (see Figures 1.1 and 1.2). In the case of filling with organic filler specimens become more rigid; at the introduction of inorganic filler even at a high concentration (60 wt. %) specimens preserve a high plasticity (the breaking strain is 400%).

Considering three-component systems (SEVA/inorganic filler/organic filler, Figure 1.3), it should be noted that in this case too an increase in the content of inorganic filler leads to more plastic specimens.

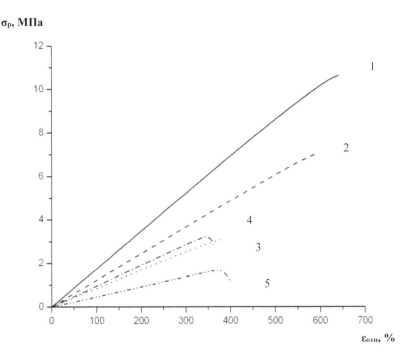

FIGURE 1.1 Change of the physicomechanical properties of a two-component system (SEVA/inorganic filler, wt. %): 1, 100/0; 2, 80/20; 3, 60/40; 4, 50/50; 5, 40/60.

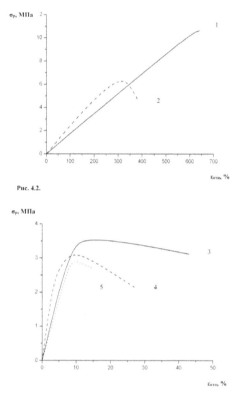

Рис. 4.2.

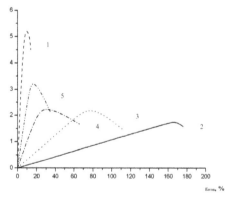

FIGURE 1.2 Change of the physicomechanical properties of a two-component system (SEVA/organic filler, wt. %): 1, 100/0; 2, 80/20; 3, 60/40; 4, 50/50; 5, 40/60.

FIGURE 1.3 The physicomechanical properties of a three-component system based on SEVA-113, wt. %: 1, SEVA/organic filler, 50/50; 2, SEVA/inorganic filler, 50/50; 3, SEVA/inorganic filler/organic filler, 50/37/13; 4, SEVA/inorganic filler/organic filler, 50/25/25; 5, SEVA/inorganic filler/organic filler, 50/13/37.

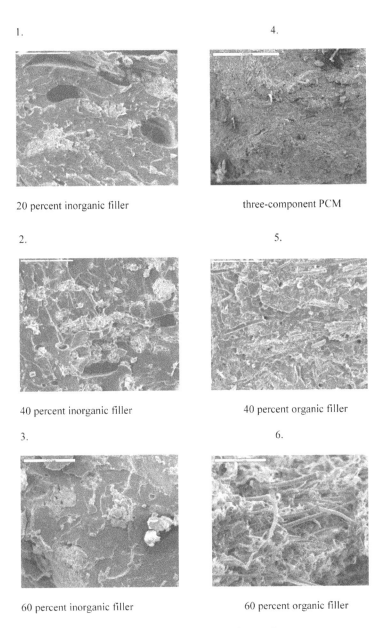

FIGURE 1.4 Electron micrographs of the fracture faces of two-component systems of polymer—composite material based on sevilene and inorganic filler (1, 2, 3); sevilene and organic filler (5, 6); three-component PCM (4): 15 percent sevilene, 50 percent organic filler, 35 percent inorganic filler (magnification ×200 μm).

Figure 1.4 presents electron micrographs of the fracture faces of composites. The systems are seen to be heterogeneous. The systems with inorganic filler include both crystallites of filler salts and their more complex formations. Defects in the system are "pressed into" the homogeneous structure of the composite. For this reason, a decrease of strength of polymer–composite material occurs owing to a decrease of the content of polymer in the composition.

The EPR method supports the data obtained (Table 1.1): The higher the concentration of organic wastes in a hybrid composite, the greater the correlation time, that is, the more rigid the system is and the smaller the deformability of the system is. At the same time, the higher the concentration of inorganic filler, the more plastic the systems are.

Noteworthy is the behavior of two ternary systems: 50 percent sevilene at 25/25 of organic and inorganic filler; and 15 percent sevilene at 50/35 of organic and inorganic filler, for which the correlation time of all ternary systems studied is minimal. The systems are similarly plastic and similarly capable of being processed by commercial equipment into particular articles.

TABLE 1.1   Correlation time of the paramagnetic probe in PCM specimens depending on the extent of filling with various fillers

Extent of Filling with Organic Filler (%)	τ	Extent of Filling with Inorganic Filler (%)	τ	Extent of Filling with Organic/Inorganic Filler (%)	τ
10	$6.1 \cdot 10^{-10}$	20	$7.7 \cdot 10^{-10}$	18.75/18.75	$6.1 \cdot 10^{-10}$
15	$6.8 \cdot 10^{-10}$	40	$7.4 \cdot 10^{-10}$	**25/25**	$5.1 \cdot 10^{-10}$
20	$7.1 \cdot 10^{-10}$	50	$7.1 \cdot 10^{-10}$	50/25	$6.2 \cdot 10^{-10}$
35	$6.3 \cdot 10^{-10}$	60	$6.8 \cdot 10^{-10}$	60/25	$6.5 \cdot 10^{-10}$
40	$7.0 \cdot 10^{-10}$	80	$6.4 \cdot 10^{-10}$	30/35	$6.0 \cdot 10^{-10}$
50	$7.1 \cdot 10^{-10}$	-	-	**50/35**	$6.2 \cdot 10^{-10}$
60	$8.0 \cdot 10^{-10}$	-	-	50/35	$5.7 \cdot 10^{-10}$

Given that inorganic filler is a mixture of salts (potassium, magnesium chlorides; magnesium sulfates; potassium and ammonium nitrates), it was shown (Table 1.2) that exactly nitrogen-containing salts led to a significant increase of the melt flow index in sevilene-based PCMs.

TABLE 1.2 Effect of the composition and concentration of inorganic salts on the melt flow index (g/10 min) of sevilene-based composites

Content (wt. %)	Unfilled Specimen	MgCl₂	KCl	MgSO₄	KNO₃	NH₄NO₃
1	7.8	7.5	6.7	8.1	2.9	4.3
2	7.5	7.1	6.3	7.1	Too high flow rate made its measurement impossible	

To assess the possibility of chemical interaction of fillers with polymers, we used the method of IR spectroscopy. The following IR bands were found to be characteristic of SEVA: 1,740 cm^{-1} (vibrations of the aldehyde group); 1,380 cm^{-1} (vibrations of the vinyl acetate group); 1,240 cm^{-1}; 1,050 cm^{-1}; 797 cm^{-1}; and 607 cm^{-1} [3].

According to [3], vibrations in the region of 1,440 cm^{-1} can be assigned to—C–OCOCH$_3$, the methylene groups -CH$_2$- in SEVA have characteristic valent vibrations at 2,850 cm^{-1}. The group NH$_4^+$ has the absorption band in the region of 3,300–3,030 cm^{-1}; the group P-O-C$_{alkyl}$; 1,180–1,150 cm^{-1}; the groups SO$_4^{2-}$, 1,130–1,080 cm^{-1}; S-CH$_2$, 2,700–2,630 cm^{-1}; the groups PO$_4^3$, HPO$_4^{2-}$, H$_2$PO$_4^-$, in the region of 1,100–1,000 cm^{-1}; NO$_3^-$, 1,380 cm^{-1}.

Changes in the ratio of the optical density of the—C(OCOCH$_3$) – (D_{1440}) band, of the optical density of the vibration bands of the vinyl acetate group (D_{1380}) to the optical density of the bands, assigned to vibrations of the methylene groups (D_{2850}), were used to assess the changes in the composition of polymer at temperatures of producing PCMs together with salts (Table 1.3). It follows from the table that at a heating with the salts KNO$_3$, NH$_4$NO$_3$, and K$_2$HPO$_4$ there is an insignificant splitting-off of the vinyl acetate group from the main chain. Heating of polymer with K$_2$HPO$_4$ produces a band responsible for the interaction of P-O-C$_{alkyl}$ (D_{1180} = 0.62).

In the case of the other salts (KH$_2$PO$_4$, (NH$_4$)$_2$HPO$_4$, (NH$_4$)$_2$SO$_4$, and MgSO$_4$), such an effect could not be identified due to the coincidence of the spectra of the characteristic absorption bands.

TABLE 1.3 Ratios of the optical densities of the bands 1,440 cm^{-1} and 1,380 cm^{-1} to the optical density of the band 2,850 cm^{-1} taken as the internal standard of the specimens of mixtures of SEVA with the salts KNO$_3$, NH$_4$NO$_3$, and K$_2$HPO$_4$.

	Sevilene	Sevilene + KNO₃	Sevilene + NH₄NO₃	Sevilene + K₂HPO₄	Sevilene + Inorganic Filler
D_{1440}/D_{2850}	0.66	0.47	0.46	0.42	0.41
D_{1380}/D_{2850}	0.59	0.42	0.43	0.31	0.36

Table 1.4 presents the melting heats of polymers and polymer–composite materials at temperatures of solid-phase transitions of nitric salts, coinciding with sevilene fluidity temperature of 80°C. According to [4], ammonium nitrates have solid-phase transitions at temperatures of 80°C and 130°C with ΔH_m equal to 0.32 ccal/mol and 1.01 ccal/mol, respectively; the salt KNO_3 has the solid-phase transition at 130°C with ΔH_m equal to 1.3 ccal/mol.

TABLE 1.4   Melting heats of polymers and their composites at temperatures of solid-phase transitions of nitric salts (J/g)

Material	80°C	130°C
Sevilene unfilled ($T_t = 80°C$)	49.9	-
KNO_3	-	54.2
NH_4NO_3	10.1	53.9
Sevilene filled with 20% NH_4NO_3	49.9	9.8
Sevilene filled with 20% KNO_3	44.9	7.6
Sevilene filled with 2% KNO_3	65.1	1.1
Sevilene filled with 40% inorganic filler	50.4	-
Sevilene/organic filler/inorganic filler (15:50:35)	-	94.9

At 80°C, for polymer–composite material based on sevilene an increase of melting heat at the introduction of 2 percent KNO_3 is due to the fact that densely packed structures are formed in polymer's liquid phase at a small content of solid filler at the interface with solid particles of filler KNO_3 [5]. At an increase of the content of salt up to 20 percent the decrease of the melting heat is due to a decrease of the content of polymer per 1 g of specimen studied.

At the introduction of NH_4NO_3 into this system, there is no change of melting enthalpy as compared with pure polymer. At 130°C, when the temperature of the solid-phase transition of the salts is higher than the melting temperature of sevilene, the enthalpy of the composite decreases, which is due to a decrease of the content of salt per 1 g of specimen studied.

1.5   CONCLUSIONS

Thus, the thermal effect of two-component hybrid systems formed from the melt is minimal when T_m (T_t) of the polymer is equal to or lower than the solid-phase transition temperature of the salts. By the extent of a decrease of enthalpy of the measured component, we can make an indirect judgment on the looseness of the filled system as compared with the individual components of the composite.

The process properties of hybrid composites are related to not only the decrease of the concentration of the matrix in the bulk of the composite, but also to the plasticizing action of inorganic filler.

KEYWORDS

- **Correlation time**
- **Hybrid composites**
- **Industrial wastes**
- **Inorganic fillers**
- **Melt flow index**
- **Organic fillers**
- **Physical and chemical properties**
- **Polymers**
- **Structure**
- **Technological properties**
- **Utilization**

REFERENCES

1. Legonkova, O. A.; "Biodeterioration of Polymer-based Composite Materials in the Environment." *Mater. Sci.* **2008,** *2*, 50–55.
2. Antsyferova, L. I.; Wasserman, A. M.; Ivanova, A. I.; Lifshits, V. A.; and Nazemets, N. S.; "Atlas of the Spectra." Moscow: Nauka Publishers; **2012,** 160 p (In Russian).
3. Nakanishi, K.; Infrared Spectra and Structure of Organic Compounds. Moscow: Mir Publishers; **2012,** 200 p.
4. Chemist's Reference Book Moscow–Leningrad: Khimiya Publishers; **2011,** *1*, 1070 p (In Russian).
5. Bryk, M. T.; "Utilization of Filled Polymers." Moscow: Khimiya Publishers; **2012,** 190 p (In Russian).

CHAPTER 2

PREDICATION OF PHOTOELECTROCHEMICAL PROPERTIES OF SELECTED MOLECULES BY THEIR STRUCTURE

V. A. ILATOVSKY, G. V. SINKO, G. A. PTITSYN, and G. G. KOMISSAROV

N.N. Semenov Institute of Chemical Physics, Russian Academy of Sciences, 4 Kosygin str., 119991 Moscow , Russia; E-mail: gkomiss@yandex.ru; komiss@chph.ras.ru

CONTENTS

2.1 AIM AND BACKGROUND

Primary goal of our study was to provide objective criteria to predict theoretically, at least qualitatively, photoelectrochemical properties of the molecules by their structure. In our opinion, the most important factor in the selection process is the electron density distribution in molecules.

2.2 INTRODUCTION

With the improvement of photovoltaic and photoelectrochemical solar energy converters interest in organic semiconductors is becoming increasingly applied nature inherent to the transition to industrial development. At this stage it is especially important to select correctly the most promising classes of organic semiconductors, and even more important to provide objective criteria to predict theoretically, at least qualitatively, photoelectrochemical properties of the molecules by their structure. In our opinion, the most important factor in the selection process is the electron density distribution in molecules, as it determines both the individual properties of the specific compound (organic semiconductor, pigment, and dye), and the intermolecular interaction, that is the character of arrangement of molecules in a solid at condensation. Good examples of powerful rearrangement of electronic structure and corresponding changes of physical and chemical properties of molecules can be seen in the well-known tetrapyrrole compounds (TPC), which, of course, belong to the group of the most promising pigments. This is largely due to the enormous variability of the structure of the TPC, typified by porphyrins and their derivatives, which are cyclic aromatic polyamines, conjugated with multiloop system containing a 16-membered macrocycle with a closed π-conjugation system including 4-8 nitrogen atoms. By replacement of hydrogen atoms in the pyrrole rings and mesopositions for a variety of donor-acceptor groups it has already been produced more than 1000 porphyrins. To this is added the variations due to the formation of different metal complexes and the introduction of extraligands. Substitution of CH-bridges by nitrogen atom (aza-substitution) in tetrabenzoporphyrin (TBP) gives a representative group of phthalocyanines (Pc) (otherwise tetrabenzo-porphyrazines/ tetra-(butadiene-1,3-ylene-1,4) tetraazaporphin/) and their derivatives.

To choose the most effective working pigment from that abundance is extremely difficult, especially as the theoretical assumptions for this choice are almost none. At certain stages one has to use intuitively guided screening to detect certain patterns that will eventually lead to the creation of a more or less acceptable theory. Thus, in comparison of the photocatalytic activity (the photovoltaic Becquerel effect) of thin films of TPC with different macrocycle structure it was noted that the maximum photocurrent (I_{ph}) and photopotential (U_{ph}) are given by pigments, which macrocycle's electronic structure modification is caused by exposure to a carbon atom in mesoposition [1]. For example, in tetraphenyl porphyrin (TPP) substitution of hydrogen to phenyl groups leads to a strong donation of electron density into the π- conjugation circuit (and hence on the pyrrole rings) and significantly increases the photoactivity of TPP. Additional polarization arises due to direct dipole interaction of atoms in the β-positions with the phenyl group. Slightly higher photocatalytic activity was shown by phthalocyanines [2], but, unlike TPP, in Pc nitrogen directly substitutes carbon atoms in mesoposition. Furthermore, Pc has more substituents—benzene rings conjugated with the pyrrole ones—so, aromaticity of this compound is much higher, and it is difficult to determine which substitutions (aza- or benzo-) causes an increase in photoactivity.

As the nitrogen atoms are likely to be key points of the TPC macrocycle structure, there was an intention to conduct a sequential aza-substitution of all four CH groups in the TBP molecule to form phthalocyanine, with a fairly representative group of metal complexes with different degrees of the ligand bond ionicity, and explore changing of the photoelectrochemical characteristics of the films of produced pigments. Hoping to get a large enough material for reflection, in parallel with the experiments there were carried out quantum-chemical calculations of the changes in the distribution of electron density, bond order, the energies of the orbitals in molecules of TPC. The calculations were performed by the program GAMESS, using Rutan molecular orbitals, by restricted Hartree-Fock method (not taking into account the correlation effects) and by density functional theory method (approximately takes into account the correlation effects). Both, experimental and calculated data have shown a good correlation with phenomenological assumptions.

2.3 EXPERIMENTAL

In setting up experiments comparing the PEC properties of various aza-derivatives of TBP and their metal complexes we focused on the adequacy of the conditions of measurement, reproducibility and statistical significance of the results. For each modification of a pigment there were measured parameters of 30 electrodes and performed statistical analysis. The mean square variation of parameters in the cited experimental data $\sigma = 5.6$ percent. Moreover, given the strong dependence of photocurrents I_{ph} (pH) and photopotentials U_{ph} (pH) of pigmented electrodes on the pH of the electrolyte [3, 4], for all kinds of pigments there were determined extreme points of the maximum values of these parameters. Figure 2.1 shows the complete dependence of the potentials and currents of Pt-ZnPc electrode plotted from the average data for the 30 electrodes. For other pigments there were obtained similar dependences, by which it was compared their photocatalytic activities in the reaction of oxygen reduction, wherein the first single-electron step is endothermic:

$$H^+ + O_2 + e^- \rightarrow HO_2, \ E_o = -0.32 \ V,$$

and requires a large activation energy. Excitons, photogenerated in the bulk of the film, migrate to the pigment—electrolyte interface. High concentrations of oxygen adsorbed on the film ($\sim10^{15}$ cm^{-3}) provides necessary conditions for separation of electron—hole pairs arising from the collapse of the excitons due to efficient trapping of electrons with formation of the charged form of adsorption O_2^-. The increase in the concentration of hydrogen ions promotes the completion of the reduction process and the transfer of the reaction products in the electrolyte. However, as can be seen from Figure 2.1, there is an optimal pH zone for I_{ph} (pH), U_{ph} (pH), beyond which further increase the concentration of H$^+$ leads to a decrease in the photoresponse. This is due to the different photoelectrochemical stability of pigments, particularly—the possibility of protonation in the acidic environment, which is largely determined by the redox potential and ionization potential. In this regard, the comparison was performed by photoactive maxima of I_{ph} (pH), U_{ph} (pH).

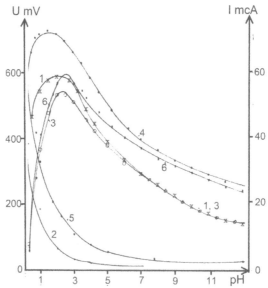

FIGURE 2.1   The dependence of currents and potentials of the electrode with Zn-Pc film on the Pt substrate on the pH of the electrolyte. 1—I_l, light-current; 2—I_d, dark current; 3—I_{ph}, photocurrent; 4—U_l, light potential; 5—U_d, dark potential; 6—U_{ph}, photopotential.

Films of TBP and its metal derivatives with thickness of 50 nm were applied to the polished platinum substrates with a diameter of 11 mm by thermal vacuum sublimation in quasi-closed volume [5–7], using a turbo-molecular vacuum system Varian Mini-TASK (vacuum upto 10^{-9} Torr) with original design of evaporation chamber. Pigments used were previously purified twice by sublimation in vacuum 10^{-7} Torr. The amounts of impurities did not exceed 10^{-4} percent. The film thickness control was carried out by frequency change of a quartz resonator located near the substrates in the evaporator system. Calibration curves were obtained by determining the thickness of pigmented films on transparent quartz substrates placed in the pigments deposition zone, with a micro-interferometer MII-11. Control of spectral parameters of the films confirmed integrity of the structure of pigment molecules. Thanks to the carefully refined method we obtained highly reproducible parameters of the films on platinum substrates. In particular, in the study of PEC parameters variations in the photocurrents did not exceed 15 percent and the photopotentials—20 percent.

Photoactivity of the films were compared by Becquerel effect value, measuring I_{ph}, U_{ph} relatively a saturated Ag/AgCl electrode in aerated 1.0 N KCl electrolyte, changing the pH from 1 to 14 by titration with 1.0 N KOH or HCl solution. Pigmented electrodes were illuminated with an arc xenon lamp DKSSh-120 with a stabilized power supply, the light output in the plane of the electrode was 100 mW/cm^2. For measurement of the quantum yield on a current η pigment films with thickness d from 2 to 300 nm were applied on polished quartz substrates coated with a conductive layer of platinum. The η value was determined from a ratio

$$\eta = 6.25 \cdot 10^7 \cdot [hc\, I_{ph}/W\lambda]\ (\%),$$

where h—Planck constant, I_{ph}—photocurrent in μA, W—absorbed light power W, λ—wavelength m, c—speed of light in m/s. Measurements were carried out in the region of maximum absorption of the pigments, separating from the spectrum of the lamp DKSSh-120 band width of 10 nm using interference filters at luminous flux in the plane of the electrode 10 mW/cm^2.

2.4   RESULTS AND DISCUSSION

The results of measurement of photoelectrochemical activity of the films of 45 metal complexes of derivatives of tetraazabenzoporphyrin with varying degrees of azasubstitution are shown in Table 2.1.

TABLE 2.1 Variations of photopotentials, photocurrents, quantum yield on a current of thin film electrodes based on the azasubstituted metal complexes of tetrabenzo-porphyrin depending on the degree of substitution and the type of central atom

No.	TPC	U_{ph}	pH	I_{ph}	pH	η
1.	Mn -TBP	102	4	0.7	4	2.0
2.	Mn -MATBP	145	3	2.8	3	4.0
3.	Mn -DATBP	178	3	4.1	3	4.5
4.	Mn -TATBP	198	2	5.5	2	5.0
5.	Mn -Pc	230	1	6.5	1	6.0
6.	Ni -TBP	53	3	0.1	3	1.1
7.	Ni -MATBP	67	2	0.2	3	1.3
8.	Ni -DATBP	72	1	0.2	2	1.3
9.	Ni -TATB	78	1	0.3	1	1.4
10.	Ni -Pc	80	0	0.3	0	1.4
11.	Co -TBP	71	2	0.1	2	1.2
12.	Co -MATBP	92	1	0.2	1	1.5
13.	Cu -DATBP	105	1	0.3	1	1.5
14.	Co -TATBP	108	1	0.4	1	1.6
15.	Co -Pc	110	0	0.4	0	1.6
16.	Zn -TBP	308	5	5.2	5	9.0
17.	Zn -MATBP	431	4	20.0	4	13.0
18.	Zn -DATBP	500	3	32.0	3	15.0
19.	Zn -TATBP	540	2	46.0	2	17.0
20.	Zn -Pc	570	2	54.0	2	18.0
21.	Mg -TBP	160	4	4.0	3	4.0
22.	Mg -MATBP	280	4	9.0	3	6.0
23.	Mg -DATBP	320	3	14.0	2	7.0
24.	Mg -TATBP	370	2	17.0	1.5	9.0
25.	Mg -Pc	400	2	21.0	1.5	10.0
26.	H_2 -TBP	162	4	1.0	3	2.0
27.	H_2 -MATBP	227	3	4.0	2	4.0
28.	H_2 -DATBP	260	2	6.2	2	5.0

TABLE 2.1 *(Continued)*

No.	TPC	U_{ph}	pH	I_{ph}	pH	η
29.	H_2-TATBP	285	1	7.7	1	6.0
30.	H_2-Pc	300	1	8.2	0	6.0
31.	Fe -TBP	85	3	0.4	2	1.0
32.	Fe -MATBP	140	1	1.1	1	3.0
33.	Fe -DATBP	155	1	1.2	0	3.2
34.	Fe -TATBP	170	0	1.3	0	3.5
35.	Fe -Pc	180	0	1.4	0	4.0
36.	ClFe -TBP	105	4	1.7	3	2.5
37.	ClFe -MATBP	180	3	3.5	2	3.8
38.	ClFe -DATBP	210	2	4.6	2	4.5
39.	ClFe -TATBP	240	1	5.0	1	5.0
40.	ClFe -Pc	250	1	5.4	1	5.0
41.	VO -TBP	240	4	11.0	3	5.0
42.	VO -MATBP	290	2	17.0	2	9.0
43.	VO -DATBP	330	1	21.0	1	13.0
44.	VO -TATBP	350	1	26.0	1	15.0
45.	VO -Pc	350	1	28.0	0	16.0

U_{ph}—maximum values of the photopotential in mV; I_{ph}—maximum photocurrent in μA; pH—value of pH at which the maximum of U_{ph},I_{ph}; η—quantum efficiency of a current. TBP—tetrabenzoporphyrin, MATBP—mono-tetraazabenzoporphyrin, DATBP—di-tetraazabenzoporphyrin, TATBP—tri-tetraazabenzoporphyrin, Pc—phthalocyanine (tetra-tetraazabenzoporphyrin).

For all studied compounds the same type effect of aza-substitution is observed, the most dramatic change of photocurrent (2–5 times) and photopotential (1.3–1.7 times) is the first step—the transition from the TBP to a mono-aza-substituted derivatives. Each subsequent step makes a smaller contribution, but the overall increase in the transition to the structure of phthalocyanine for U_{ph} reaches 185 percent, and for I_{ph}—1,000 percent. The stability of pigments changes significantly—the pH value, at which protonation of the molecules begins, decreases by four to five units as the proportion of nitrogen atoms in the macrocycle grows, which is associated with the increase of redox potential with aza-substitution. Changing in stability of pigments is seen in the stability of the electrodes. Photoactivity of the TBP films lost in 10–12 hr of light saturating light. At the same time an output of dications of the pigments into solution was spectrally recorded. Phthalocyanine films (regardless of the nature of the central atom) did not change their characteristics after hundreds of hours of light in the same conditions. Basically, the differences in the photoactivity of pigments are caused by three factors—the efficiency of energy (charge) transfer, the band gap, the spectral range, and extent of light absorption. All three are ultimately determined by the distribution of electron density in the molecules. So, for all of aza-derivatives of TBP, as a result of the transition from the structure of pure porphyrin to the phthalocyanine, there is observed an increase of photochromic properties of molecules and molecular interaction. For example, Figure 2.2 shows the change in absorption spectrum of zinc TBP at the sequential aza-substitution: Absorption bandwidth expansion occurs simultaneously with the increase of the extinction coefficient and bathochromic shift, leading to a decrease in gap width.

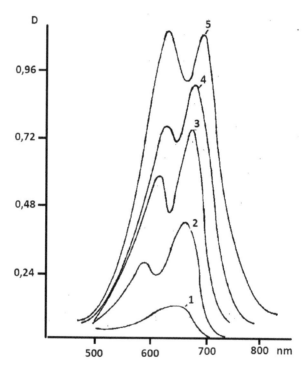

FIGURE 2.2 The absorption spectra of the films of aza-derivatives of Zn-TBP. 1—Zn-TBP, 2—Zn-MATBP, 3—Zn-DATBP, 4—Zn-TATBP, and 5—Zn-Pc.

Quantitative changes in the quantum yield on a current at sequential aza-substitution are less significant than those for the photocurrent, but nevertheless, the total increase of η reaches 300 percent, the greatest increase in efficiency is observed for the first step—a mono-aza-substitution. For the photocurrents that can be explained by a sharp increase in the extinction coefficient, that is, photochromic properties at the molecular level. In measurements of η this property has less influence on the final result, and increased quantum yield can be attributed to the high efficiency of energy and charge transfer between the molecules, as well as to a more advantageous arrangement of the band structure energy levels.

Lattice constants in a preferred crystal orientation plane (parallel to the substrate) are 1.98 nm for H_2-Pc and 2.19 nm for H_2-TBP, which corresponds to the axis "a" of the unit cell, connecting centers of molecules lying in one plane. Even on electron micrographs of the films pigments thickness of 5 nm (Figure 2.3), obtained with an electron microscope JEM-100B with the ultimate resolution of 0.2 nm at the same scale is a clear difference in the interatomic distances and the changes in the lattice. This difference in interatomic distances and changes in the lattice constants are evident even in electron micrographs of pigment films with thickness of 5 nm (Figure 2.3) received by an electron microscope JEM-100B with maximum resolution of 0.2 nm at the same scale.

 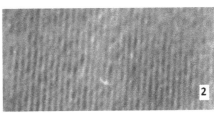

FIGURE 2.3 The change in the lattice with fourfold azazameschenii in metal-free tetrabenzporfirine. (**a**)—tetrabenzporfirin (**b**)—phthalocyanine. Resolution of 0.2 nm.

Given the almost identical molecular size (distance between the outer atoms of the benzene rings due to compression of the coordination sphere as a result of aza-substitution in Pc is of 0.13 nm smaller), the minimum distance between the nearest atoms for H_2-Pc is ~0.34 nm, and for H_2-TBP ~0.42 nm. This increases the electron affinity, decreases both ionization potential and the energy difference between the highest occupied and lowest unoccupied molecular orbitals, leading for Pc to a bathochromic shift of the first absorption band. Simultaneously, there are observed strengthening of the bond to the metal Me^{2+} at the ion radii less than 1.4 A, the appearance of dative π-bonds Me-N, change in the acid-base properties, in particular, a considerable reduction of protonation even in a strongly acidic medium. Thus, the interatomic distances are reduced by about 20 percent, which certainly contributes to the ring currents in the conjugation macrocycle, respectively, to strengthening of intermolecular interactions and increase of the efficiency of energy and charge transfer (TPC are characterized by "hopping" charge-transfer mechanism in which intermolecular distances are particularly important) and to an increase of the quantum yield on a current, a maximum value in the investigated group of compounds is 18 percent. Characteristically, the net effect depends on the nature of the central atom, whose interaction with the ligand significantly changes the electron density distribution in the macrocycle and its inner diameter.

Figure 2.4 shows the position of the energy bands for the limiting cases of substitution (no substitution—TBP, complete replacement—Pc) and the position of the acceptor level of oxygen in accordance with its redox potential.

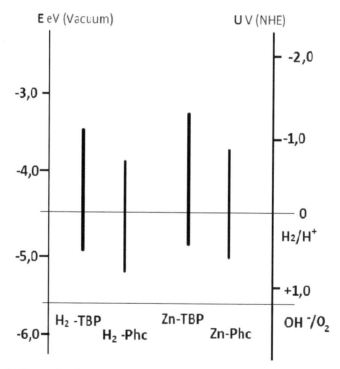

FIGURE 2.4 Energy band diagram of Me-TBP and Me-Pc.

If the surface layer is filled, that is, oxygen captures an electron to form a charge-transfer complex:

$$O_2 \rightarrow O_{2(a)} \rightarrow [O_2^{-d}]_{(a)} + p^{+d},$$

then the concentration of adsorbed charged molecules is determined by conventional expression [8]:

$$\vartheta^- = \vartheta \, [1 + 1/2 \cdot exp(w/kT)] = \vartheta \, [1 + 1/2 \cdot exp((E_a - F_s)/kT)],$$

where ϑ—total concentration of adsorbed molecules, w—activation energy of the transition to the charged form, F_s—the distance from the Fermi level at the surface to E_v, E_a—the distance from the surface acceptor level to the E_v (E_v—the ceiling of the valence band). Consequently, the energy diagrams shown are initially determine increased population of acceptor levels in the dark and less photoresponse of TBP than Pc.

However, arguments based on the general concepts and physicochemical properties of the molecules do not provide sufficient certainty in assessing the impact of the electronic structure of molecules on the properties of their solid agglomerates. In connection with this there was attempted mathematical modeling of transformation of the structures by means of quantum-chemical calculations of the electron density distribution and energy characteristics of the orbitals.

The calculation of the spatial and electronic structure of molecules was accomplished by the method of molecular orbitals on the program GAUSSIAN 03. In finding the molecular orbitals there were used density functional theory and the approach of Rutan in which the molecular orbitals are defined in the class of functions of the form

$$\varphi_p(\vec{r}) = \sum_{k=1}^{N_{atom}} \sum_{s=1}^{M(k)} c_{sp}^k \eta_s^k(\vec{r} - \vec{s}_k)$$

Here $\eta_1^k(\vec{r}), \eta_2^k(\vec{r}), ..., \eta_{M(k)}^k(\vec{r})$—a given set of functions for the k-th atom in the molecule, the position of which is determined by the vector $\vec{s}_k$. These functions, called atomic orbitals, are not assumed to be linearly independent or orthogonal, but are chosen normalized. c_{sp}^k—varying coefficients. The index s is a set of three indices (n, l, m), and the functions $\eta_s^k(\vec{r})$ are:

$$\eta_s^k(\vec{r}) = R_{n\ell}^k(r) Y_{\ell m}(\vec{\Omega}).$$

Accordingly, the sum over s is a triple sum:

$$\sum_{s=1}^{M(k)} \equiv \sum_{n=1}^{N(k)} \sum_{\ell=0}^{L(k,n)} \sum_{m=-\ell}^{\ell}$$

The set of radial parts of all atomic orbitals, used in the construction of the molecular orbitals, is the atomic basis of calculation. The basis was taken by a linear combination of Gaussian orbitals

$$R_{n\ell}(r) \approx \sum_{q=1}^{Q'} d_q^{n\ell} r^{k_q^{n\ell}} e^{-\alpha_q^{n\ell} r^2} \, ,$$

in the form of a standard correlation consistent basis 6-31G**, which is denoted also as *6-31G (1d1pH)* or *cc-pVDZ*. In the calculations we used two forms of the exchange-correlation functional: form PBEPBE, described in [9, 10], and the form PBE1PBE, described in [11, 12]. To determine the value of the spin in the ground state of the molecule there were carried out calculations of molecules with different spins and the total energies compared.

The results of calculation by the limited Hartree-Fock method (not taking into account correlation effects) and by the density functional theory (approximately taking into account correlation effects) are shown in Tables 2.2 and 2.3, and the corresponding numbering of atoms—in Figure 2.5.

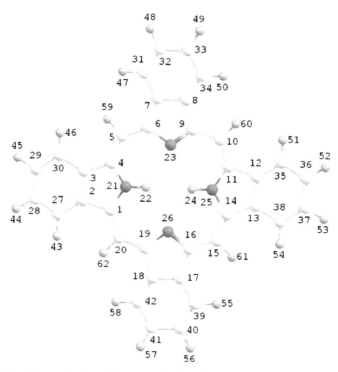

FIGURE 2.5   The numbering of the atoms in the molecules under study.

TABLE 2.2 Changes in the distribution of electron density in the molecule of tetrabenzoporphyrin with sequential aza-substitution (Hartree-Fock)

Atom Number	Before Substitution	0	1	2-orto	2-para	3	4	After Substitution
1.	C	0.321051	0.437988	0.274334	0.305926	0.286533	0.584203	C
2	C	−0.08527	−0.04998	−0.03553	−0.0736	−0.06077	−0.10211	C
3	C	−0.08527	−0.12067	−0.14554	−0.11234	−0.12495	−0.1021	C
4	C	0.321052	0.750765	0.654557	0.599145	0.610537	0.584205	C
5	C	−0.16368	−0.73145	−0.68168	−0.64345	−0.65047	−0.63157	N
6	C	0.297132	0.660397	0.638233	0.592743	0.60889	0.588361	C
7	C	−0.01972	−0.12967	−0.06071	−0.05498	−0.04915	−0.04479	C
8	C	−0.10452	−0.01325	−0.10769	−0.09229	−0.11678	−0.11989	C
9	C	0.389694	0.316014	0.688339	0.411316	0.677646	0.664767	C
10	C	−0.24887	−0.17681	−0.74176	−0.24582	−0.71878	−0.70297	N
11	C	0.462781	0.336566	0.76172	0.46861	0.756837	0.746162	C
12	C	−0.07667	−0.09999	−0.13756	−0.06601	−0.10803	−0.0971	C
13	C	−0.07667	−0.07001	−0.03211	−0.10721	−0.0843	−0.0971	C
14	C	0.462781	0.298136	0.397134	0.73886	0.725921	0.746162	C
15	C	−0.24887	−0.14239	−0.18628	−0.71159	−0.68705	−0.70297	N
16	C	0.389694	0.27503	0.335136	0.643082	0.630436	0.664767	C
17	C	−0.10452	−0.01414	−0.10075	−0.13073	−0.13214	−0.11989	C
18	C	−0.01972	−0.10653	−0.0199	−0.01065	−0.00717	−0.04479	C
19	C	0.297132	0.373251	0.247478	0.291599	0.273391	0.588361	C
20	C	−0.16368	−0.2212	−0.11956	−0.15497	−0.13735	−0.63157	N
21	N	−0.80003	−0.86562	−0.78381	−0.78313	−0.78259	−0.76738	N
22	H	0.392221	0.402829	0.400495	0.397033	0.399423	0.402385	H
23	N	−0.77395	−0.77932	−0.80634	−0.78307	−0.78431	−0.77959	N
24	H	0.393755	0.394374	0.409868	0.409103	0.418961	0.425386	H
25	N	−0.88939	−0.7999	−0.84798	−0.87062	−0.85203	−0.85387	N
26	N	−0.77395	−0.76184	−0.75895	−0.77188	−0.76373	−0.77959	N
27	C	−0.0997	−0.12647	−0.12085	−0.10631	−0.11073	−0.09636	C
28	C	−0.16758	−0.14571	−0.14872	−0.16291	−0.15832	−0.16731	C
29	C	−0.16758	−0.16069	−0.18104	−0.1718	−0.17595	−0.16731	C
30	C	−0.0997	−0.09793	−0.08089	−0.09017	−0.08606	−0.09636	C
31	C	−0.13039	−0.10677	−0.11961	−0.11728	−0.12169	−0.1229	C
32	C	−0.14944	−0.16613	−0.14993	−0.14985	−0.14807	−0.14769	C

TABLE 2.2   *(Continued)*

33	C	−0.16614	−0.1469	−0.15724	−0.15952	−0.15966	−0.16182	C
34	C	−0.11626	−0.13576	−0.11295	−0.12166	−0.11153	−0.11097	C
35	C	−0.1158	−0.09454	−0.09165	−0.12122	−0.10319	−0.10546	C
36	C	−0.1545	−0.17267	−0.16575	−0.14896	−0.15422	−0.15053	C
37	C	−0.1545	−0.16223	−0.14114	−0.15581	−0.14808	−0.15053	C
38	C	−0.1158	−0.10477	−0.12963	−0.10095	−0.10982	−0.10546	C
39	C	−0.11626	−0.13104	−0.1141	−0.10608	−0.10508	−0.11097	C
40	C	−0.16614	−0.14869	−0.17055	−0.16857	−0.17003	−0.16182	C
41	C	−0.14944	−0.16774	−0.15142	−0.14685	−0.1462	−0.14769	C
42	C	−0.13039	−0.11528	−0.12971	−0.13627	−0.1366	−0.1229	C
43	H	0.151604	0.168197	0.151732	0.149808	0.149478	0.167836	H
44	H	0.146236	0.16209	0.148571	0.144667	0.145562	0.143205	H
45	H	0.146236	0.162386	0.14674	0.14434	0.144719	0.143205	H
46	H	0.151604	0.193695	0.1753	0.16955	0.170926	0.167835	H
47	H	0.156107	0.178144	0.185657	0.185897	0.185685	0.18475	H
48	H	0.152006	0.150813	0.157821	0.158502	0.158321	0.157229	H
49	H	0.150338	0.151751	0.156449	0.155772	0.156378	0.154787	H
50	H	0.155859	0.156186	0.183034	0.160131	0.181867	0.179148	H
51	H	0.168911	0.153006	0.193815	0.170583	0.195648	0.197801	H
52	H	0.160866	0.146531	0.162843	0.164413	0.166934	0.16954	H
53	H	0.160866	0.147148	0.163625	0.165534	0.167665	0.16954	H
54	H	0.168911	0.151038	0.169237	0.196189	0.196229	0.197801	H
55	H	0.155859	0.155679	0.149019	0.174765	0.174001	0.179148	H
56	H	0.150338	0.152402	0.146604	0.148928	0.149169	0.154787	H
57	H	0.152006	0.15058	0.148584	0.150422	0.150828	0.157229	H
58	H	0.156107	0.154874	0.151201	0.154902	0.154583	0.18475	H
59	H	0.16157	0.161812	-	-	-	-	-
60	H	0.150037	-	-	0.160002	-	--	-
61	H	0.150037	0.165883	0.164818	-	-	-	-
62	H	0.16157	0.158493	0.168946	0.16471	0.168241	-	-

TABLE 2.3 Changes in the distribution of electron density in the molecule of tetrabenzoporphyr in sequential aza-substitution (by density functional theory on the grid)

Atom Number	Before Substitution	0	1	2-orto	2-para	3	4	After Substitution
1.	C	0.318705	0.335082	0.324168	0.322311	0.336346	0.492694	C
2.	C	0.079445	0.072616	0.082218	0.080802	0.07099	0.075752	C
3.	C	0.064812	0.083751	0.072442	0.075919	0.086511	0.090177	C
4.	C	0.34823	0.481363	0.490204	0.489379	0.477113	0.47302	C
5.	C	−0.17246	−0.57271	−0.56948	−0.56449	−0.56676	−0.56427	N
6.	C	0.297463	0.441181	0.443079	0.438523	0.443982	0.442716	C
7.	C	0.083174	0.091614	0.100919	0.087589	0.098647	0.105369	C
8.	C	0.100962	0.100847	0.100646	0.106798	0.099981	0.094572	C
9.	C	0.286769	0.291985	0.443657	0.291054	0.442895	0.443515	C
10.	C	−0.18365	−0.1751	−0.56978	−0.17955	−0.56456	−0.56052	N
11.	C	0.318659	0.335855	0.489895	0.322308	0.488486	0.492695	C
12.	C	0.079431	0.069906	0.072524	0.080799	0.079997	0.075719	C
13.	C	0.064842	0.075938	0.081677	0.075952	0.087275	0.090224	C
14.	C	0.348227	0.329671	0.324412	0.489344	0.480765	0.472982	C
15.	C	−0.17247	−0.1744	−0.17218	−0.56452	−0.56436	−0.56428	N
16.	C	0.297455	0.289554	0.289232	0.43853	0.439488	0.442713	C
17.	C	0.08318	0.093555	0.09381	0.087571	0.092703	0.105405	C
18.	C	0.100975	0.093156	0.094067	0.106829	0.100872	0.094569	C
19.	C	0.286773	0.288464	0.289497	0.291053	0.293325	0.443511	C
20.	C	−0.18371	−0.16973	−0.17306	−0.17962	−0.17148	−0.56055	N
21.	N	−0.65905	−0.62733	−0.62116	−0.62168	−0.6195	−0.59016	N
22.	H	0.294134	0.2974	0.300256	0.301075	0.30412	0.310317	H
23.	N	−0.68786	−0.67021	−0.64064	−0.66216	−0.64477	−0.64205	N
24.	H	0.294135	0.297583	0.300498	0.30107	0.304198	0.310326	H
25.	N	−0.65905	−0.65586	−0.62216	−0.62166	−0.58796	−0.59017	N
26.	N	−0.68785	−0.68467	−0.68606	−0.66224	−0.66394	−0.64213	N
27.	C	−0.13023	−0.12928	−0.13376	−0.13117	−0.13039	−0.12134	C
28.	C	−0.06303	−0.06185	−0.06018	−0.06119	−0.06131	−0.0614	C
29.	C	−0.06419	−0.06384	−0.06413	−0.06341	−0.0622	−0.0601	C
30.	C	−0.12825	−0.12248	−0.11767	−0.11866	−0.12307	−0.12376	C
31.	C	−0.13672	−0.13249	−0.13799	−0.13083	−0.13473	−0.13856	C

TABLE 2.3   *(Continued)*

32.	C	−0.0652	−0.06233	−0.06072	−0.06316	−0.06073	−0.05982	C
33.	C	−0.0628	−0.06243	−0.06059	−0.06106	−0.06048	−0.06057	C
34.	C	−0.13974	−0.14349	−0.13721	−0.14275	−0.13808	−0.1338	C
35.	C	−0.13025	−0.12735	−0.11733	−0.13118	−0.12374	−0.12133	C
36.	C	−0.06303	−0.06337	−0.06442	−0.06118	−0.0611	−0.0614	C
37.	C	−0.06419	−0.06322	−0.0601	−0.06343	−0.06033	−0.0601	C
38.	C	−0.12825	−0.12971	−0.13326	−0.11867	−0.12509	−0.12376	C
39.	C	−0.13672	−0.13633	−0.14028	−0.13082	−0.13341	−0.13856	C
40.	C	−0.0652	−0.06374	−0.06313	−0.06316	−0.06204	−0.05982	C
41.	C	−0.0628	−0.06362	−0.06298	−0.06105	−0.0615	−0.06058	C
42.	C	−0.13974	−0.1404	−0.13993	−0.14276	−0.14191	−0.13379	C
43.	H	0.065096	0.063859	0.064969	0.065592	0.066454	0.084349	H
44.	H	0.064287	0.064673	0.065905	0.066126	0.067747	0.06954	H
45.	H	0.064374	0.065828	0.067117	0.067146	0.06879	0.069374	H
46.	H	0.065148	0.081908	0.083745	0.082705	0.084124	0.08418	H
47.	H	0.056666	0.073646	0.073892	0.074406	0.07482	0.076813	H
48.	H	0.058693	0.059998	0.061032	0.061571	0.062358	0.063934	H
49.	H	0.05859	0.058996	0.061	0.060343	0.062486	0.063877	H
50.	H	0.056809	0.056007	0.073897	0.05794	0.07516	0.076201	H
51.	H	0.065097	0.066542	0.083793	0.065592	0.083283	0.084348	H
52.	H	0.064288	0.066032	0.067165	0.066134	0.068048	0.06954	H
53.	H	0.064374	0.066174	0.065889	0.067143	0.068044	0.069375	H
54.	H	0.065148	0.066017	0.064993	0.082704	0.082491	0.084174	H
55.	H	0.056666	0.056817	0.059009	0.074391	0.076035	0.076817	H
56.	H	0.058693	0.059954	0.061288	0.061565	0.063045	0.063933	H
57.	H	0.05859	0.060174	0.061263	0.06035	0.061879	0.063876	H
58.	H	0.056808	0.058152	0.058946	0.057926	0.058215	0.076205	H
59.	H	0.064099	0.067717	-	-	-	-	-
60.	H	0.065772	-	-	0.070909	-	-	-
61.	H	0.064098	0.066716	0.070527	-	-	-	-
62.	H	0.065775	0.067212	0.070551	0.070928	0.07273	-	-

In addition, there were calculated spectrum of single-particle states and the spatial distribution of electrons in the highest occupied state (HOMO) and the lowest free state (LUMO) (Figure 2.6). The spatial distribution of electron states in the HOMO and LUMO is illustrated in Figures 2.7 and 2.8. These figures show the surface level of functions $|\Psi(\vec{r})|^2$, that is surfaces satisfying the condition $|\Psi(\vec{r})|^2 = const$. Here $\Psi(\vec{r})$ is the one-particle wave function of the electron in the states HOMO and LUMO of the corresponding molecules, const values are the same for all molecules (Figures 2.9, 2.10, 2.12, 2.13 and 2.14).

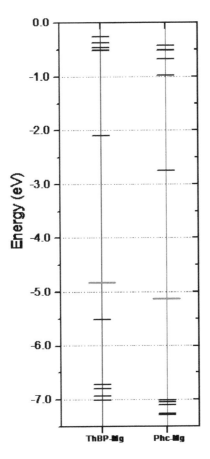

FIGURE 2.6   Spectrum of one-electron energies in the ground state of molecules Mg-tetrabenzoporphyrin and Mg-phthalocyanine (the spin of the ground state S = 0).

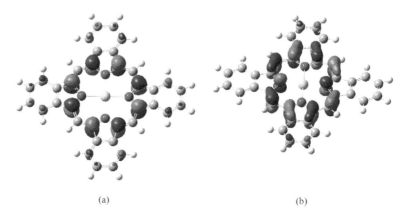

FIGURE 2.7   Spatial distribution of the electron in the one-particle states of HOMO (a) and LUMO (b) for the ground state of Mg-tetrabenzoporphyrin.

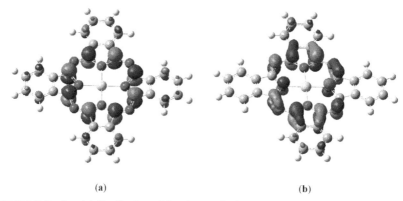

FIGURE 2.8   Spatial distribution of the electron in the one-particle states of HOMO (a) and LUMO (b) for the ground state of Mg-phthalocyanine.

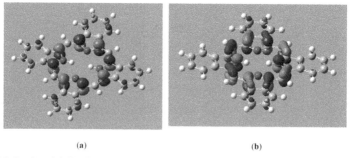

FIGURE 2.9   Spatial distribution of the electron in the one-particle states of HOMO (a) and LUMO (b) for the singlet state of the metal-free tetrabenzoporphyrin

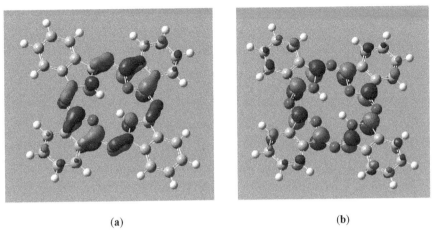

(a)   **(b)**

FIGURE 2.10   Spatial distribution of the electron in the one-particle states of HOMO (a) and LUMO (b) for the singlet state of the metal-free phthalocyanine.

In addition, changes of the energy in sequential aza-substitution were assessed (Figures 2.11, 2.12, and 2.13).

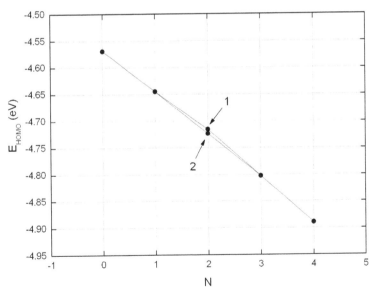

FIGURE 2.11   Dependence of the energy EHOMO on the number of substitutions in the molecule of tetrabenzoporphyrin.
1—molecule shown in Figure 2.14(c);
2—molecule shown in Figure 2.14(d).

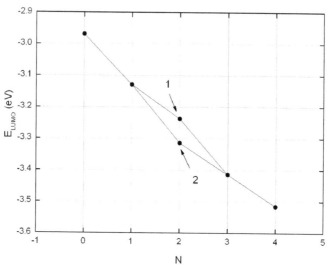

FIGURE 2.12 Dependence of the energy E_{LUMO} on the number of substitutions in the molecule of tetrabenzoporphyrin.
1—molecule shown in Figure 2.14(c);
2—molecule shown in Figure 2.14(d).

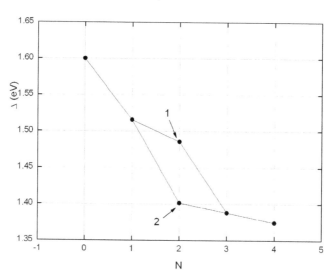

FIGURE 2.13 Dependence of the energy difference $\Delta E = ELUMO - EHOMO$ on the number of substitutions in the molecule of tetrabenzoporphyrin. 1—molecule shown in Figure 2.14(c);
2—molecule shown in Figure 2.14(d).

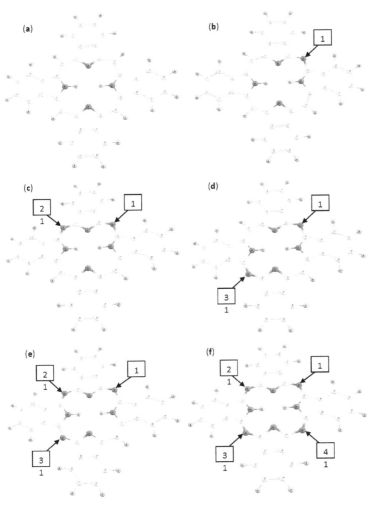

FIGURE 2.14   Molecule tetrabenzoporphyrin, arrows mark places of aza-substitution. (**a**) tetrabenzoporphyrin (symmetry D2h), (**b**) mono-substitution (symmetry C1), (**c**) disubstituted (symmetry D1h), (**d**) disubstituted (symmetry C2h), (**e**) triple substitution (symmetry C1), and (**f**) phthalocyanine (symmetry D2h).

As can be seen from the above calculations, the most revealing changes occur, naturally, in the areas of substitution, with the apparent concentration of the charge in the most likely coordination of potential electron acceptor. Characteristically, the improved method of density functional on the grid clearer tracks -orto and -para substitution options. In similar calculations for other metal derivatives of tetrabenzo-porphyrin there was no significant difference, however, it should be noted a significant effect of extraligand in the case of Fe-TBP/FeCl-TBP. According to Table 2.1, the ratio of their photocurrents (0.4/1.7 = 0.24) after full aza-substitution (Fe-Pc/ClFe-Pc) is almost unchanged (1.4/5.4 = 0.26), that is a significant advantage of the pigment having extraligands on the central atom is saved. The quantum yield of the chlorinated form is five times higher. Therefore, looking more forward to the next subject of research, we present some of the data on these compounds (Figures 2.15, 2.16, 2.17, and 2.18).

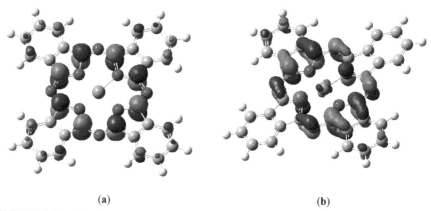

(a)   (b)

FIGURE 2.15   Spatial distribution of the electron with spin "up" in the one-particle states of HOMO (**a**) and LUMO (**b**) for the ground state of Fe-phthalocyanine. Ground-state spin $S = 1$.

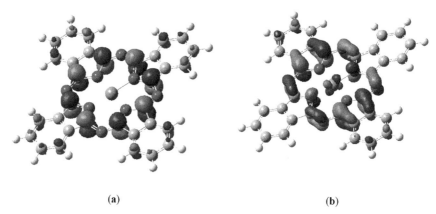

(a)   (b)

FIGURE 2.16   Spatial distribution of the electron with spin "down" in the one-particle states of HOMO (**a**) and LUMO (**b**) for the ground state of Fe-phthalocyanine. Ground-state spin $S = 1$.

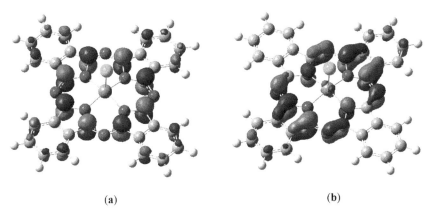

FIGURE 2.17 Spatial distribution of the electron with spin "up" in the one-particle states of HOMO (**a**) and LUMO (**b**) for the ground state of FeCl-phthalocyanine. Ground-state spin $S = 3/2$.

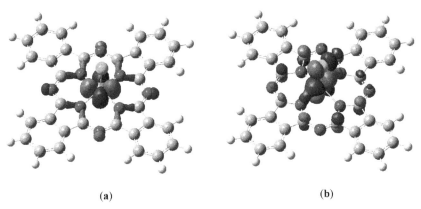

FIGURE 2.18 Spatial distribution of the electron with spin "down" in the one-particle states of HOMO (**a**) and LUMO (**b**) for the ground state of FeCl-phthalocyanine. Ground-state spin $S = 3/2$.

The picture of changes in the distribution of electron density is so obvious, that practically does not require any comment. We note only that the most impressive version of Figure 2.18 is unlikely, since it corresponds to the band gap of about 0.8 eV, and the experimental values (about 1.5 eV) is more consistent with the spin-up states.

2.5 CONCLUSION

Thus, the set of considered measurements on films of aza-substituted metal derivatives of tetrabenzoporphyrin in combination with the data of quantum-chemical calculations of the electron density and energy characteristics of the pigments, leads to several conclusions:

(i) Sequential substitution of carbon atoms in mesoposition by nitrogen, that leads to a transition from the porphyrin structure to a phthalocyanine structure, increases the electron density in the area under consideration with the corresponding increase of charges in the π-electronic bonds.

(ii) The obvious consequence of the increase of electron density in the important area of structural conjugation of macrocycle is a ring currents increase and compression of the cycle, which leads to a considerable increase of the intermolecular interaction.

(iii) Consequences of the strengthening of the intermolecular interactions are bathochromic red shift of the absorption maxima, the increase in the extinction coefficients and the broadening of the absorption bands in the visible part of the spectrum, as well as the hypsochromic shift of the Soret band.

(iv) Accordingly with the changes in the optical characteristics of the pigments it takes place a reduction of the band gap, lowering of the energy level of the valence band, increasing of the depth of acceptor levels formed by the adsorbed oxygen. In this case, it can be seen as a positive change, since it brings pigments to the value of the band gap of 1.5 eV, which is ideal for the converter of solar energy on the ground level.

(v) From a practical point of view, the redistribution of electron density in the very structurally similar molecules leads to a significant increase in chemo- and light fastness of pigments and oxidation potentials, to an 8–10 fold increase in the photocurrent, 1.6–1.8 fold increase in photopotentials, 2–5 fold—in the quantum yield on a current.

KEYWORDS

- **Aza-substitution**
- **Electron density distribution**
- **Organic semiconductors**
- **Porphyrins**
- **Tetrapyrrole compounds**

REFERENCES

1. Rudakov, V. M.; Ilatovsky, V. A.; and Komissarov, G. G.; Photoactivity of metal derivatives of tetraphenylporphyrin. *Khim. Fizika.* **1987,** *6(4),* 552–554, (In Russian).

2. Apresyan, E. S.; Ilatovsky, V. A.; and Komissarov, G. G.; Photoactivity of thin films of metal derivatives of phthalocyanine. *Zh. fiz.himii.* **1989,** *63(8),* 2239–2242, (In Russian).

3. Ilatovsky, V. A.; Shaposhnikov, G. P.; Dmitriev, I. B.; Rudakov, V. M.; Zhiltsov, S. L.; and Komissarov, G. G.; Photocatalytic activity of thin films of azasubstituted tetrabenzoporphy-rins. *Zh. fiz.khimii.* **1999,** *73(12),* 2240–2245, (In Russian).

4. Ilatovsky, V. A.; Dmitriev, I. B.; Kokorin, A. I.; Ptitsyn, G. A.; and Komissarov, G. G.; The influence of the nature of the coordinated metal on the photoelectrochemical activity of thin films of tetrapyrrole compounds. *Khim.Fiz.* **2009,** *28(1),* 89–96, (In Russian).

5. Ilatovsky, V. A.; Ptitsyn, G. A.; and Komissarov, G. G.; Influence of molecular structure of the films of tetrapyrrole compounds on their photoelectrochemical characteristics at the various types of sensitization. *Khim.Fiz.* **2008,** *27(12),* 66–70, (In Russian).

6. Ilatovsky, V. A.; Sinko, G. V.; Ptitsyn, G. A.; and Komissarov, G. G.; Structural sensitization of pigmented films in the formation of nano-sized monocrystal clusters. In: Collection: The Dynamics of Chemical and Biological Processes, XXI Century. Moscow: Institute of Chemical Physics RAS; **2012,** 173–180.

7. Ilatovsky, V. A.; Apresyan, E. S.; and Komissarov, G. G.; Increase in photoactivity of phthalocyanines at structural modification of thin film electrodes. *Zhurn.fiz.khimii.* **1988,** *62(6),* 1612–1617, (In Russian).

8. Wolkenstein, F. F.; Physical Chemistry of Semiconductor Surfaces. Moscow: Nauka; **1973,** 398.

9. Perdew, J. P.; Burke, K.; and Ernzerhof, M.; *Phys. Rev. Lett.* **1996,** *77,* 3865.

10. Perdew, J. P.; Burke, K.; and Ernzerhof, M.; *Phys. Rev. Lett.* **1997,** *78,* 1396.

11. Ernzerhof, M.; Perdew, J. P.; and Burke, K.; *Int. J. Quantum Chem.* **1997,** *64,* 285.

12. Ernzerhof, M.; and Scuseria, G. E.; *J. Chem. Phys.* **1999,** *110,* 5029.

CHAPTER 3

INVESTIGATION ON PERFORMANCE OF TURBOCHARGED SPARK-IGNITION ENGINE EQUIPPED WITH ANTISURGE VALVE AND BYPASS FLOW CONTROL MECHANISM AT VARIOUS WORKING CONDITIONS

M. M. DOUSTDAR, A. GOUDARZI, and
M. GHANBARNIA SOOTEH

I.H.U, Thermal Engine Research Center

CONTENTS

3.1   INTRODUCTION

Turbocharging history backs to invention of internal combustion engines. Alfred Buchi invented the first practical version of turbochargers for internal combustion engines which was moved by engine exhaust gases. Brown and Baveri Company started commercial producing of turbochargers in 1923 [1].

Designers use turbochargers, to achieve an engine with smaller dimension and preserving power production and efficiency. The basis of this approach is to drive a compressor by using a turbine. Thereupon pressure and density of inlet air to engine increase and the inlet air flow rate will increased too. This approach provides more inlet fuel to engine and more power of engine as well [1, 2, 3].

For reduction of fuel consumption and preserving low pollution in spark-ignition engines, the designers use some methods like; low displacement volume of engine, turbochargers, direct fuel injection and various performance of valves. This approach will increase the potential of higher engine efficiency up to 30 percent [4]. Silva et al. show the pollution of carbon dioxide decreases 15–49 percent, by reduction of engine dimension and using turbochargers and flowing tow low-cost strategies of engine stop-start and fuel cutoff [5].

Many researches on effects of turbochargers on internal combustion engines have been done. In accordance with these studies, turbochargers increase the power of engines, ofcourse, if appropriate turbocharger is chosen [1, 2, 3]. Krakianitis and Sadoi presented a method for selecting an appropriate turbocharger for a special engine. However, they found a range of turbochargers for a special engine by using theoric relations, but final selection only obtained by experimental tests.

Developments in manufacturing of turbocharger system components, have led to a series of studies on the impact of these changes. Kesgin investigated the effect of turbocharging system on the performance of a natural gas engine. He showed the efficiency of turbochargers has a direct effect on engine efficiency. Whatever back pressure of turbine and pressure drop of compressor be lower, the efficiency of engine will be further. He showed, however, if the turbine inlet and compressor outlet are connected to the center of the related manifold, engine efficiency is greater than the case of connection from one end, but due to the high costs of construction and maintenance, it is not commonly.

Some examples of high-tech turbocharger are bi-inlet and variant blades turbines. Westin, exhibited turbine with variable geometry improves turbocharger efficacy up to 60 percent, whereas, twin-entry turbines increases up to 24 percent [8]. Utilizing a turbine with variable geometry decreases 18 percent diameter of turbine wheels, on the other hand it increases efficiency and outlet power. Adaptation of engine with unstable performance nature and turbine and compressor with stable performance nature requires to use accurate control mechanisms. Common case of these mechanisms is wastegate. Thomasson et al. have investigated pressure strengthen control mechanism by wastegate in steady state and transient flow [9]. As another control mechanism we can mention to antisurge valve. In order to control compressor performance and avoid the occurrence of surge, the antisurge valve has been employed usually. Dimitrios and George have inspected on the various mechanisms of antisurge valves. They reduced the standard deviation of pressure fluctuations and provided safe operation of compressor away from surge area as well as by control algorithm of PID.

In this study we have investigated the effects of altitude on engine performance due to geometrical characteristic and performance condition. Then selection criterion of appropriate engine because of improvement of engine performance has been studied. In the next step, to avoid surge occurrence, we have introduced suitable control mechanism and investigated its effects on safe performance of system. Finally, by using bypass flow rate mechanism we have controlled effects of turbocharger and avoided mechanical and thermal stresses.

3.2 TEST CASE

In this experiment we have used a LYCOMING engine model of HIO-360-D which is a four-cylinder spark-ignition engine with air cooling system. Locating pattern of cylinders is horizontally opposed. More information of specific engine has been shown in Table 3.1.

TABLE 3.1   Engine specifications

HIO-360-D Engine	
No. of cylinder	4
Bore	83(mm)
Stroke	81.4(mm)
Displacement	5.91(L)
Compression ratio	10:1
Rated power	141 kW at 3,200 rpm

Note that the fuel is aviation 100 LL gasoline with higher Octane number than regular one. This selection makes more resistance against knock phenomena.

In order to test the engine, we have used an eddy current dynamometer which is made by API company model of FR 400 BRL. The maximum amount of power absorption, torque and measurable rotation are 400 hp, 850 Nm, and 12,000 rpm, respectively. In Figure 3.1, test room and connecting posture of engine to dynamometer has been illustrated. The measurements have been done at rotation of 1,800–3,200 rpm. In each cycle the throttle valve has varied from lowest to maximum values (full load) and the results has been recorded. Power and fuel consumption graphs of engine are shown in Figures 3.2 and 3.3 correspondingly. It should be noted that these graphs have been obtained at full load mode to brief presentation.

FIGURE 3.1   A schematic view of test room.

3.3   ENGINE PERFORMANCE SIMULATION

In this part we have simulated the engine by GT-POWER software and compared the results with experimental data.

3.3.1   SIMULATION AND VERIFICATION

Today, CAD/CAM simulations as a tool to estimate performance of various systems, reduce costs and testing time is taken into consideration. In this study specified engine and its components have been simulated by GT-POWER software carefully. Modeling of software is based on fluid mechanics laws, thermodynamic laws, mass conservation law, equilibrium reactions, and chemical kinetics. In particular, combustion simulation of this software is based on one region and multiregions approaches. In accordance with unification of thermo-dynamic properties in each region, pressure unification developed in all volume of cylinder [11]. In this investigation we have used multiregions approach. Bor-der detection of regions is based on flame turbulent level. Flame propagation speed is different depend on fuel and thermodynamic properties of mixture in combustion chamber [12]. To obtain accurate results, it is necessary to calibrate empirical coefficients of the combustion model by test data [13]. Hence, we have paid special attention to calibrate the model coefficient by experimental data, in order to understand the behavior of engine different operation condi-tions.

Experimental data and numerical results of power and fuel consumption are compared in Figures 3.2 and 3.3.

In according to Figure 3.3, good agreement between experimental and nu-merical results about fuel consumption can be observed. The averaged error between numerical and experimental results is about 7 percent which shows verification of the selected model.

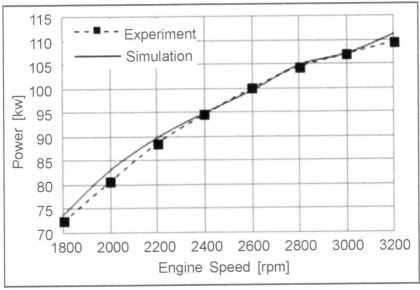

FIGURE 3.2   Brake power variations versus engine speed.

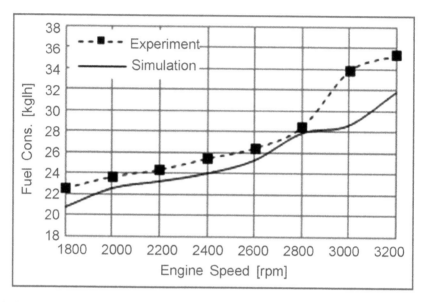

FIGURE 3.3   Fuel consumption variations versus engine speed.

3.3.2   Performance Simulation in Altitude

According to our desire, the engine should work at high altitude. As altitude increases air pressure drops and this reduction has a destructive effect on engine performance. It is difficult to provide engine performance conditions at high altitude in a lab. So we have to use software to obtain influence of altitude on the engine.

The maximum operation altitude of the above mentioned engine is 4.5 km from sea level. Equation (3.1) shows the relation between temperature, altitude and pressure.

$$P = P_{S.L.}\, \exp\!\left(\frac{-gh}{RT}\right) \tag{3.1}$$

Where T is temperature at desired altitude.

By using Eq. (3.1) we have investigated engine performance condition at altitude of 2.7 km from sea level. A comparison between power at desired altitude and sea level is shown in Figure 3.4.

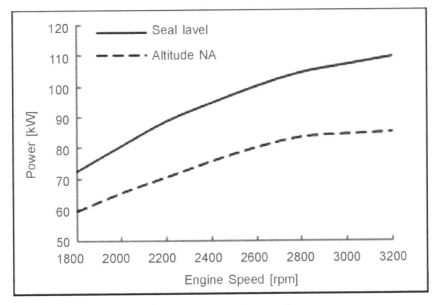

FIGURE 3.4 Variation of brake power versus engine speed.

According to Figure 3.4, by increase of altitude the engine power drops dramatically. The approach of this study is to recoup the losses by using an appropriate turbocharger.

3.4 TURBOCHARGER

In this part, we introduce general description about turbocharger system and its performance. Generally turbo machines are not suit to work with a piston machines. Thus the composition of the gasoline engine and the turbocharger must be carefully designed. General purpose of this part is to find good agreement between engine and turbocharger for best performance.

3.4.1 Investigation of Components and Match Logic

The amount of passing air based on geometrical characteristics, engine performance condition and amount of amplification in compressor specify by Eq. (3.2).

$$\dot{m}_a = \eta_v \frac{N}{2} \rho_{a,i} V_d \tag{3.2}$$

Compressor power needed to create the desired pressure ratio will be obtained from Eq. (3.3).

$$\dot{W}_C = \frac{1}{\eta_C}(\dot{m}_a + \dot{m}_B)h_{0i}\left[\left(\frac{P_e}{P_{0i}}\right)^{\frac{\gamma-1}{\gamma}} - 1\right] \tag{3.3}$$

where $\dot{m}_B$ is outlet mass flow rate from antisurge valve to atmosphere.

Equation (3.4) by consideration of shaft efficiency which has connected compressor and turbine determines compressor power needed.

$$\dot{W}_C = \eta_{mech}\dot{W}_T \tag{3.4}$$

The amount of turbine power through passing fluid can be calculated by following equation:

$$\dot{W}_T = \eta_T\left(\dot{m}_a + \dot{m}_f - \dot{m}_W\right)h_{0i}\left[1 - \left(\frac{P_e}{P_{0i}}\right)^{\frac{\gamma-1}{\gamma}}\right] \tag{3.5}$$

In Eqs. (3.3) and (3.5) we should note that the efficiency of turbine and compressor in turbochargers system is based on the stagnation to static logic. This is because of less usage of flow velocity and less waste in this system [1]. In Eq. (3.5) $\dot{m}_f$ is flow rate of injection fuel in manifold and will be computed according to $\dot{m}_f = \dot{m}_a[FAR]$ and fuel to air ratio. $\dot{m}_w$ is passing flow rate of wastegate and will be calculated based on geometrical characteristic and movement mechanism of wastegate to achieve functional specifications and optimal performance at desired area. Mass flow rate through a valve is established upon by passing of a compressible flow through an obstacle. This relation is based on the analysis of one-dimensional compressible flow and considering effects of actual flow as a discharge factor. In accordance to choking or not choking in manifold, mass flow rate will be calculated from Eqs. (3.6) or (3.7), respectively. The criteria of choking occurrence is a pressure ratio greater than $\left(\frac{2}{\gamma+1}\right)^{\frac{\gamma}{\gamma-1}}$ on each side of manifold [7].

$$\dot{m} = \frac{C_D A_R P_0}{\sqrt{RT_0}}\left(\frac{P}{P_0}\right)^{\frac{1}{\gamma}}\left\{\frac{2\gamma}{\gamma-1}\left[1 - \left(\frac{P}{P_0}\right)^{\frac{\gamma-1}{\gamma}}\right]\right\}^{\frac{1}{2}} \tag{3.6}$$

$$\dot{m} = \frac{C_D A_R P_0}{\sqrt{RT_0}} \sqrt{\gamma} \left(\frac{2}{\gamma+1} \right)^{\frac{\gamma+1}{2(\gamma-1)}} \tag{3.7}$$

Where T_0 and P_0 in these equations are stagnation temperature and pressure of upstream and P is downstream pressure.

The amount of required air by the engine determines the initial size of turbocharger. Looking at catalogs which provided by the manufactures, the initial size of turbocharger can be selected. On the other hand, turbocharger's engine can operate in a wider range of mass flow than compressor. It shows the importance of drawn air flow on the compressor map than turbine map. According to full line of engine operation which is overlapped in all range of velocities and loads to compressor characteristics, the final selection of turbocharger will be done. Figure 3.5 shows this overlapping.

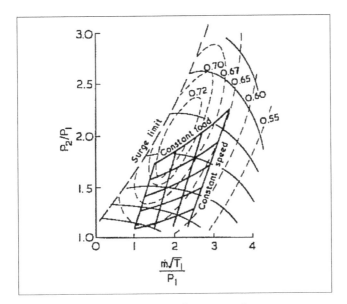

FIGURE 3.5 Overlapping curve of engine performance and compressor.

By increasing of inlet density of compressor, mass flow rate rises gradually, if the engine operates at constant speed but the load grows gradually. Air flow of engine can overlaps on compressor characteristic curve (e.g., Figure 3.5). The goal is to select a compressor which constant speed line of engine be at the middle level of higher efficiency part of compressor's map. If engine works at constant load, this situation is similar to fix Brake Mean Effective Pressure (BMEP), while the engine speed increases. Volume ratio of air flow because of growth of engine speed will be increased also. Effective surface of turbo-charger turbine flow is constant. So the inlet pressure of turbine enlarges. It is clear that increase in pressure leads to an increase in available energy for expansion through the turbine and increase amplification pressure in compressor also. Thus, the engine constant load line is not on compressor characteristic curve horizontally. It increases by growing of engine speed. It is obvious on constant load lines in Figure 3.5.

General characteristics curve of engine should be between choke (right side of compressor curve in Figure 3.5) and surge (left side of compressor curve in Figure 3.5) lines at the upper level of compressor efficiency curve.

The margin between surge and nearest point of the engine performance should be enough large to stay away from surge area.

3.4.2   Selection of an Appropriate Turbocharger

In agreement with pervious parts, in order to achieve a proper adoption for engine, all performance points of engine which is equipped to turbocharger should be obtained. As mentioned above, performance points include constant speed lines and constant load which are overlapped on compressor curve. If engine performance condition be at safe region and upper level of compressor efficient curve, it ensures suitable adaptation of engine and compressor.

In this study we have used GT-POWER software to obtain performance conditions of an engine which equipped with a turbocharger. In spark-ignition engines, the throttle controls load at constant rotation. This process performs by changing the density of suction air and the consumption fuel of engine [6].

Since the simulation mode requires control mechanisms that are abstract in vitro condition, we have used opening changes in throttle. Thus, the constant speed lines have been acquired by preserving of engine speed and changes in opening of throttle and constant load lines simulated by impounding of throttle position and changing in engine speed. So engine performance condition in conjunction with turbocharger may be obtained. The process repeated for several of turbochargers and engine curves adopted on compressor map. Compressor map includes compressor efficient contours in term of reduced mass flow through the engine and pressure ratio in two sides of compressor. In this experiment in order to adaptation of engine and turbocharger, different turbochargers of Garrett Company are used. Figures 3.6, 3.7, and 3.8 show overlapping of engine performance condition on compressor performance maps in types of GT32, GT35, and GT37 [15]. Hajilo et.al showed characteristic curve of twin-entry turbines-pulse turbocharger, according to flow rate through each duct will change. It shows pulse turbochargers need to be more studied.

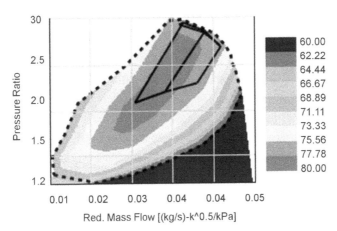

FIGURE 3.6 Adaptation of engine performance condition with turbocharger of GT32.

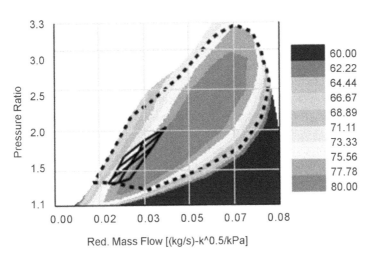

FIGURE 3.7 Adaptation of engine performance condition with turbocharger of GT35.

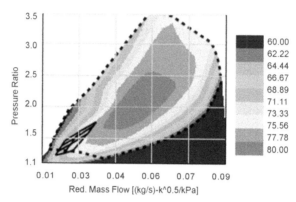

FIGURE 3.8   Adaptation of engine performance condition with turbocharger of GT37.

According to Figure 3.6, engine performance condition is so close to choke region of compressor and as a result turbocharger of GT-32 is small for our desired engine. It should be noted that approaching of engine performance condition to choke region is because of software extrapolation features [13]. The engine performance condition actually choke off, but the compressor performance condition is not to define outside of choke region.

The adaptation of engine performance condition with turbocharger of GT37 has been illustrated in Figure 3.8. According to this figure, performance condition of engine is close to surge region and is at the down level of compressor map.

But in accordance with Figure 3.7, it is obvious that the conjunction of engine with turbocharger of GT35 is simulated better than two other turbochargers. Because engine performance condition is located at upper region of compressor efficient map and the closest performance point to surge region is about 20 percent of recommended mass flow rate [1].

It should be noted that the choice of turbocharger has been carried out without considering antisurge valve and wastegate.

3.5   ANTISURGE VALVE

Surge is an instability of flow through compressor which appears as a periodical sweep. This phenomena causes a lot of noise, a sudden increase in temperature, an increase in blades stress and mechanical damages. The phenomena is due to a decrease in flow rate at specific speed and as a result of inability of fluid to pass over from current pressure ratio. In this situation return flow is appeared also. It usually starts with period stall and with further reduction in flow rate which leads to surge. Necessary condition for occurrence of surge is to be at positive slope in constant speed lines. These points cannot self-compensate probably disturbances. So surge character line at compressor character curve will be obtain by $\dfrac{dPR}{d\dot{m}} = 0$

In order to avoid surge occurrence in compressor, axial or radial one, antisurge valve is used usually. Various mechanisms have been presented for these valves which all of them are based on reduction of compressor downstream pressure and according to surge control line (to avoid surge occurrence). We use this line to compensate uncertainly in the measurements and delay in valve performance. There are some descriptions for this parameter that one of them has been shown in Eq. (3.8).

$$SM = \frac{PR_s - PR_{sc}}{PR_{sc}} \tag{3.8}$$

where PR_s is pressure ratio at surge line, PR_{sc} is pressure ratio at surge control line, and SM is surge control limitation. Some amounts have been suggested for surge control line. Among them we have used 0.25 for axial multistage compressor of Komposty and 0.15 for radial compressor of Henderson [17]. We should note that high amount of this parameter causes compressor performance drop because of proximity of high efficiency lines and surge line. Figure 3.9 shows, compressor characteristic curve of turbocharger GT35 and surge control line.

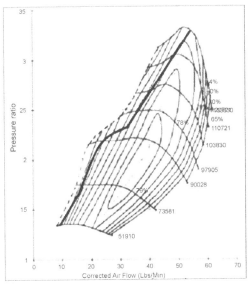

FIGURE 3.9 Compressor characteristic curve of turbocharger GT35 and surge control line [15].

Figure 3.10 presents some antivalves which have been used in this examination. This mechanism calculates the downstream pressure leads to surge by reading of mass flow rate and compressor upstream pressure and according to surge control line. In next stage, read upstream pressure has been compared with calculated upstream pressure. If the pressure exceeds the valve opening command is issued. In this study because of computer simulation we can remove delay of valve performance and consider lower amount for surge control limitation.

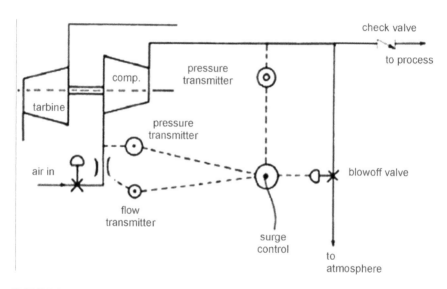

FIGURE 3.10 Antisurge valve and its control mechanism.

Outlet fluid of valve chooses one of the two selectable paths according to performance condition. First path is entering to atmosphere and second is return to the inlet flow of compressor. In this study, outlet flow of valve is exhausted to atmosphere because of nonharmful of fluid and greater control of inlet flow to the compressor. We use check valve in main path of engine to prevent backflow to engine and instability of engine.

In this part we have investigated the effect of this mechanism on performance of system. For this purpose we compared pressure ratio versus modified flow rate curve of a compressor with compressor characteristic curve for totally open wastage in 1,800; 2,200; and 3,000 rpm. The comparisons have been illustrated in Figures 3.11 and 3.12. According to Figure 3.11, the performance of compressor in rotation of 3,200 rpm is in safe region and faraway of surge and choking. But in rotation of 1,800 and 2,200 rpm, compressor performance curve is in the surge region. To avoid surge occurrence we have used an antisurge valve. By using this valve, compressor performance curve will be modified to Figure 3.12. This modification is happened by opening of the valve and reduction pressure ratio at low flow rate regions.

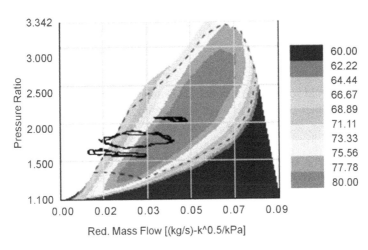

FIGURE 3.11   Comparison of compressor performance curve in an engine cycle.

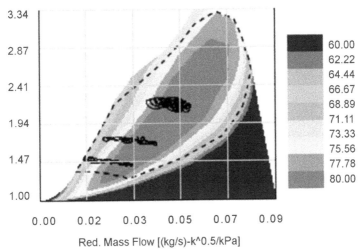

FIGURE 3.12   Comparison of compressor performance curve in an engine cycle equipped to an antisurge valve.

3.6 EFFECTS OF TURBOCHARGERS ON ENGINE PERFORMANCE IN ALTITUDE

The goal of using turbochargers in this investigation is to solve engine performance problem at high altitude. We have introduced the bad effects of altitude
on engine performance by using GT-software. Therefore, in this section we have
tried to compensate power drop by using of turbocharger GT35 and more entrance air flow. Engine turbocharger circuit is equipped to antisurge valve and
there is no wastegate also.

Figure 3.13 shows, engine power at sea level and at high altitude without/
with turbocharger for completely open throttle. In accordance with this figure,
turbocharger could solve engine power drop even it works better than sea level.
In Figure 3.14, fuel consumption with and without turbocharger has been compared. As it clear, because of more entrance air flow to engine, the fuel consumption with turbocharger is more than without it. By a comparison between
Figures 3.13 and 3.14 it is obvious that fuel-specific consumption by using of
turbocharger has improved.

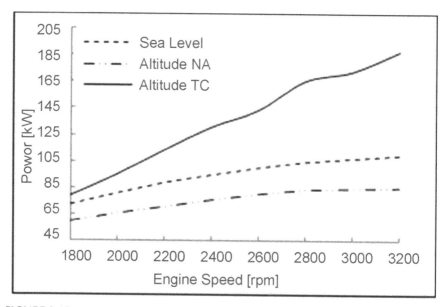

FIGURE 3.13 Break power variation versus engine speed.

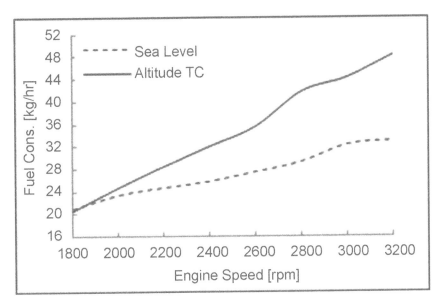

FIGURE 3.14   Variation of fuel consumption versus engine speed.

The analysis of antisurge valve performance shows, the effects of this valve is up to rotation of 2,600 rpm and after that it does not have significant effect on system performance. This deficiency is because of receding from surge region. Figure 3.15 illustrates effects of antisurge valve regions at rotation of 1,800 rpm. According to this figure, opening valve command has been sent at specific times because of growth of pressure after compressor. The opening of valve will be continued when the pressure after compressor drop under surge control pressure. According to results, about 18 percent of flow through compressor has been bypassed to prevent surge by antisurge valve.

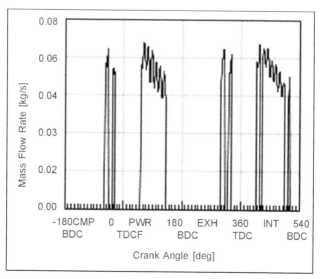

FIGURE 3.15 Mass flow rate variation of antisurge valve versus crank angle in engine speed of 1,800 rpm.

Pressure variation after compressor and before turbine as a cycle average has been shown in Figure 3.16.

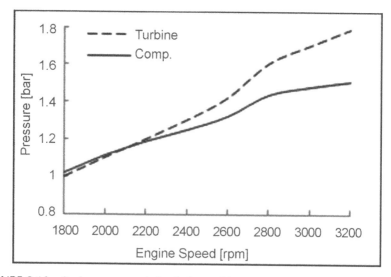

FIGURE 3.16 Cycle average variation before turbine and after compressor versus engine speed.

In pervious part, engine performance simulation based on constant performance equivalence ratio for different altitudes and rotations have been done. The result of this simulation is increase of production power and fuel consumption of engine, as it clears in Figure 3.15. This growth is because of the lack of control mechanism on inlet flow rate. We use control mechanism like wastegate to control engine fuel consumption and production power to avoid thermal and mechanical stress. This study is an attempt to use developed surge control mechanism. It works by setting the effective surface and constant discharge coefficient for valve in different rotations. Thus, the occurrence of surge will be decreased. If surge phenomena happen, surge control mechanism works and increases valve effective surface by more opening and estranges system from surge range. Figure 3.17 shows engine production power with inlet flow control mechanism for completely open throttle. In this situation, flow control criteria is amount of inlet fuel to engine at sea level (test conditions) according to Figure 3.18. For practical application, pressure amplitude by compressor can be criteria and acts as surge control mechanism.

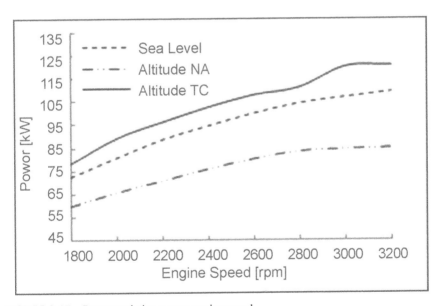

FIGURE 3.17 Power variation versus engine speed.

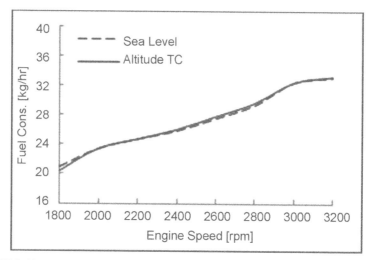

FIGURE 3.18   Fuel consumption variation versus engine speed.

Figure 3.19 shows, mass flow rate through the compressor, engine and anti-surge valve. By this graph we can investigate the influence and quality of valve performance in different rotations. According to this figure, at high speed rotation, about 25 percent of flow rate through compressor is bypassed by valve. Flow rate through turbine and compressor are equal because we do not have wastegate in this system. Also we can investigate presented equation on session four by curves of Figure 3.19.

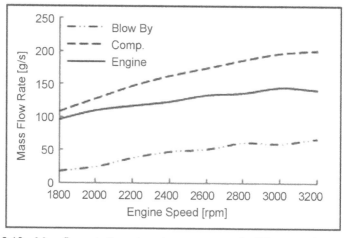

FIGURE 3.19   Mass flow rate through system components in different engine speeds.

The variation of pressure before turbine and after compressor has been exhibited in Figure 3.20. By a comparison between Figures 3.16 and 3.20, we find that valve control performance on flow rate can also control production and consumption power of turbine and compressor respectively.

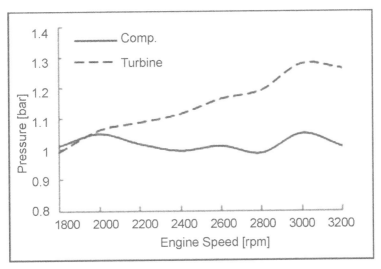

FIGURE 3.20   Variation of pressure before turbine and after compressor versus engine speed.

3.7   CONCLUSION

In this study GT-POWER software have been used to model the performance of spark-ignition internal combustion engines. In accordance with experimental and simulation results and average error of 2 and 7 percent for power and fuel consumption respectively, a validation of the study has obtained. Simulation results indicate engine power drop at altitude. To address this shortcoming, the engine has been equipped to a turbocharger selected by surge and choke criteria. To avoid surge occurrence in compressor, antisurge valve have been used. In the final step, another control mechanism on the antisurge valve to control amount of inlet fuel to engine have been utilized. This mechanism bypasses extra flow rate of compressor. The results show that by using turbocharge and the mentioned control mechanism even at altitude of 2,700 m achievement of a higher power than that of sea level with the same amount of fuel consumption is possible.

KEYWORDS

- **Antisurge valve**
- **GT-software**
- **Internal combustion engine**
- **Turbocharging**

REFERENCES

1. Watson, N.; and Jonata, M. S.; Turbocharging the Internal Combustion Engine. New York: MacMillan Press Ltd; **1982**.
2. Heywood, J. B.; Internal Combustion Engine Fundamentals. New York: McGraw-Hill; **1988**.
3. Garrett, T. K.; Newton, K.; and Steeds, W.; The Motor Vehicle. 13th ed., Butterworth-Heinemann; **2001**.
4. Renberg, U.; 1D Engine Simulation of a Turbocharged SI Engine with CFD Computation on Components. PhD diss., Linköping; **2008**.
5. Silva, C.; Ross, M.; and Farias, T.; Analysis and simulation of "low-cost" strategies to reduce fuel consumption and emissions in conventional gasoline light-duty vehicles. *Energy Convers Manage.* **2009**, *50(2)*, 215–222.
6. Korakianitis, T.; and Sadoi, T.; Turbocharger-design effects on gasoline-engine performance. *J. Eng. Gas Turbines Power.* **2005**, *127*, 525–530.
7. Kesgin, U.; Effect of turbocharging system on the performance of a natural gas engine. *Energy Conver. Manag.* **2005**, *46*, 11–32.
8. Westin, F.; Accuracy of Turbocharged si-Engine Simulations. PhD diss., KTH; **2002**.
9. Thomasson, A.; Eriksson, L.; Leufvén, O.; and Andersson, P.; Wastegate actuator modeling and model-based boost pressure control. *IFAC Workshop Engine Powertrain Control, Simul. Model.* Parice; **2009**.
10. Dimitrios, D.; and George, P.; Industrial compressor anti-surge computer control. *Int. J. Comput. Inform. Eng.* **2007**, *1*, 220.
11. Benson, R. S.; and Rowland, S.; Internal Combustion Engine. Oxford: Pergamon; **1979**.
12. Pourkhesalian, A. M.; Shamekhi, A. H.; and Salimi, F.; Alternative fuel and gasoline in an SI engine: A comparative study ofperformance and emissions characteristics. *Fuel.* **2010**, *89(5)*, 1056–1063.
13. *Manual, Engine Performance Application.* Version 7.1, Westmont, IL: Gamma Technologies Inc.; **2010**.
14. Berberan-Santos, M. N.; Bodunov, E. N.; and Pogliani, L.; On the barometric formula. *Am. J. Phys.* **1997**, *65*, 404–412.
15. www.TURBOBYGARRETT.com
16. Hajilouy-Benisi, A.; Rad, M.; and Shahhosseini, M. R.; Modeling of twin-entry radial turbine performance characteristics based on experimental investigation under full and partial admission conditions. *ScientiaIranica, Transact. B: Mech. Eng.* **2009**, *16(4)*, 281–290.
17. Eftari, M.; Shahhosseini, M. R.; Ghadak, F.; and Rad, M.; Study of Surge phenomena in compressor and its control. *J. Mech. Eng.* **2011**, *76*, (In Persian).

CHAPTER 4

THE DEPENDENCE OF SOME THERMODYNAMIC CHARACTERISTICS UPON INITIAL SPATIAL-ENERGY PARAMETERS OF FREE ATOMS

G. A. KORABLEV[1], N. G. KORABLEVA[1], and G. E. ZAIKOV[2]

[1]Izhevsk State Agricultural Academy, Russia, Izhevsk 426000, E-mail: korablev@udm.net.

[2]N.M. Emanuel Institute of Biochemical Physics, RAS, Russia, Moscow 119991, 4 Kosygina St., E-mail: chembio@sky.chph.ras.ru

CONTENTS

4.1   INTRODUCTION

Thermodynamic parameters (enthalpy, entropy, thermodynamic potential) allows describing various physical, chemical and other processes and explicitly assess the possibility of their flow without vividly using physical models. The application of reliable values of formation enthalpies is required for searching new perspective materials and compounds, assessment of molecule kinetic properties, analysis mechanisms of rocket propellant combustion, etc. [1, 2, 3].

The establishment of connection between the structure and thermochemical parameters, as well as kinetic and thermochemical characteristics of interacting systems is of great importance [4, 5]. But the analysis of dependences between main parameters of chemical thermodynamics and spatial-energy characteristics of free atoms is still topical.

For this purpose the methodology of spatial-energy parameter (P-parameter) has been used in this research [6].

4.2   RESEARCH TECHNIQUE

The comparison of multiple regularities of physical and chemical processes allows assuming that in many cases the principle of adding reciprocals of volume energies or kinetic parameters of interacting structures is fulfilled.

Some examples are: ambipolar diffusion, cumulative rate of topochemical reaction, change in the light velocity when moving from vacuum into the given medium, resultant constant of chemical reaction rate (initial product—intermediary activated complex—final product).

Lagrangian equation for the relative motion of isolated system of two interacting material points with masses m_1 and m_2 in the coordinate x with acceleration α can be as follows:

$$\frac{1}{1/(m_1 a\Delta x)+1/(m_2 a\Delta x)} \approx -\Delta U \quad \text{or} \quad \frac{1}{\Delta U} \approx \frac{1}{\Delta U1} + \frac{1}{\Delta U2} \tag{4.1}$$

where ΔU_1 and ΔU_2—potential energies of material points on elemental interaction section, ΔU—resultant (mutual) potential energy of these interactions.

The atom system is formed by oppositely charged masses of nucleus and electrons. In this system the energy characteristics of the subsystems are: orbital energy of electrons and effective energy of nucleus considering screening effects (by Clementi). Either the bond energy of electrons (W) or the ionization energy of an atom (E_i) can be used as the orbital energy.

Therefore, assuming that the resultant interaction energy in the system orbital-nucleus (responsible for interatomic interactions) can be calculated following

the principle of adding reciprocals of some initial energy components, the introduction of P-parameter [6, 7] as averaged energy characteristic of valent orbitals
based on the following equations has been substantiated:

$$\frac{1}{q^2/r_i} + \frac{1}{W_i n_i} = \frac{1}{P_э}$$
(4.2)

$$P_E = \frac{P_0}{r_i}$$
(4.3)

$$\frac{1}{P_0} = \frac{1}{q^2} + \frac{1}{(wrn)_i}$$
(4.4)

$$q = \frac{Z^*}{n^*}$$
(4.5)

where, W_i—bond energy of an electron [8]; n_i—number of elements of the given
orbital; r_i—orbital radius of i—orbital [9]; Z^* and n^*—effective charge of a
nucleus and effective main quantum number [10, 11].

The value of P_0 will be called spatial-energy parameter, and the value of
P_E—effective P-parameter. The effective P_E-parameter has a physical sense of
some averaged energy of valent electrons inside the atom and is measured in
energy units (e.g., in electron-volts (eV)).

The calculations demonstrated that the values of P_E-parameters numerically
equal (in the limits of 2%) the total energy of valent electrons (U) by statistic
model of an atom. Using the known relationship between the electron density
(b) and interatomic potential by statistic model of an atom, the direct dependence of P_E-parameter upon the electron density on the distance r_i from the
nucleus can be obtained:

$$\beta_i^{2/3} = \frac{AP_0}{r_i} = AP_E$$

where A—constant.

Based on the Eqs. (4.2), (4.3), (4.4), and (4.5) P_E and P_0—parameters of free
atoms for the majority of elements of the periodic system have been calculated
[6, 7]. Some of these calculations are given in Table 4.1.

Modifying the rules of addition for reciprocals of energy values of subsystems with reference to complex structures, the formula for calculating P_C-
parameter of a complex structure is obtained:

$$\frac{1}{P_c} = \left(\frac{1}{NPE}\right)_1 + \left(\frac{1}{NPE}\right)_2 + ...$$
(4.6)

where N_1 and N_2—number of homogeneous atoms in subsystems.

4.3   CALCULATION OF DISSOCIATION ENERGY OF BINARY MOLECULES VIA AVERAGE VALUES OF P_0-PARAMETERS

The application of methods of valent bond and molecular orbitals to complex structures faces significant obstacles for prognosticating energy characteristics of the bonds formed. Based on the Eq. (4.6), the application of the formula for calculating the dissociation energy has turned out to be practically suitable [12]:

$$\frac{1}{D_0} = \frac{1}{P_C} = \frac{1}{\left(P_E \frac{N}{K}\right)_1} + \frac{1}{\left(P_E \frac{N}{K}\right)_2} \tag{4.7}$$

where N—bond order, K—maxing or hybridization coefficient that usually equals the number of valent electrons considered, and the value of P_E (N/K) has a physical sense of the averaged energy of spatial-energy parameter accrued to one valent electron of the orbitals registered. For complex structures P_E-parameter was being averaged by all main valent orbitals.

For binary molecules the dissociation energy (D_o) corresponds to the value of chemical bond energy: $D_o = E$.

The calculation results of dissociation energy by the Eq. (4.7), given ion Table 4.2, demonstrated that $P_C=Д_0$. For some molecules containing such elements as F, N and O, the values of ion radius have been applied to register the bond ionic character for calculating P_E-parameter (in Table 4.2 marked with *). For such molecules as C_2, N_2, O_2 the calculations have been made by multiple bonds. In other cases the average values of bond energy have been applied. The calculated data do not contradict the experimental ones [2, 3].

4.4   EQUATION OF DEPENDENCE OF THERMODYNAMIC CHARACTERISTICS UPON SPATIAL-ENERGY PARAMETERS OF FREE ATOMS

The dissociation energy (E) of the molecule breakage into two parts numerically equals the difference of the heat produced during the formation of dissociation products and during the initial molecule formation:

$$E = D_0 = \left[\Delta H_0(R_1) + \Delta H_0(R_2)\right] - \Delta H_0(R_1 R_2) \tag{4.8}$$

where D_0—dissociation energy of a molecule (rupture energy of its bond), $\Delta H_0(R_1)$ and $\Delta H_0(R_1)$—formation enthalpy at 0°K respectively of dissociation products R_1 and R_2, $\Delta H_0(R_1\,R_1)$—initial energy formation enthalpy.

Since for binary molecules $E=D_0$, the Eq. (4.7) gives:

$$E = \frac{\left(P_E\dfrac{N}{K}\right)_1\left(P_E\dfrac{N}{K}\right)_2}{\left(P_E\dfrac{N}{K}\right)_1+\left(P_E\dfrac{N}{K}\right)_2} \tag{4.8a}$$

From the formulas (4.8) and (4.7a) we have:

$$\frac{\left(P_E\dfrac{N}{K}\right)_1\left(P_E\dfrac{N}{K}\right)_2}{\left(P_E\dfrac{N}{K}\right)_1+\left(P_E\dfrac{N}{K}\right)_2} = \Delta H_0(R_1)+\Delta H_0(R_2)-\Delta H_0(R_1R_2) \tag{4.9}$$

This is the equation of direct dependence of thermodynamic values and initial spatial-energy characteristics of a free atom.

The relationship obtained Eq. (4.9) allows defining the corresponding dependences with the experimental thermodynamic characteristics of chemical reactions.

Thus, in thermodynamic experimental methods for determining the formation enthalpy, the following equation is used:

$$\Delta H_0 = T\left(\Delta\Phi_T^* - R\ln K_E\right)$$

Where, K_E—equilibrium constants of chemical reaction; T—thermodynamic temperature of the process; $\Delta\Phi_T^*$—change in the reaction of considered reduced thermodynamic potential; R—universal gas constant.

4.5 EVALUATION OF SINGLE-ATOM GAS FORMATION ENTHALPY (ΔH_G^0)

In the experimental evaluation method of single-atom gas formation enthalpy from solids, a known thermodynamic identity can be used:

$$\Delta H_G^0 \equiv \Delta H_{SD}^0 + \Delta H_{S,298} \tag{4.10}$$

Where, ΔH_G^0—formation enthalpy of gaseous substance; ΔH^0_{SD}—formation enthalpy of solid in nonstandard state; $\Delta H_{S,298}$—sublimation enthalpy (substance transition from solid into gaseous state).

The heats of formation of chemical elements in standard state are taken as zero. The list of such elements is known, for example, [13] and mainly incorporates elements in solid state. Thus the single-atom gas formation enthalpy in these cases equals the sublimation enthalpy and is determined by physical and chemical criteria of the process. The sublimation comes to diffusion movements of heated particles into the surface layer with further effusion (outflow). The diffusion activation energies in external and internal regions are quite different [14].

It is usually believed that sublimation flows by the particle migration from more strongly bonded state with the biggest number of neighbors to less strongly bonded and further—to the adsorbed surface layer [15]. By analogy, the number of interacting molecules in liquid decreases, when molecules move from the lower part of the surface layer to its upper part.

The diffusion activation energy is defined by the values of electron densities of migrating particle and particles surrounding it at the distance of atom-molecular interaction radius (R), that is, such interaction is carried out via the power field of particles evaluated by the values of their P-parameters [16]. In such a model the diffusion process has basically the similar nature in any aggregate state and its energy and directedness are defined by three main factors:

(i)   value of P-parameters of structures;
(ii) number of particles;
(iii) radius of atom-molecular interaction.

In advanced researches carried out for solid solutions in the frames of generalized grid model it has been found out that "effective diffusion coefficient depends on local composition, own volumes of component atoms and potentials of paired interactions" [17]. At the same time two types of diffusion are also distinguished: "normal" and "bottom-up" [18], this apparently agrees with the notion of volume (internal) and surface diffusion.

TABLE 4.1 P-parameters of atoms calculated via bond energy of electrons

Atom	Valent Electrons	W (eV)	r_i (Å)	q^2_0 (eVÅ)	P_0 (eVÅ)	R (Å)	P_0/R (eV)	$P_0/R(n^*+1)$	r_i (Å)	P_0/r_i (eV)	$P_0/r_i(n^*+1)$
1	2	3	4	5	6	7	8	9	10	11	12
Li	2S¹	5.3416	1.506	5.86902	3.475	1.55	2.2419	0.7473	0.68	5.1103	1.7034
Na	3S1	4.9552	1.713	10.058	4.6034	1.69	2.4357	0.60892	0.96	4.6973	1.1743
K	4S1	4.0130	2.612	10.993	4.8490	2.36	2.0547	0.4372	1.33	3.6459	0.7757
Rb	5S1	3.7511	2.287	14.309	5.3630	2.48	2.1625	0.43250	1.49	3.5993	0.71986
Cs	6S1	3.3647	2.516	16.193	5.5628	2.68	2.0757	0.3992	1.65	3.3714	0.6483
Mg	3S13S2	6.8859 6.8859	1.279 1.279	17.501 17.501	5.8568 8.7787	1.60 1.60	3.6616 514667	0.91544 1.3717	0.74 0.74	7.9173 11.863	1.9793 2.9658
Ca	4S14S2	5.3212 5.3212	1.690 1.690	17.406 17.406	5.929 6.6456	1.97 1.97	3.00964.902	0.64035 0.95535	1.331.04	8.5054	1.8097
Sr	5S15S2	4.8559	1.836	21.224	6.790 9.6901	2.15 2.15	2.9205 4.5070	0.56409 0.90140	1.20	8.0751	1.6150
Ba	6S16S2	4.2872	2.060	22.950	6.3768 9.9812	2.21 2.21	2.6854 4.5164	0.55490.8685	1.38	7.2328	1.3909

TABLE 4.1 (*Continued*)

1	2	3	4	5	6	7	8	9	10	11	12
Sc	4 S 1 4 S - 23d14S2 + 3d1	5 . 7 1 7 4 5 . 7 1 7 4 9.3532	1.570 1.570 0.539 0.539	19.311 19.311 81.099 81.099	6.1279 9.3035 4.7463 14.050	1.64 1.64 1.64	3.73652.89418 .5671	0.79500.61581 .8228	0.83	16.928	3.6016
Y	5 S 1 5 S - 24d14d1 + 5S2	6.33766.7965	1.6930.856	22.540229.18 5.675615.706	6.4505 10.030	1.81	3.5638 5.5417 3.13578.6771	0.71276 1.1083 0.627141.7354	0.97	16.192	3.2384
La	6S16S24f- 15P16S2 + 5P16S2 + 4f1	4 . 3 5 2 8 1 0 .30225.470	1.9150.423 40.827 145.53	34.68117870 65915.517	6.720311.2594. 257618.40029.	1.87	3.59376.02092 .2 7 6 8 9 0 .8 3 9 6 1 5 . 8 6 0 8.2976	0.6911 1.1579 0.43781.8922 3.0501 1.5957	1.041.04	28.518 14.920	5.48432 .8693
Ti	4 S 1 4 S - 23d13d24S2 + 3d24S1 + 3d1	6 . 0 0 8 2 6 . 0 0 8 2 11.99011.990	1.477 1.477 0.496 0.469	20.879 20.879 106.04 106.04	6.2273 9.5934 5.556 10.558 20.151	1.46 1.461.46	4.2653 6.5708 3.805313.802	0.90751.3960 0.809652.93661 .7172	0.780.64 .486	12.29931 .486	2.61696 .6981
Zr	5 S 1 5 S - 24d14d25S2 + 4d2	5.54149.1611	1.5930.790	23.926153.76 6.9121 23.492	6.5330 10.263 13.229	1.601.6	4.0831 6.4146 4.3201 8.2681 14.683	0.8166 1.2829 0.8640 1.6536 2.9365	0.82	28.640	5.7298

TABLE 4.2 Dissociation energies of two-atom molecules—$D_0 \left(\dfrac{KJ}{mol} \right)$

Structure	First Atom				Second Atom				P_C (eV)	D_0 cal-cul.	D_0 exper-im.
	Orbitals	N/k	P_E (eV)	$P_E \dfrac{N}{k}$	Orbitals	N/k	P_E (eV)	$P_E \dfrac{N}{k}$			
1	2	3	4	5	6	7	8	9	10	11	12
CCl	2P^1	1/1	7.6208	7.6208	3P^1	1/1	8.5461	8.5461	4.0209	388.9	393.3
CBr	2P^1	1/1	7.6208	7.6208	4P^1	1/1	8.0430	8.0430	3.9130	377.7	364
CJ	2P^1	1/1	7.6208	7.6208	5P^1	1/1	7.2545	7.2545	2.2523	217.4	209.2
CN	2P^2	2/2	13.066	13.066	2S^{2}2P^3	2/5	47.413	18.965	7.7358	746.7	755.6
CN	2P^2	2/2	14.581	14.581	2P^3	2/3	25.127	16.751	7.796	752.5	755.6
C-O	2P^2	1/2	13.066	6.533	2P^2	1/2	17.967	8.984	3.782	365	356
NO	2P^1	1/1	9.2839	9.2839	2P^2	2/2	20.048	20.048	6.346	612.5	626.8
CH	2P^2	1/2	13.066	6.533	1S^1	1/1	9.066	9.066	3.7969	366.5	333±1
OH	2P^2	1/2	17.967	8.9835	1S^1	1/1	9.066	9.066	4.5118	435.5	423.7
ClF	3S^{2}3P^5	1/7	29.391*	4.1987	2S^{2}2P^5	1/7	38.202*	5.4574	2.5579	246.9	229.1
ClO	3S^{2}3P^5	1/7	29.391*	4.1987	2P^2	2/2	8.7191*	8.7191	2.8337	273.5	264
ClO	3P^1	1/1	4.7216*	4.7216	2S^{2}2P^4	1/6	30.738*	5.123	2.450	237.2	264
FO	2P^1	1/1	4.9887*	4.9887	2P^2	1/2	8.7191*	4.3596	2.327	224.6	219.2
NF	2P^3	1/3	10.696*	3.5653	2P^1	1/7	38.202*	5.4574	2.486	239.5	298.9
NCl	2P^3	1/3	22.296	7.432	3P^1	1/1	8.5461	8.5461	3.9751	383.7	384.9
H$_2$	1S^1	1/1	9.0624	9.0624	1S^1	1/1	9.066	9.066	4.533	437.5	432.2
Li$_2$	2S^1	1/1	2.2419	2.2419	2S^1	1/1	2.2419	2.2419	1.121	108.2	98.99
B$_2$	2P^1	1/1	5.4885	5.4885	2P^1	1/1	5.4885	5.4885	2.744	264.9	276±21
C-C	2P^1	1/1	7.6208	7.6208	2P^1	1/1	7.6208	7.6208	3.810	367.8	376.7
C=C	2P^2	2/2	13.066	13.066	2P^2	2/2	13.066	13.066	6.533	630.6	611
N-N	2P^3	1/3	10.696*	3.5653	2P^3	1/3	10.696*	3.5653	1.783	172.1	161
N=N	2S^{2}2P^3	2/5	22.745*	9.098	2S^{2}2P^3	2/5	22.745*	9.098	4.549	439	418
O-O	2P^2	1/2	8.7191	4.3596	2P^2	1/2	8.7191	4.3596	2.1798	210.4	213.4
O=O	2S^{2}2P^4	2/6	30.738*	10.246	2S^{2}2P^4	2/6	30.738*	10.246	5.123	494.5	498.3

In a liquid the radius of the sphere of molecular interaction is $R \approx 3r$, where r—radius of a molecule. Liquids are mainly formed by the elements of first and second periods of the system. For the second period we can write down: $R \approx 3r = (n + 1)r$, where n—main quantum number. For both periods (first and second) we have $R = (\langle n \rangle + 1)r \approx 2.5\ r$.

Let us assume that this principle with definite approximation can be extended to various aggregate states of elements of all other periods but taking into consideration screening effects introducing the value of effective main quantum number (n^*) instead of n. These values of n^* and $n^* + 1$ taken by Slater [19] are shown in Table 4.3.

TABLE 4.3   Effective main quantum number

n	1	2	3	4	5	6
n^*	1	2	3	3.7	4	4.2
$n^* + 1$	2	3	4	4.7	5	5.2

Thus we assume that the radius of sphere of atom-molecular interaction during the particle diffusion in the sublimation process is defined as follows:

$$R = (n^* + 1)r,$$

where, r—dimensional characteristic of atom structure. Total change of R is from 3r to 5.2r.

When forming single-atom gases, the sublimation process is accompanied by the rupture of paired bond of atoms of near surroundings. The averaged value of structural P_S-parameter of interacting atoms can be the assessment of formation enthalpy and numerically equals the value of P_0-parameter falling on the radius unit of atom-molecular interaction but taking into consideration the relative number of interacting particles by the equation:

$$P_S = \frac{P_0}{R}\gamma = \frac{P_0\gamma}{r(n^* + 1)} \approx \Delta H_\Gamma^o$$

where g—coefficient taking into consideration the relative number of interacting particles and equaled to (as the calculations demonstrated):

$$\gamma = \frac{N_0}{N}$$

Here N_0—number of particles in the sphere volume of the radius R, N—number of particles of realized interactions depending on the process type (internal or surface diffusion).

Inside the liquid below the top layer 2R thick, the resultant force of molecular interaction equals zero.

Applying the initial analogy to internal diffusion and sublimation we can consider that such equilibrium state corresponds to the equality $N_0 = N$, and then g = 1.

On the top part of liquid surface layer the volume of the sphere of atom-molecular interaction and the number of particles in it is practically two times lower in comparison with internal layers under 2R, that is, $\frac{N_0}{N} \approx \frac{1}{2}$ and $g = \frac{1}{2}$—for surface diffusion and sublimation.

In case of volume diffusion even a more extreme option is possible, when the number of particles of realized interactions two times exceeds N_0.

Thus two-valent elements magnesium and calcium form either two or even four covalent bonds (in chelate compounds): two covalent bonds and two—by donor-acceptor mechanism. For such and analogous cases with volume diffusion $g = 2$.

4.6 CALCULATIONS AND COMPARISONS

Based on such initial statements and assumptions the value of P_s-parameter was calculated (and DH^0_r) by the Eq. (4.12) for the majority of elements in periodic system—Table 4.4. The values of covalent, atom and ion radii are basically taken by Belov-Bokiy and partially by Batsanov [20].

The structural coefficient g was taken as equaled to two for covalent spatial-energy bonds at the distances of only atom or covalent radii (surface diffusion). Coefficient g was taken as equaled to one with ion spatial-energy bonds of elements from subgroups «a» and group eight in periodic system (internal diffusion). In all other cases this coefficient equals 1 or 1/2 (internal or volume diffusion).

In some cases the values of DH^0_r of the given element was found as average value by its two possible valent states (marked with <...>). The calculations are carried out practically for all elements of six periods irrespectively of their aggregate state that definitely could give higher accuracy than reference data.

Experimental and reference data [2, 3] have a relative error within 0.5–1.5 (%), but in the given calculations such average error was about 5 percent. Probably the search for a more rational technique of registering screening effects for clarifying the effective main quantum number can eventually bring more reliable results for evaluating DH^0_r.

4.7 CONCLUSIONS

The dependences of dissociation energy of binary molecules and formation enthalpies of single-atom gases upon initial spatial-energy characteristics of free atoms have been determined.

The corresponding equations, the calculations on which basically agree with the experimental and reference data, have been obtained.

TABLE 4.4 Calculation of formation enthalpy of single-atom gases—$\Delta H_{298 \; 1 \, eV} = 96.525 \; kJ/mol$

Group	Atom	Orbitals	P_o (eVA^0)	γ	r (A°)	$\Delta H = \dfrac{P_o\gamma}{r(n^*+1)}$ (eV)	r_u (A°)	$\Delta H = \dfrac{P_o\gamma}{r_u(n^*+1)}$ (eV)	ΔH calculation (kJ/mol)	ΔH°_{298} ref. data (kJ/mol)
1	2	3	4	5	6	7	8	9	10	11
	Li	$2S^1$	3.4750	1			0.68	1.703	164.4	159.3
	Na	$3S^1$	4.6034	1			0.98	1.1743	113.3	107.5
1a	K	$4S^1$	4.8490	2	2.36	0.874			84.36	88.9
	Rb	$5S^1$	5.3630	2	2.48	0.865			83.49	80.9
	Cs	$6S^1$	5.5628	2	2.68	0.7984			77.07	77
	Be	$2S^2$	7.512	½			0.34	3.682	355.4	326.4
	Mg	$3S^2$	8.7787	½			0.74	1.483	143.1	147.1
2a	Ca	$4S^2$	8.8456	1			1.04	1.8097	174.7	177.8
	Sr	$5S^2$	9.6901	1			1.20	1.6150	155.9	160.7
	Ba	$6S^2$	9.9812	2	2.21	1.737			167.7	179.1
	Sc	$4S^2 3d^1$	14.050	1			0.74	4.040	389.96	379.1
3a	Y	$5S^1 4d^2$	17.527	2	1.62	4.342			419.2	423
	La	$6S^2 5p^1$	29.659	2	1.87	6.1002			<448.4>	429.7
		$6S^2 4f^1$	15.517	2	1.87	3.1915				
	Ti	$4S^1 3d^1$	11.7853	2	1.46	3.4344			<449.2>	468.6
		$4S^2 3d^2$	20.151	2	1.46	5.8732				
4a	Ti	$4S^1 3d$	11.7853	1			0.78	3.214	<478.4>	468.6
		1^1	20.151	1			0.64	6.6981		
		$4S^2 3d^2$								
	Zr	$5S^2 4d^2$	23.492	2	1.60	5.873			566.9	600
	Hf	$6S^2 5d^2$	24.498	2	$r_g \approx 1.44$	6.543			631.6	620.1

1	2	3	4	5	6	7	8	9	10	11
	V	$4S^2 3d^1$	15.776	1			0.67	5.6010	<512.5>	514.6
		$4S^1 3d^2$	17.665	1			0.67	5.6090		
	Nb	$5S^2 4d^2$	25.577	1			0.67	7.6349	736.9	722.6
	$(5S^2 4d^3)$									
5a	Nb	$5S^1 4d^2$	20.805	1			0.767	5.425	<709.4>	722.6
	$(5S^1 4d^4)$	$5S^1 4d^4$	30.607	1			0.65	9.2748		
	Nb			1					<723.2>	722.6
	(3, 4, 5)									
	Ta	$6S^2 5d^2$	26.314	1			0.737	6.8662	<791.5>	786.1
		$6S^2 5d^3$	32.722	1			(0.66)	9.5344		
	Cr	$4S^2$	10.535	1			0.83	2.7006	<406.1>	397.5
		$4S^2 3d^1$	17.168	1			0.64	5.7074		
6a	Mo (2)	$5S^1 4d^1$	19.574	1			0.737	5.312		
	$(5S^2 4d^4)$								<658>	656.5
	Mo (4)									
	$(5S^1 4d^4)$	$5S^1 4d^3$	28.293	1			0.68	8.3215		

1	2	3	4	5	6	7	8	9	10	11
	W(4)	$6S^2 6S^1$	27.879	2	1.40	3.8295			<831.5>	856.9
	W(5)	$6S^2 5d^3$	34.828		1.40	4.7841			:	
	Mn (2)	$4S^1 3d^1$	12.924	1			0.91	3.0217	291.7	284.5
	Tc (3)	$5S^1 4d^2$	23.866	2	1.36	7.0194			674.1	657
7a	Re (4)	$6S^2 5d^2$	29.806	1			0.72	7.961	768.4	
	Re (6)	$6S^2 5d^4$	44.519	½			0.52	8.232	794.6	775.7
	Re (4,6)								<781>	
	Fe (2)	$4S^1 3d^1$	12.717	1			0.80	3.3822	<418.6>	417.1
	Fe (3)	$4S^2 3d^1$	16.664	1			0.67	5.2918		

	Co (2)	$4S^1 3d^1$	12.707	1			0.78	3.4662	<436.5>	428.4
	Co (3)	$4S^2 3d^1$	16.680	1			0.64	5.5785		
	Ni (2)	$4S^1 3d^1$	12.705	1			0.74	3.6530	<465.5>	429.3
	Ni (3)	$4S^1 3d^2$	16.897	1			0.60	5.992		
8	Ni (2)	$4S^1 3d^1$	12.705	2	1.24	4.360			420.8	429.3
	Ru (3)	$5S^1 4d^2$	23.636	1			0.68	6.982	671	656.8
	Rh (3)	$5S^2 4d^1$	21.114	1			0.75	5.6304	543.5	556.5
	Pd (2)	$5S^2$	12.057	1			0.64	3.7681	363.7	372.3
	Os (3)	$5d^3$	25.986	2	$r_k = 1.26$	7.922			766	790.2

	Ir (2)	$6S^1 5d^1$	17.631	2	1.35	5.023			<665.8>	669.5
8	Ir (4)	$6S^2 5d^2$	30.790	2	1.35	8.772				
	Pt (2)	$6S^1 5d^1$	17.381	1			0.6	5.971	576.4	565.7
	Cu (2)	$4S^1 3d^1$	13.242	1			0.80	3.5218	339.9	337.6
	Ag	$4d^1$	9.7843	2	$r_k = 1.25$	3.131			<282.3>	284.9
16	Ag (1)	$4d^1$	9.7843	2	1.44	2.718				
	Au (3)	$6S^1 5d^2$	27.536	1	1.44	3.6774			355.0	368.8
	Zn (1)	$3d^1$	6.1153	1	1.39	0.9361			<127.1>	130.5
	Zn (2)	$4S^2$	11.085	1	1.39	1.6968				
	Cd (2)	$5S^2$	11.839	½			0.99	1.1959	115.4	111.8
26	Cd (2)	$5S^1 4d^1$	17.145	½	1.56	1.0991			106.1	111.8
	Cd (1, 2)								<110.8>	111.8
	Hg (1)	$5d^1$	11.266	½	1.60	0.6771			65.35	61.40
	B	$2S^2 2p^1$	16.086	1	0.91	5.892			568.7	561
36	Al	$3S^2 3p^1$	18.093	1	1.43	3.163			<327.3>	329.3
					$r_k = 1.25$	3.619				

1	2	3	4	5	6	7	8	9	10	11
	Ga	$4S^2 4p^1$	20.760	1	1.39	3.178			<284.8>	273.0
	Ga		8.8961	2	1.39	2.723				
36	In	$5S^2 5p^1$	21.841	1	1.66	2.6314				238.1
	In	$5S^2 5p^1$	21.841	½			0.92	2.374	<241.5>	
	Tl	$6S^2 6p^1$	22.012	½			1.05	2.015	194.5	181.0
	C	$2p^2$	10.061	1	0.77	4.3554			<723.9>	716.7
		$2S^2 2p^2$	24.585	1	0.77	10.643				
	Si	$3p^2$	10.876	2	1.17	4.648			448.6	452
46	Ge	$4p^2$	12.072	2	1.39	3.3656			<378.3>	376.6
		$4p^2$	12.072	2	1.24	4.1428				
	Ge	$4p^2$	12.072	1			0.65	3.9516	381.43	376.6
	Sn	$5p^2$	13.009	2	1.58	3.2934			317.9	302.1
	Pb	$6S^2 6p^2$	32.526	1	1.50	2.0952			202.2	195.1
	Pb	$6p^2$	13.460				1.26	2.0543	198.3	195.1
	N	$2p_r^5$	21.966				1.48	4.947	477.5	472.7
	P	$3p^1$	7.7864	2	$r_к = 1.16$	3.3562			323.9	316.3
56	As (5)	$4S^2 4p^3$	40.232	½	1.40	3.057			295.1	
	As (3)	$4p^3$	18.645	1	$r_к = 1.21$	3.2785			316.5	301.8
	As (3 ,5)								<305.8>	

1	2	3	4	5	6	7	8	9	10	11
56	Sb	$5p^3$	20.509	1	r_K −1.39	2.9509			<265.2>	268.2
		$5p^3$	20.509	1	1.61	2.5477				
	Bi	$6p^3$	21.919	1	1.82	2.3160			<207.3>	209.2
	Bi	$6p^3$	21.919	1			2.13	1.9790		
	O (2)	$2p^2$	11.858	1			1.40	2.823	<253.1>	249.2
	O (4)	$2p^4$	20.338	½			1.40	2.421		
66	S (2)	$3s^1 3p^1$	21.673	2	r_K ~0.94	2.882			<280.8>	277.0
	S (4)	$3p^4$	21.375	1			1.82	2.9360		
	Se (4)	$4p^4$	24.213	½	r_K =.17	2.2032			<226.7>	223.4
	Se (2)	$4s^1 4p^1$	23.283	1			1.98	2.495		
	Te	$5s^1 5p^1$	23.882	1			2.11	2.264	218.5	215.6
	Po	$6s^1 6p^1$	23.664	½	r_K = 1.50	1.516			146.4	146
76	F	$2p^1$	6.635	½			1.33	0.83145	80.3	79.5
	Cl	$3p^1$	8.5461	1			1.81	1.2804	123.6	121.3
	Br	$4p^1$	9.3068	1			1.79	1.1062	106.8	111.8
	I	$5p^1$	9.9812	½			r^{5+} = 0.94	1.062	102.5	106.8

KEYWORDS

- **Dissociation energy**
- **Formation enthalpy**
- **Spatial-energy parameter**
- **Sublimation**

REFERENCES

1. Lebedev, Yu. A.; and Miroshnichenko, E. A.; Thermal chemistry of steam-formation of organic substances. M.: "Nauka"; **2012,** 216 p.
2. Properties of Inorganic Compounds. Reference-Book. Efimov, A. I.; et al. L.: "Khimiya"; **2012,** 392 p.
3. Rupture energy of chemical bonds. Ionization Potentials and Affinity to an Electron. Reference-Book. Kondratyev, V. I.; et al. M.: "Nauka"; **1974,** 351p.

4. Benson, S.; Thermochemical Kinetics. M.: Mir; **2012**, 308 p.
5. Berlin, A. A.; Wolfson, S. A.; and Enikolopyan, N. S.; Kinetics of Polymerization Processes. M.: "Khimiya"; **2012**, 319 p.
6. Korablev, G. A.; Spatial-Energy Principles of Complex Structures Formation. Netherlands: Brill Academic Publishers and VSP; **2005**, 426 p (Monograph).
7. Korablev, G. A.; Spatial-Energy Criteria of Phase-Formation Processes. Izhevsk: Publishing House "Udmurt University"; **2008**, 494 p.
8. Fischer, C. F.; Average-energy of configuration Hartree-Fock results for the atoms helium to radonV. *Atomic Data.* **2012**, *4*, 301–399 pp.
9. Waber, J. T.; and Cromer, D. T.; Orbital radii of atoms and ions. *J. Chem. Phys.* **1965**, *42(12)*, 4116–4123.
10. Clementi, E.; and Raimondi, D. L.; Atomic screening constants from S.C.F. functions, 1. *J. Chem. Phys.* **1963**, *38(11)*, 2686–2689.
11. Clementi, E.; and Raimondi, D. L.; Atomik screening constants from S.C.F. functions, 2. *J. Chem. Phys.* **1967**, *47(4)*, 1300–1307.
12. Korablev, G. A.; and Zaikov, G. E.; Energy of chemical bond and spatial-energy principles of hybridization of atom orbitalls. *J. Appl. Polym. Sci.* USA, **2006**, *101(3)*, 283–293.
13. Thermodynamic Properties of Individual Substances. Reference-book of Academy of Science of the USSR, Glushko, V. P.; ed. M.: "Science"; **1978**, *1*, 496 p.
14. Kofstad, P.; Deviation from Stoichio, Diffusion and Electrical Conductivity in Simple Metal Oxides. M.: "Mir"; **2012**, 398 p.
15. Pound, G. M.; *S. Phys. Chem. Ref. Data.* **1972**, *1*, 135–146.
16. Korablev, G. A.; and Solovyev, S. D.; Energy of diffusion activation in metal systems. *Bull. ISTU.* **2007**, *4*, 128–132.
17. Zaharov, M. A.; Grid models of multi-component solid solutions: statistic thermodynamics and kinetics. Abstract of doctoral thesis. Novgorod State University; **2008**, 36 p.
18. Geguzin, Ya. E.; Ascending diffusion and diffusion aftereffect. *UFN.* **2012**, *149(1)*, 149–159.
19. Batsanov, S. S.; and Zvyagina, R. A.; Overlap integrals and problem of effective charges. "Nauka", Novosibirsk: Siberian Branch; **1966**, 386 p.
20. Batsanov, S. S.; Structural chemistry. Facts and Dependencies. M.: MSU; **2012**, 292 p.

AUTHORS

1. Korablev Grigory Andreevich, Doctor of Chemical Science, Professor, Head of Department of Physics at Izhevsk State Agricultural Academy. Izhevsk 426052, 30 let Pobedy St., 98-14, tel.: +7(3412) 591946, e-mail: biakaa@mail.ru, korablev@udm.net.
2. Zaikov Gennady Efremovich, Doctor of Chemical Science, Professor of N.M. Emanuel Institute of Biochemical Physics, RAS, Russia, Moscow 119991, 4 Kosygina St., tel.: +7(495)9397320, e-mail: chembio@chph.ras.ru.

CHAPTER 5

FIRE RETARDANT COATINGS BASED ON PERCHLOROVINYL RESIN WITH IMPROVED ADHESIVE PROPERTIES TO PROTECT FIBERGLASS PLASTICS

G. E. ZAIKOV

Russian Academy of Sciences, Moscow, Russia

CONTENTS

5.1 INTRODUCTION

In many cases polymer construction materials are a good alternative to metals and reinforced concrete. Still, it is known that the majority of such materials are combustible. That is why implementation of the materials to the building industry is associated with solving a range of engineering problems; one of them is providing the materials with the required fire safety. The fire hazard of polymers and composite materials is understood as a complex of properties which along with combustibility includes the ability for ignition, lighting, flame spreading, quantitative evaluation of smoke generation ability, and toxicity of combustion products.

Fiberglass plastics find ever-widening applications in different industrial fields. The main benefit of fiberglass plastics is higher strength and lower density compared to metals; they are not subjected to corrosion.

However, together with the valuable property complex of fiberglass plastics, they also have a significant drawback that is a low resistance to open flame.

The sufficient increase in fire safety of fiberglass constructions may be achieved by using the passive protection measures—applying flame retardant intumescent coatings.

Under the influence of high temperatures on a surface of the object protected from fire an intumescent surface appears which obstructs penetration of heat and fire spread over the surface of the material.

For effective fire protection it is necessary to apply compounds which components inhibit combustion comprehensively: In a solid phase it is carried out by transforming the destruction process in a material, in a gaseous phase—preventing the oxidation of the degradation products [1, 2].

A standard formulation of a fire retardant coating includes an oligomer binder as also fire retarding nitrogen, phosphorus and/or halogen containing inorganic and organic compounds. The fire retarding effect is enhanced at the combination of different heteroatoms in an antipyrene [3, 4].

Previously, it was found that phosphorus boron containing compounds are effective antipyrenes in a fire retardant composition [5–7].

In the investigation we developed a new phosphorus-boron-nitrogen-containing oligomer (PEDA). The oligomer has a good compatibility to a polymer binder, slightly migrates from a polymer material, and is an effective antipyrene when has the lower phosphorus content [6, 7].

Phosphorus-boron-nitrogen-containing oligomer and polymer products comprising —P=O, —P—O—B—, —B—O—C—, and —C—N—H— bonds are not studied enough. IR spectroscopy showed that these groups are a part of the PEDA macromolecule composition.

To improve physical and mechanical properties of coatings and characteristics of fire protection efficiency, the fire retardant coatings including PEDA as a modifier and based on perchlorovinyl resin (PVC resin) were obtained. The coatings are used for fiberglass plastics.

5.2 EXPERIMENTAL

With a purpose of defining the efficiency of the developed fire retardant coatings for fiberglass, a set of experiments was conducted.

The experiments on fire retardant properties were carried out on the developed technique by exposure of a coated fiberglass plastic sample to open flame. Time-temperature transformations on the nonheated surface of the fiberglass test sample was registered using a pyrometer measuring the moment of achieving the limit state—a temperature of the fiberglass destruction beginning (280–300°C).

Then, the intumescence index of the coating was calculated. The intumescence index was determined by a relative increase in height of the porous coke layer compared to the initial coating height.

The coke residue was estimated by the relative decrease in the sample weight after keeping it in the electric muffle during 10, 20, and 30 min at 600°C.

For a possibility to use the fire retardant coatings, we need to solve a problem concerned with providing the required adhesion between the coating and protected material. The adhesive strength of the coatings to fiberglass plastic defined as the shear strength of a joint.

During the work, the studies on the combustibility and water absorption of the coatings were conducted.

The combustibility was evaluated by exposure of a sample to the burner flame (temperature peak 840°C) and fixing the burning and smoldering time after fire source elimination.

The experiments on the water absorption were performed in distilled water at temperature $23 \pm 2°C$ for 24 hr. The water absorption was estimated by a sample weight change before and after exposure to water.

The coke microstructure formed after a test sample burning was studied as well.

5.3 RESULTS AND DISCUSSION

As part of the research, the investigation of the coatings based on perchlorovinyl resin and containing the developed intumescent additive PEDA on fire protective properties was conducted. The results are presented in Table 5.1.

TABLE 5.1   Influence of PEDA content on fire resistance of the coatings based on PVC resin

Parameter	Without Coating	PEDA Content (wt. %)							
		0		**2.5**		**5.0**		**7.5**	
Coating Thickness (mm)		0.7	1.0	0.7	1.0	0.7	1.0	0.7	1.0
Intumescence Index	-	1.55	2.7	4.89	5.55	5.12	6.0	5.64	6.47
Time to the Limit State (sec)	18	29	32	44	52	48	57	55	63
Temperature of the Non-Heated Sample Side in 25 sec (°C)	-	247	223	131	115	116	108	109	102

When a coating of 1 mm in thickness containing 7.5 percent PEDA (percent of the initial composition weight) is used, the peak time to the limit state is established, and the intumescence index reaches 6.47 at that.

The temperature dependence of the nonheated sample side on flame exposure time at different PEDA content is shown in Figure 5.1.

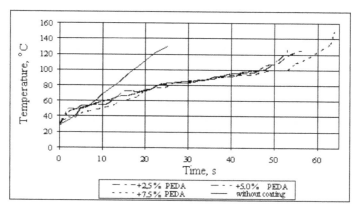

FIGURE 5.1   Dependence of temperature on the nonheated sample side on flame exposure time.

As it is seen in Figure 5.1, the studied coatings allow keeping temperature on the nonheated sample side within the range 80–100°C for quite a long time; time to the limit state of test samples increased by 2–2.5 times.

Coke formation is an important process in fire and heat protection of a material. Achieving a higher intumescence ratio for carbonized mass, lower heat conductivity of coke, and its sufficient strength, all these characteristics are the necessary conditions for effective fire protection.

The dependence of PEDA content influence on ability to form coke is presented on Figure 5.2.

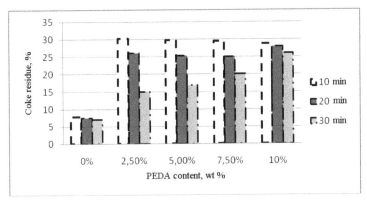

FIGURE 5.2   Effect of PEDA content on the coke residue values at 600°C.

As the diagram shows, with the growth of PEDA content the coke residue increases as well. It can be explained by the catalytic processes in coke formation caused by phosphorus boron containing substances [8].

In the experiments on combustibility it was found that the coatings containing PEDA are resistant to combustion and can be assigned the fire reaction class 1 as nonflammable (see Table 5.2).

TABLE 5.2   Influence of PEDA content on combustibility of the coatings based on PVC resin

PEDA Content (wt. %)	Combustibility of a Coating
0	Burning
2.5	self-extinguishing in 2 sec
5.0	self-extinguishing in a second
7.5	Not burning

The combustibility tests demonstrate that introducing PEDA into the compositions based on PVC resin promotes formation of a large coke layer; the coating film does not burn, because the presence of nitrogen in the modifying additive enables an enhancement of the fire and heat resisting effect.

TABLE 5.3   Influence of PEDA content on water absorption of the coatings based on PVC resin

PEDA Content (wt. %)	Extent of Change in Sample Weight	pH
0	0.02	7
2.5	−0.05	5
5.0	−0.6	5
7.5	−0.05	5
10.0	−0.07	4

The results on determining the water absorption of the modified samples revealed an insignificant washout of PEDA that takes place through the slight diffusion of the modifying additive to the film surface that is evidenced by a change in pH in 24 hr (Table 5.3). Nevertheless, this has no effect on fire resistance of the coatings.

As noted above, intumescent coatings should have good adhesion to the protected material; therefore, the studies on the influence of PEDA content on the adhesion strength of the coatings based on PVC resin to the fiberglass plastics were carried out while researching. The test results are illustrated by Figure 5.3.

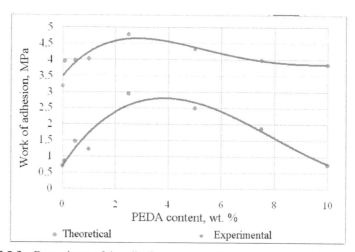

FIGURE 5.3   Dependence of the adhesion strength of the coatings on PEDA content.

So, it was established that the introduction of PEDA to the coating composition in amounts of 2.5–7.5 percent provides an increase in the adhesion strength by 1.5–4 times.

For confirmation of the experimental data, the work of adhesion was calculated according to the Young-Dupre equation. The surface tension was observed on the duNouy tensiometer. The contact angles of wetting were defined using a goniometric method.

The calculated values of the work of adhesion are well correlated with experimental data.

In order to improve intumescence and fire protection, the effect of introduced thermal expanded graphite (TEG), which served as a filler, on coke formation and physical and mechanical characteristics of the coatings was also studied in the work.

In a course of the study an optimal graphite amount was chosen so that the adhesion characteristics of the coatings would not become worse and allow obtaining a sufficiently hard coke.

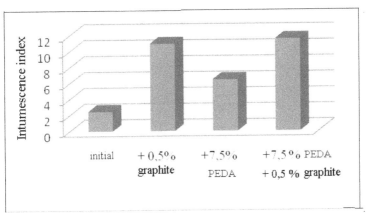

FIGURE 5.4   Influence of PEDA and TEG presence on the intumescence index of coatings.

The best results were achieved when applying PEDA and filler TEG. In this case the intumescence ratio reached 11.6 (see Figure 5.4). The results on the influence of the coating modification and filler presence on the coke structure are presented in Figures 5.5, 5.6, and 5.7.

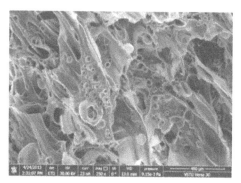

FIGURE 5.5 Micrograph of a coke structure on the initial coating at 250-fold magnification.

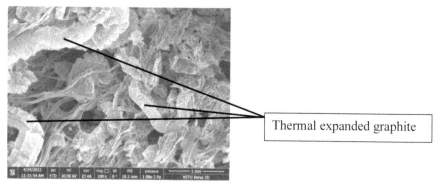

Thermal expanded graphite

FIGURE 5.6 Micrograph of a coke structure on the coating containing TEG at 100-fold magnification.

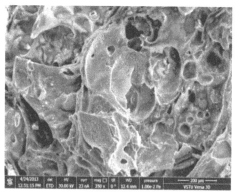

FIGURE 5.7 Micrograph of a coke structure on the coating containing PEDA and TEG at 350-fold magnification.

The foamed coke formed at the composition testing and not containing modifying additives and fillers has a coarse amorphous structure (Figure 5.5); there are foamed globular formations of 10–100 μm in the coke volume grouping to associates.

In the compositions containing TEG only (Figure 5.6), the coke structure is mainly determined by graphite that is present in the form of extended structures longer than 1,000 μm, 50–100 times greater than the pore size of the foamed phase. The presence of these structures leads to increased friability of the foamed mass, coke has low strength.

In the coke structure of the coating containing PEDA and TEG (see Figure 5.7) the extended structures, formed by graphite, disappear, and there are only short fragments of these formations. A consolidation of the carbon layers is observed which, probably, takes place due to formation of the high temperatures of polyphosphoric acids on the surface and between the layers of expanded graphite sites that solder layers, and, thereby, impede TEG intumescence; there is a slight shrinkage of the intumescent layer. As a result, the intumescence index of this composition is not substantially exceed the intumescence index of the composition containing only filler, but with a more ordered structure of coke and a sufficiently high strength and hardness of the composition the fire resistance increases. Such a coke can withstand more intense combustion gas streams.

5.4 CONCLUSION

Thus, the fire retardant coatings based on the developed phosphorus-boron-nitrogen-containing oligomer have high fire and heat protective and adhesive properties. The structure and presence of phosphorus, boron and nitrogen heteroatoms promotes an enhancement of the film-forming polymer carbonization and increase in the intumescence ration of the coatings. In addition, the definite advantage of PEDA application is that it is slightly washed out of a coating when exposed to water.

Introduction of the modifying additive PEDA in combination with a filler—thermal expanded graphite—permits to increase the coating intumescence by 11 times, resulting in improved fire and heat protective properties of the coatings and reduced destruction of fiberglass plastic.

The research has been done with financial support from Ministry of Education and Science of the Russian Federation under realization of the federal special-purpose program Scientific, academic and teaching staff of innovative Russia for 2009–2013 years: The Grant Agreement No. 14.B37.21.0837 "Development of active adhesive compositions based on element organic polymers and vinyl monomers."

KEYWORDS

- **Fire retardant coatings**
- **Perchlorovinyl resin**
- **Phosphorus-boron-nitrogen-containing oligomer (PEDA)**

REFERENCES

1. Berlin, Al. Al.; Combustion of polymers and polymer materials of reduced combustibility. *Soros Edu. J.* **1996,** *9,* 57–63.
2. Shuklin, S. G.; Kodolov, V. I.; and Klimenko, E. N.; Intumescent coatings and the processes that take place in them. *Fibre Chem.* **2004,** *36(3),* 200–205.
3. Nenakhov, S. A.; and Pimenova, V. P.; Physical Transformations in fire retardant intumescent coatings based on organic and inorganic compounds. *Pozharovzryvobezopasnost—Fire Exp. Safety.* **2011,** *20(8),* **17–24.**
4. Balakin, V. M.; and Polishchuk, E. Yu.; Nitrogen and phosphorus containing antipyrenes for wood and wood composite materials. *Pozharovzryvobezopasnost—Fire Exp. Safety.* **2008,** *17(2),* 43–51.
5. Shipovskiy, I. Ya.; Bondarenko, S. N.; and Goryainov, I. Yu.; Fire protective modification of wood. Proceedings of the International Scientific and Practical Conference. Dnepropetrovsk. **2005,** *47,* 20.
6. Gonoshilov, D. G.; Keibal, N. A.; Bondarenko, S. N.; and Kablov, V. F.; Phosphorus Boron Containing Fire Retardant Compounds for Polyamide Fibers. Proceedings of the 16th International Scientific and Practical Conference Rubber industry: Raw Materials. Manufactured Materials. Technologies. Moscow; **2010,** 160–162.
7. Lobanova, M. S.; Kablov, V. F.; Keibal, N. A.; and Bondarenko, S. N.; Development of Active Adhesive Fire and Heat Retardant Coatings for Fiber-Glass Plastics. All the Materials. Encyclopaedic Reference Book. **2013,** *4,* 55–58.
8. Korobeynichev, O. P.; Shmakov, A. G.; and Shvartsberg, V. M.; The Combustion Chemistry of Organophosphorus Compounds. *Uspekhi khimii.* **2007,** *76(11),* 1094–1121.

CHAPTER 6

MODIFICATION OF PECULIARITIES OF MICROCRYSTALLINE CELLULOSE (MCC) AND ITS OXIDIZED FORM (DIALDEHYDE CELLULOSE DAC) GUANIDINE-CONTAINING MONOMERS AND POLYMERS OF VINYL AND DIALLYL SERIES

AZAMAT A. KHASHIROV[1], AZAMAT A. ZHANSITOV[1],
G. E. ZAIKOV[2], and SVETLANA YU. KHASHIROVA[2]

[1]Kabardino-Balkarian State University a. Kh.M. Berbekov, 173 Chernyshevskogo st., 360004, Nalchik, Russian Federation, new_kompozit@mail.ru

[2]N.M. Emanuel Institute of Biochemical Physics of Russian Academy of Sciences, 4, Kosygin St., 119991, Moscow, Russian Federation

CONTENTS

6.1   INTRODUCTION

Formation and research of systems "polymeric carrier"—biologically active substance have lately got great importance.

Such systems find an application as immobilized biocatalysts, bioregulators and an active form of medicinal substances of the prolonged action.

In this work for the first time have been studied modification peculiarities of microcrystalline cellulose (MCC) and its oxidized form (dialdehyde cellulose DAC) guanidine-containing monomers and polymers of vinyl and diallyl series [1–3]. The structure and some characteristics of used guanidine-containing modifiers are shown in Table 6.1.

TABLE 6.1   Structure and some characteristics of guanidine-containing modifiers of cellulose and dialdehyde cellulose

Modifier	Molecular Weight	Melting Point (°C)	Structure
Acrylate guanidine (AG)	131,134	175–176	R=H
Methacrylate guanidine (MAG)	145,160	161–163	R=CH3
N,N-Diallylguanidine acetate (DAGA)	199,253	211–212	IV V X = H or F
N,N-Diallylguanidine trifluoroacetate (DAGTFA)	253,224	157–158	IV V X = H or F

6.2   RESULTS AND DISCUSSION

Composite materials were received by treating MCC or DAC, water-soluble monomeric guanidine derivatives (Table 6.1), with subsequent polymerization. Quantity of the monomer/polymer zwitterionic quaternary ammonium cations acrylate and diallyl guanidine derivatives were included in MCC or DAC determined by nitrogen content using elemental analysis.

The results of IR spectroscopic studies show the structural differences of cellulose samples and its modified forms (Figure 6.1).

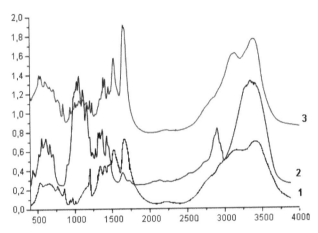

FIGURE 6.1   Comparison of the spectra PMAG (1), DAC-MAG in situ (2), DAC (3).

For example, the polymerization in DAC-MAG in situ (Figure 6.1, curve 3) in the spectra varies the ratio of the intensity of stripes both cellulose (within $1,000-1,100$ cm^{-1}) and MAG, moreover, the stripe disappears within 860 cm^{-1} indicating the presence of a double tie. Splitting of the tie C = O ties PMAG within 1,250 cm^{-1} is taking place that clearly shows strong mutual influence of the DAC and MAG/PMAG and formation of bimatrix systems. There is an increase in the intensity of the peak 1,660 cm^{-1} in the spectrum of the DAC-MAG formation aldimine connection and increase in the width of the characteristic absorption stripes in DAC-MAG within $1,450-1,680$ cm^{-1}, probably due to the formation of relatively strong ties with the active MAG DAC centers.

Comparison of the IR spectra of the composites and DAZ DAZ-DAG and DAZ-DAGTFA demonstrates differences in the spectra of these compounds and indicates the formation of new structures (Figure 6.2).

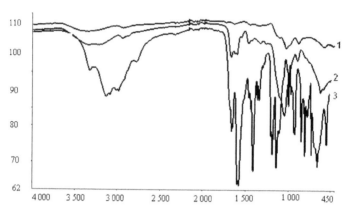

FIGURE 6.2   IR-spectra DAC (1), DAC-DAGTFA (2); DAGTFA (3).

In the spectra of the modified DAC v_{OH} observed increase in absorption from the high-frequency, especially in the spectrum of DAC modified DAG. This is due to increasing of the hydroxyls involved in weak hydrogen bonds. Stretching vibrations of C-H bonds of methane and methylene groups DAC appear in $3{,}000$–$2{,}800$ cm^{-1}. In the spectra of the modified DAG and DAGTFA DAC, these valence oscillations are superimposed on the absorption of the CH$_2$ groups that are part of diallyl compounds.

This leads to an increase in the intensity of the absorption bands with a frequency of ~$2{,}900$ cm^{-1}. In the area of ~$1{,}650$ cm^{-1} peaks appear adsorbed water. Raising polar amino groups comprising the modifier that increases the polarity of the substrate that assists in keeping the surface modified pulp samples more water adsorption due to hydrogen bonds. Increasing the intensity of the peak $1{,}655$ cm^{-1} in the spectrum of DAC-DAG and DAC-DAGTFA may indicate formation aldimine communication giving a signal in this area. In the polymerization in the DAG and DAGTFA, DAC in situ peak at $1{,}140$ cm^{-1} present in the DAC disappears. Obviously, the terminal CHO groups DAC and guanidine-containing diallyl modifier reacted with each other.

Thus, immobilization AG, MAG, DAG, and DAGTFA, DAC between components in the formation of various types of bonds: due to van der Waals forces, intra-and intermolecular coordination and hydrogen bonds, C-C bonds formed during the free radical polymerization in situ immobilized AG, MAG, DAG, and DAGTFA, bonds formed during the graft copolymerization of monomer radical salts with DAC and labile covalent aldimine C = N bonds formed by reacting aldehyde groups with amino DAC guanidine-containing compounds.

Composite materials obtained by polymerization and DAG, DAGTFA in situ in the inter- and intra-fibrillar DAC pores, dissolve well in water. We can assume that the action of pulp and DAGTFA, DAG, a major role in breaking the inter- and intra-molecular hydrogen bonds play anions CH_3COO^- and SF_3COO^- which form a DAC stable complexes through its hydroxyl groups. Simultaneously, the esterification may occur partly sterically more accessible alcohol group's DAC. SEM method shows that the dissolution of cellulose sphere-like type of complex is formed between the components of the solution in which the cellulose macromolecules have coil conformation (Figure 6.3).

(a)

(b)

FIGURE 6.3   SEM images dialdehyde cellulose (**a**) and complex of dialdehyde cellulose with N,N-diallylguanidine trifluoroacetate (**b**).

In the process of dissolution of dialdehyde cellulose and DAG, DAGTFA also act as acceptors of hydrogen bonds and associated solvent molecules, thereby preventing reassociation of cellulose macromolecules. The biological activity of the synthesized composite materials was investigated and shown that the composite synthesized materials are quite active and have a biocidal effect against Gram-positive (St. Aureus) and Gram (E.coli) microorganisms. Being the most expressed biocidal activity is shown in the composites with diallyl derivatives of guanidine.

6.3   CONCLUSION

Studies of the structure of the composites by SEM showed that DAG and DAGTFA localized in the surface layers of the composite, which increases the availability of biocidal centers and explains a higher relative activity of these composites.

In the case of dialdehyde cellulose, modified acrylate derivatives, guanidine antibacterial active groups are in the deeper layers of the interfibrillar dialdehyde cellulose, which reduces their bioavailability and therefore, the DAC-PAG and DAC-PMAG and start to show the bactericidal activity of only 48 hr. Slowing down the release rate of the bactericidal agent opens prospect of long-acting drugs with a controlled-release bactericide.

KEYWORDS

- **Acrylate guanidine**
- **Cellulose**
- **IR spectroscopy**
- **methacrylate guanidine**
- **N,N-diallylguanidine acetate**
- **N,N-diallylguanidine trifluoroacetate**
- **SEM**

REFERENCES

1. Khashirova, S. Yu.; Zaikov, G. E.; Malkanduev, U. A.; Sivov, N. A.; and Martinenko, A. I.; Synthesis of new monomers on diallyl gyanidine basis and their ability to radikal (co)polymerization. J. Biochemistry and Chemistry: Researsh and Developments. New York: Nova Science Publishers Inc.; **2003**, 39–48.
2. Sivov, N. A.; Martynenko, A. I.; Kabanova, E. Yu.; Popova, N. I.; Khashirova, S. Yu.; and Ésmurziev, A. M.; Methacrylate and acrylate guanidines: Synthesis and properties. *Petrol. Chem.* **2004**, *44(1)*, 43–48.
3. Sivov, N. A.; Khashirova, S. Yu.; Martynenko, A. I.; Kabanova, E. Yu.; Popova, N. I.; NMR[1]H spectral characteristics of diallyl monomers derivatives. Handbook of Condensed Phase Chemistry. Nova Science Publishers; **2011**, 293–301.

CHAPTER 7

RESEARCH PROGRESS ON CARBON NANOTUBE–POLYMER COMPOSITES

SHIMA MAGHSOODLOU and AREZOO AFZALI

University of Guilan, Rasht, Iran

CONTENTS

7.1   INTRODUCTION

Nowadays, many researchers have interested on the properties and applications of carbon nanotubes (CNTs) and related materials [1–5].

CNTs show, through experiments, many special properties such as high flexibility, low mass density, large aspect ratio (typically >1,000), extremely high moduli and strengths for these materials [6–7]. Therefore, various studies have been focused great attention to utilize these noteworthy characteristics for engineering applications such as polymeric composites, hydrogen storage [8], actuators [9], field emission materials [10], chemical sensors [11], and nano-electronic devices[12].

For instance, CNTs behave as metallic or semiconductors, depending upon their dimensions, electronic structure, and topological defects present on the tube surface. Individual single-walled carbon nanotubes (SWCNTs) can be metallic or semiconducting [1]. They can transport electrons over long lengths without considerable interruption. It makes them more conductive than copper [2]. This combination of mechanical and electrical properties of individual nanotubes makes them the ideal reinforcing agents in a number of applications. In 1994, Ajayan et al. introduced the first ever polymer nanocomposites using CNTs as fillers [3]. Since then, there have been many papers dedicated to processing and resulting mechanical and/or electrical properties of fabricated polymer nanocomposites. Although, CNTs are normally mixtures of various chiralities , diameters, and lengths, they have some defects like ignoring the presence of impurities. In addition to, CNT aggregation has been found to dramatically hamper theme chanical properties of fabricated nanocomposites. CNTs are normally curled and twisted due to their small size so some of individual CNTs, which are embedded in a polymer, only exhibit a fraction of their potential [3–5]. The splendid properties of CNTs cannot as yet be fully translated into high strength and stiffness finished products.

Due to the lack of solubility in either aqueous or non-aqueous solvents because of CNTs nature network, they often need to be prior functionalized for possible applications [6].

Surface functionalization of CNTs can be applied improve the compatibility of structure. A variety of methods based on covalent attachment and non-covalent interactions have been developed to achieve effective dispersions of CNTs. For obtaining the effective dispersions of CNTs, it can be utilized the covalent attachment and non-covalent interactions.

Many polymers have been successfully used for the coating of CNTs via these approaches and in general, presynthesized polymers react with the surface of pristine or oxidized CNTs [7–8]. Modification of CNTs with polymers not only improves dispersion properties of CNTs but also provide the combination of physical and chemical properties of CNTs and polymers. This combination of properties also makes CNTs ideal reinforcing agent for the polymer composite materials [9].The incorporation of only a few percentages of carbon nanotubes (CNTs) into many polymers has led to a significant improvement in mechanical and electrical properties [10].

Various polymer matrices are used for composites, including thermoplastics, thermosetting resins, liquid crystalline polymers, water-soluble polymers, conjugated polymer, among others [11].

The CNTs/polymer composites are commonly prepared by solution casting [12], melt processing [13], wet spinning [14–15] and electrospinning [15] are used to fabricate CNT-composite materials. To fully explore their reinforcing properties, a uniform dispersion, exfoliation, and orientation of CNTs are important [16]. Because they are present in the form of bundles and ropes due to Vander wall's forces. Several approaches are used in order to obtain a good dispersion and alignment of the CNTs, such as chemical functionalization [17], wrapping [18], and the synthesis of aligned nanotubes by the deposition of nanotubes onto chemically modified substrate [19]. The properties of polymer composites that can be improved due to presence of CNTs include tensile strength, tensile modulus, toughness, glass transition temperature, thermal conductivity, electrical conductivity, solvent resistance, optical properties , etc. [11].

7.2   FUNCTIONALIZATION OF CARBON NANOTUBES FOR POLYMER-COMPOSITE PREPARATION

The chemical functionalization of CNTs has been a subject of several studies. Here, we are going to focus only on recent developments of functionalization of CNTs for polymer composite formation. It is known that nanotube solubility, dispersion, and stress transfer must all be maximized to reach optimum mechanical properties. Unless the interface between nanotube and polymer is carefully engineered, poor load transfer between nanotubes, when in bundles, and between nanotubes and surrounding polymer chains may result in interfacial slippage [20–21]. Therefore, functionalization of nanotubes is extremely important for their processing and potential applications in polymer composites. In general, composites based on chemically modified nanotubes show the best mechanical results because functionalization enables a significant improvement in both dispersion and stress–strain transfer. The treatment of CNTs by chemical functionalization and/or ultrasonication is widely used to increase the dispersion of nanotubes in solvents. In particular, a better dispersion of nanotubes in solvents overcomes the problems of nanotube aggregation in polymer composites processed by the solution casting technique [20, 22–24].

7.2.1 DEFECT FUNCTIONALIZATION

CNTs are purified by oxidative methods to remove metal particles or amorphous carbon from the raw materials. In these methods, defects are preferentially observed at the open ends of CNTs. The purified SWNTs contain oxidized carbon atoms in the form of –COOH group. In this oxidizing method, SWNTs are broken to very short tubes (pipes) of lengths 100–300 nm [25–27]. However, the defective sites created at the CNT surfaces by this method are extremely sparse, and cannot promote good dispersion in the polymer/CNT composites [27–28]. However, they can be used for covalent attachment of organic groups by converting them into acid chlorides and subsequently reacting with amines to give amides [5].

The functionalized CNTs are more soluble in organic solvents than the raw CNTs. Most of the SWNT bundles exfoliate to give individual SWNT macromolecules, if the reaction time of acid chloride group with amines is at elevated temperature. These SWNTs are soluble in organic solvents [11].

7.2.2 COVALENT FUNCTIONALIZATION

Covalent functionalization of CNTs can be achieved by either direct addition reactions of reagents to the sidewalls of nanotubes or modification of appropriate surface-bound functional (e.g., carboxylic acid) groups on the nanotubes. An example of direct functionalization of nanotubes with phenol groups is shown in Figure 7.1. This functionalization provided stable dispersions of CNTs in a range of polar solvents, including water. An advantage of the phenolic functionalities is that they allow post-functionalization of the MWNTs with other molecules that can be employed in preparing customized products. In fact, the functionalized CNTs were found to be compatible with polymers or layered aluminosilicate clay minerals, giving homogeneous, coherent, transparent CNT thin films and/or gels [29].

Because of the π-orbitals of the sp^2-hybridized C atoms, CNTs are more reactive than those with a flat graphene sheet and they have an enhanced tendency to covalently attach with chemical species. In the case of covalent functionalization, the translational symmetry of CNTs is disrupted by changing sp^2 carbon atoms to sp^3 carbon atoms, and the properties of CNT, such as electronic and transport are influenced. But this functionalization of CNT can improve solubility as well as dispersion in solvents and polymer. Covalent functionalization can be accomplished by either modification of surface-bound carboxylic acid groups on the nanotubes or by direct reagents to the side walls of nanotubes (Table 7.1). Generally, functional groups such as –COOH or –OH are created on the CNTs during the oxidation by oxygen, air, concentrated sulfuric acid, nitric acid, aqueous hydrogen peroxide, and acid mixture [30–31]. The surface of the acid treated MWNTs indicates the presence of some defects in the carbon–carbon bonding associated with the formation of carboxylic acid groups on the surface, while the raw MWNTs show uniform surfaces and a clear diffraction pattern because of their perfect lattice structure of carbon–carbon bonds .The number of –COOH groups on the surface of CNT depends on acid treatment temperature and time, increasing with increasing temperature. The extent of the induced–COOH and –OH functionality also depends on the oxidation proce-dures and oxidizing agents. Nanotube ends can be opened during the oxidation process.

TABLE 7.1   Covalent functionalization of CNTs

Nanotube type	Preparation approach	Polymer/organic molecules	Catalyst/ reagent	References
MWNT	**Grafting from (ROP)**	**Poly(-capro-**	**Sn(Oct)2**	**[32]**
MWNT	**Grafting from (ROP)**	**lactone) Poly(L-lactide)**	**Sn(Oct)2**	**[33]**
SWNT	Grafting from (ROP)	Nylon-6	Sodium	[34-35]
MWNT				
MWNT	Grafting from (ATRP)	Poly(methyl meth-acrylate)	AIBN	[36]
MWNT	Grafting from (ATRP)	Polystyrene	CuBr	[37]
MWNT	Grafting from (ATRP)	Polystyrene	Cu(I)Br/ PMDETA	[38]
MWNT	Grafting from (ATRP)	Poly(acrylic acid)	Cu(I)Br/ PMDETA	[39]
MWNT	Grafting from (ATRP)	Poly(tert-butyl acrylate)	Cu(I)Br/ PMDETA	[40]
MWNT	Grafting from (ATRP)	Poly(N-isopro-pylacrylamide)	Cu(I)Br/ PMDETA	[41]
MWNT	Grafting from (ATRP)	Glycerol Mono-methacrylate	Cu(I)Br/ PMDETA	[42]
SWNT	Grafting to	Polyethylene glycol	–	[43]
MWNT				
MWNT	Grafting to	Polyimide	–	[44]
SWNT	Grafting to	Poly(amido amine)	–	[43]
MWNT	Grafting to	Poly (-caprolactone)-diol	–	[45]

TABLE 7.1 *(Continued)*

Nanotube type	Preparation approach	Polymer/organic molecules	Catalyst/ reagent	References
SWNT	Grafting to	Poly(vinyl acetate-co-vinyl alcohol)	–	[46]
MWNT				
MWNT	Grafting to	Poly(2-vinyl pyridine)	–	[47]
SWNT	Cycloaddition of azomethine ylides	3,4-dihydroxy-benzaldehyde	N-methyl-glycine	[29]
MWNT	Cycloaddition of azomethine ylides	7-bromo-9,9-dioctyl fluorine-2-carbaldehyde	L-lysine	[48]
SWNT	Cycloaddition of azomethine ylides	Amino-acid	Parafor-maldehyde	[49]
MWNT				
SWNT	Cycloaddition of azomethine ylides	Peptides, Nucleic acids	$R-NHCH_2$ COOH	[50]
MWNT	[4+2] Cycloadditions	3,6-diaminotetra-zine	Temp.	[51]
SWNT	[4+2] Cycloadditions	Triazolinedione	Temp.	[52]
SWNT	[4+2] Cycloadditions	2,3-dimethoxy-1,3-butadiene	$Cr(CO)_6$	[53]
SWNT	[2+1] Cycloadditions	Alkyl azidoformate	Nitrene	[17]
		Dipyridyl imidaz-olidene	Carbene	
SWNT	[2+1] Cycloadditions	Dichlorocarbene	Carbene	[54]
MWNT				
SWNT	[2+1] Cycloadditions	PEG di-azidocar-bonate ester	Nitrene	[55]
SWNT	Radical additions	Polystyrene	Nitroxide	[56]
SWNT	Radical additions	Methoxyphenylhy-drazine	Microwave	[57]

The most common and facile method for the surface functionalization of CNTs is nanotube oxidation, which results in the formation of a number of carboxylic acid groups (COOH) on the surface of the nanotubes. These functionalized nanotubes are much more stable in polar solvents [27, 31]. As a result, the water-stable nanotubes can be easily embedded into water soluble polymers such as a poly (vinyl alcohol) (PVA), giving polymer–CNT composites with homogeneous nanotube dispersion. Oxidized nanotubes also show excellent stability in other solvents such as caprolactam, which is used in the production of polyamide (PA6) [42]. It has been shown that acid functionalization significantly improves the interfacial bonding properties between the CNTs and a polymer matrix. The carboxylic functional groups have been shown to give a stronger nanotube–polymer interaction, leading to enhanced values in Young's modulus and mechanical strength [58–60].

Acid-treated nanotubes can also be used to electrochemically deposit metal (e.g., gold) nanoparticles onto CNT surfaces [61–62].

A completely different strategy for the surface functionalization of CNTs with nitrogen-containing groups is the treatment of CNTs under an atomic-nitrogen flow obtained by molecular nitrogen dissociation in microwave plasma. X-ray photoelectron spectroscopy of the nanotube surface demonstrated the presence of amides, oxides and, mainly, amine and nitrile groups [63].

The presence of active functional groups such as carboxylic acids or amines allows for further covalent functionalization with polymer molecules (polymer grafting). Two main approaches for the covalent functionalization of CNTs with polymers have been reported: "grafting from" and "grafting to"[64].

The grafting-from approach is based on the initial immobilization of initiators onto the nanotube surface, followed by in situ polymerization with the formation of the polymer molecules bound to the nanotube. The benefit of this technique is that polymer-functionalized nanotubes with high grafting density can be prepared. However, this method requires a strict control of the amounts of initiator and substrate [53, 54]. The grafting- from technique is also used for the preparation of styryl-grafted nanotubes [55]. In this work, the carboxylic acid groups on the surface of oxidized MWNTs have been reacted with 4-vinylbenzyl chlorides via an esterification reaction followed by the polymerization to produce polystyrene (PS)-grafted CNTs [34, 36–38].

The grafting-to approach is based on the attachment of already preformed end-functionalized polymer molecules to functional groups on the nanotube surface via appropriate chemical reactions. An advantage of this method is that commercially available polymers containing reactive functional groups can be utilized. However, the main limitation of grafting-to technique is that initial binding of polymer chains sterically prevents diffusion of additional macromolecules to the surface leading to relatively low polymer loading [56]. The functionalization make the composites with improved dispersion of SWNTs and enhanced mechanical properties [22].

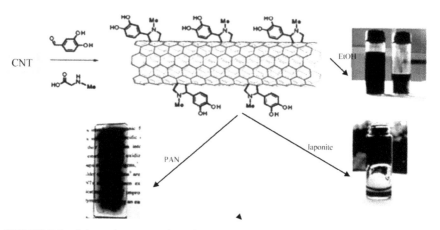

FIGURE 7.1   Schematic presentation of covalent functionalization of a CNTs.

7.2.3 *NON-COVALENT FUNCTIONALIZATION*

Non-covalent functionalization of nanotubes is of particular interest because it does not compromise the physical properties of CNTs, but improves the solubility and the process ability. This type of functionalization mainly involves surfactants, biomacromolecules or wrapping with polymers (Table 7.2). In the search for non-destructive purification methods, nanotubes can be transferred to the aqueous phase in the presence of surfactants .In this case, the nanotubes are surrounded by the hydrophobic components of the corresponding micelles. The interaction becomes stronger when the hydrophobic part of the amphiphilic contains an aromatic group [65–66].

The non-covalent functionalization of nanotubes normally involves van der Waals, p–p, CH–p or electrostatic interactions between polymer molecules and CNT surface. The advantage of non-covalent functionalization is that it does not alter the structure of the nanotubes and, therefore, both electrical and mechanical properties should also remain unchanged. However, the efficiency of the load transfer might decrease since the forces between the wrapping molecules and the nanotube surface might be relatively weak. There are several non-covalent approaches for nanotube functionalization such as surfactant-assisted dispersion, polymer wrapping, plasma polymerization-treatment, and polymerization-filling technique (PFT) [67].

TABLE 7.2 Non-covalent functionalization of CNTs

Polymer or surfactant	Nanotube type	Preparation method	References
Surfactants			
Cationic (CTAB) Cationic (CTAB) Cationic (CTVB) Cationic (MATMAC) Anionic (SDS) Anionic (SDBS) Anionic (SDBS) Non-ionic (Triton X-100) Non-ionic (Triton X-305)	SWNT SWNT SWNT MWNT MWNT SWNT SWNT MWNTSWNT MWNT	Microemulsion Ultra-sonication Sonication Emulsion polymeriza-tion Ultrasonication Ultrasonication Bath sonicaton Ultra-sonication Ultrasoni-cation	[68] [69] [70] [71-72] [66] [73-74] [75] [76-77] [78-79]
Biomacro-molecules			
Proteins/DNA Glucose (Dex-tran) β-1,3-glucans Chitosan	MWNT SWNT SWNT SWNT	Immobilization Dialysis Electroactive interac-tion Ultrasonication	[79] [80] [81] [82]
Polymers			
Poly(4-vinyl pyridine) Poly(phenyl acetylene)	SWNT MWNT	Sol–gel chemistry Solution mixing	[83] [84]
Poly(m-phenylenevinylene-co-2,5-dioctoxy-p-phenylenev-inylene)	MWNT SWNT	Solution mixing	[85] [86]
Poly(styrene)-poly(methacrylic acid)	SWNT MWNT	Solution mixing	[87]

Surfactant-assisted dispersion is a very common technique that enables to transfer nanotubes to the aqueous phase in the presence of surface-active molecules such as sodium dodecyl- sulfate (SDS) or benzylalkonium chloride. The presence of an aromatic group in the surfactant molecule allows for p–p stacking interactions with the graphitic sidewalls of the nanotubes, which results in their effective coating and dispersion. Polymer wrapping involves the utilization of conjugated and aromatic group containing polymers [e.g., polyvinyl pyrrolidone

(PVP), poly (phenylenevinylene), pyrene–poly(ethylene glycol) (PEG)], which can wrap around CNTs through p–p stacking and van der Waals interactions [88]. Coleman and Ferreira showed that polymer wrapping can minimize energy for purely geometric reasons [89]. Poly(dimethylsiloxane) (PDMS) polymers have also been used for CNT coating via non-covalent CH–p interactions [90].

Recently, Kevlar-functionalized nanotubes were prepared by heating Kevlar with MWNTs in the presence of sulfuric and nitric acids under reflux. This resulted in the partial oxidation and functionalization of CNTs with carboxylic acid groups, which formed hydrogen bonds with amido-groups, as well as terminal carboxylic acid and amino groups in Kevlar. This process produced Kevlar-coated CNTs, which have been utilized for the fabrication of MWNT–polyvinylchloride (PVC) composites. Plasma polymerization treatment enables the coating of CNTs with a very thin (3 nm) polymer layer. Polymer composites based on these coated nanotubes enhanced interfacial bonding in a PS polymer matrix [22, 91].

PFT usually involves in situ copolymerization of olefins catalyzed directly from nanotubes pretreated by a methylalumi- noxane (MAO) or highly active metallocene-based complexes (e.g., Cp_2ZrCl_2). This approach destroys nanotube bundles and results in homogeneously coated CNTs [22].

7.3 PREPARATION METHODS OF POLYMER/CNT NANOCOMPOSITES

As emphasized in the preceding, the dispersion of CNTs in polymer matrices is a critical issue in the preparation of CNT/polymer composites. Better reinforcing effects of CNTs in polymer composites will be achieved if they do not form aggregates and as such, they must be well dispersed in polymer matrixes.

Common traditional approaches for the fabrication of CNT–polymer composites include: solution processing of composites, melt spinning, melt processing, in situ polymerization, processing of composites based on thermosets, electrospinning and coagulation spinning for composite fibers and yarn production [92].

Solution casting processing of composites is one of the most usual methods for making polymer–nanotube composites on a laboratory scale. The nanotubes and polymer are mixed in a suitable solvent before evaporating the solvent to form a composite film. One of the advantages of this method is the possibility to achieve debundling and good-quality dispersion of the nanotubes in an appropriate solvent. However, solution processing techniques cannot be utilized for insoluble polymers [93]. In this case, melt processing is a good alternative technique, which is particularly useful for dealing with thermoplastic polymers

[94]. Furthermore, melt processing is the most promising approach for the production of polymer–MWNT nanocomposites on industrial scale. Normally, melt processing involves mixing of CNTs with the molten polymer by shear mixing. Bulk composites can then be prepared by compression molding, injection molding, or extrusion. Advantages of melt processing are its speed and simplicity and easy integration into standard industrial facilities (e.g., extruders, blow-molding machines). Although under high temperatures, this approach can sometimes result in unexpected polymer degradation and oxidation. Melt processing can also be used for the production of both bulk-polymer composites, composite fibers, and yarns [95].

Surface-coated CNTs produced by in situ polymerization, whether by covalent or non-covalent methods, enable the production of polymer nanocomposites displaying much-better thermomechanical, flame-retardant, and electrical conductive properties even at very-low nanotube loadings. As mentioned above, the covalent approach allows for the formation of a strong interface between the nanotube and polymer matrix due to strong chemical bonding of polymer molecules to the CNT surface. The in situ polymerization technique also enables the preparation of composites with very-high nanotube loadings. A closely related method of processing of composites is based on epoxy resins thermosets. In this approach, CNTs can be dispersed in a liquid epoxy precursor and, then, the mixtures can be cured by the addition of a hardener, such as triethylene tetramine (TETA), and the application of temperature or pressure [96].

A study of SWNT–polyethylene (PE) composites, prepared with several different methods, found that polymerization was generally preferred as the method to disperse the nanotubes. Widely used industrial approaches for the production of polymer fibers and yarns, such as coagulation spinning and electrospinning, have also been utilized for the fabrication of polymer–nanotube composites. In coagulation spinning, for example, composite fibers can be produced by an injection of a surfactant-stabilized nanotube dispersion in water into a rotating bath of polymer (e.g., PVA) dissolved in water, such that nanotube and polymer dispersions flowed in the same direction at the point of injection. In this case, polymer molecules replace surfactant molecules on the nanotube surface, thus, destabilizing the nanotubes dispersion, which collapses to form a fiber. These fibers can then be retrieved from the bath, rinsed, and dried [97].

The electrospinning technique involves electrostatically driving a jet of polymer and nanotube dispersions in an appropriate solvent out of a nozzle onto a metallic counter electrode. When the power supply is turned on, the composite solution becomes charged. This forces the solution out of the nozzle and toward the counter electrode. Charging of the solvent causes rapid evaporation, resulting in the coalescence of the composite into fibers with diameters between 10 nm and 1 mm. Yarns can also be produced by collecting the fibers on a rotating drum and twisting them [22].

The superb mechanical and physical properties of individual CNTs provide the impetus for researchers in developing high performance continuous fibers based upon carbon nanotubes. Unlike in the case of carbon fibers, the processing of CNT fibers does not require the cross-linking step of the precursor structures [98].

7.4 NEW APPROACHES FOR PREPARATION OF POLYMER–COMPOSITE FILMS AND FIBERS

7.4.1 LAYER-BY-LAYER TECHNIQUE

The layer-by-layer (LBL) approach involves building up a layered composite film by alternate dipping of a substrate into dispersions of CNTs and polyelectrolyte solutions. Additionally, to improve the structural integrity of the film, cross- linking can be induced. LBL assembly is a simple, versatile and relatively inexpensive approach, which provides multifunctional molecular assemblies of tailored architectures and material properties for various versatile reaction/sensing materials of nanometer thickness and will enable large-scale, reproducible production of membrane-based, highly integrated microsensors [99–100].

This method has significant advantages as thickness and polymer–nanotube ratio can easily be controlled and very-high nanotube loading levels can be obtained. Therefore, this has led to recent exceptional growth in the use of LBL-generated nanocomposites. This technique has been used extensively to incorporate inorganic nanoparticles, nanowires, and nanotubes into organic polymers (Figure 7.2) [101–102].

Recently, highly conductive, smart electronic yarns and textiles have been prepared by LBL coatings of cotton threads with SWNTs and appropriate polyelectrolytes. These new materials demonstrated high chemical/mechanical durability, electrical conductivity, weave ability, and wear ability, as well as interesting chemo- and bio-sensing opportunities. The high strength and conductivities of these composites are attributed to the unique homogeneity of the LBL-assembled composites [102].

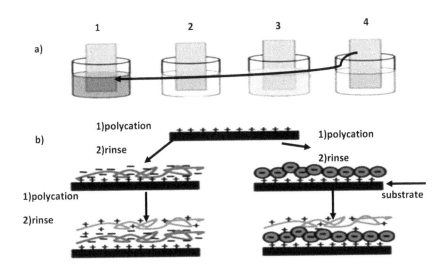

FIGURE 7.2 Principal scheme of the LBL film-deposition. (a) Steps 1 and 3 represent the adsorption steps 2 and 4 are washing steps. (b) Two adsorption routes, depicting LBL deposition for polymers and polymers with nanoparticles or nanotubes

7.4.2 SWELLING TECHNIQUE

Recently, high-strength, high-toughness Kevlar-nanotube- composite fibers have been produced by the swelling of commercially available Kevlar fibers in a suspension of MWNTs in N-methylpyrrolidone (NMP) under ultrasonication. This process allowed the nanotubes to diffuse into the interior of the fiber. The resulting composites were stronger and tougher than the original Kevlar fibers at only 1–2 wt% of nanotube loadings. There is also another very recent report on new conductive CNT–PE composites, which have been prepared similarly by swelling a thin PE film in MWNT dispersions in tetrahydrofuran (THF) under ultrasonication. This new approach is a very promising post-processing technique, which allows us to incorporate CNTs into already form polymer products, including insoluble or temperature-sensitive polymers, such as Kevlar. In addition, this technology enables the inclusion of nanotubes into a very thin (several hundred nm) top polymer layer. Therefore, only a very small percentage (1–2%) of nanotubes is needed to produce polymer composites with potentially high electrical surface conductivity and improved mechanical parameters [103–104].

7.4.3   BUCKYPAPER-BASED APPROACH

Buckypaper is a thin porous assembly of CNTs, usually formed by filtration from their dispersion in a solvent. Over the last few years, the electrical and mechanical properties of buckypaper have been studied extensively. In order to prepare buckypaper, CNTs must be dispersed in an appropriate solvent. One of the solvents used to prepare buckypaper is NMP, which enables to fabricate good quality buckypaper without the need for centrifugation [95, 105–106].

However, using an appropriate surfactant in water can be cheaper than using DMF and NMP, which also exhibit the disadvantage of high boiling points. The most common surfactants are sodium dodecyl benzene sulfonate (SDBS) and sodium dodecyl sulphate (SDS). Aqueous dispersion of SWNTs in the presence of the water-soluble perylene derivatives has also been reported. After the nanotubes are dispersed in a solvent, the buckypaper can usually be fabricated by a simple Buchner filtration[22].

Another way to produce buckypaper without the dispersion and filtration method is the "domino pushing effect" (Figure 7.3). Ding et al. present a simple and effective macroscopic manipulation of aligned CNT arrays.

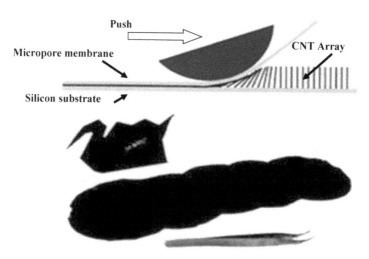

FIGURE 7.3   Domino pushing of a CNT array to produce high-quality buckypaper.

The domino pushing of the CNT arrays can efficiently ensure that most of the CNTs are well aligned tightly in the buckypaper. Initial measurements indicate that these buckypapers have better performance in thermal and electrical conductance. Overall, there is an enormous interest in developing new buckypaper-based composite materials with improved mechanical properties and conductivity for a number of potential applications from hydrogen absorption to fire retardation. Buckypaper, normally, has a laminar structure with a random orientation of the bundles of tubes in the plane of the sheet. It is therefore a porous material and, thus, liquids and gases can permeate through it [107–108].

However, the flexibility of the individual nanotubes and their affinity for each other conspire to obstruct the porosity in such films limiting the perfusion rate of liquids and gases, restricting the accessible surface area, and, thereby, limiting their utility in important applications. These pores are characterized by a diameter distribution that varies from one to well over a hundred nanometers [109].

Infusing the porous buckypaper with polymer is a facile way of improving the buckypaper mechanical properties and create polymer composites with high loadings of nanotubes (>60%). To increase the mechanical properties, sheets of buckypaper can also be inserted between laminates. Polymers or epoxies have been layered on top of buckypaper to develop a ply material that has electrochemical actuation properties [110–112].

In addition, nanotubes may also be chemically cross-linked to create stronger buckypaper. For example, using the nitrene reaction, functionalized nanotubes can be linked within a bundle as well as between bundles. By filtering and drying, a buckypaper of linked nanotubes can be obtained. This linked buckypaper has the potential to have better mechanical properties than pristine buckypaper [55, 112].

7.5   PROCESSING CNT-POLYMER COMPOSITE FILMS

CNT-polymer composites are commonly produced in the form of films for mechanical property analysis. The thicknesses of these films range from a few millimeters to electron-transparent TEM samples. Once a suitable level of CNT dispersion is achieved in the matrix, films are easily processed by pressing CNT-polymer melts, casting CNT- polymer solutions/thermosets into molds and drying/curing, or spin casting from CNT-polymer solutions. Test specimens of precise dimensions can be cut or machined from these composite films for mechanical testing. Pressing CNT-thermoplastic films is well described in the literature and is a simple method of producing composite specimens [5, 113–114].

Polymer containing dispersed CNTs is typically crushed or pelletized, placed between heated platens, and pressed at a pressure in the range of 10–30 MPa. such that molten CNT composite polymer flows outward constrained to some predetermined thickness by shims or a mold. Upon solidification, test specimens are then cut from the film. For CNT-polymer solutions and CNT-thermoset matrices, composite films are typically processed utilizing a casting technique. Well-dispersed CNT- polymer fluids are transferred to molds and heated to evaporate the solvent or cure the thermoset [115–116] .In a similar method, dispersed CNTs in PVA-water solutions are allowed to settle followed by decanting the PVA solution containing well dispersed CNTs. These suspensions are then deposited drop wise onto substrates, and the solvent is evaporated. This process is repeated to build up to the desired thickness of film [89, 117].

Very thin CNT-polymer composite films (thickness ¼ 200 nm) can be produced by means of spin casting in which suspensions of CNTs in low viscosity polymer solutions are deposited drop wise to the center of a rotating substrate [118]. Spinning for 20–30 sec at 3,000 rpm has been reported to produce electron-transparent films. The centrifugal forces induce radial flow of the drop resulting in thin films with some preferred orientation of the embedded CNTs along the direction of flow. Fabrication of continuous CNT-polymer composite films has been demonstrated by extrusion of CNT composite thermoplastic through a slit die followed by take-up onto a chilled roller [119–120].

Other techniques include resin infiltration into dry CNT preforms [121–122], polymer intercalation of aligned MWNT mats [123], SWNT "buckypaper" [124–125] and complete in situ polymerization of thermoplastic matrices [126–127].

The orientation of the embedded CNTs relative to the loading conditions, as in all fiber composites, has an effect on the properties that are measured. The process of shear- mixing or sonication results in random orientation of the CNTs the polymer matrix, which can be preserved throughout processing given the absence of strong unidirectional shear or elongation flow. For most CNT- polymer film processing, the CNT orientations are randomly arranged relative to the principal axes of the film. Exceptions occur when any degree of stretching or drawing of the film is implemented, which forces embedded CNTs into alignment with the direction of flow [128–130]. This is observed for CNT-polymer melt extrusions through slit dies [131–132] , and injection moldings of carbon nanofiber polymer composites [133-134]. Achieving the highest degrees of CNT alignment within a polymer matrix is achieved by spinning dispersions of CNTs in polymer fluids into fibers.

7.6 PROCESSING CNT-POLYMER COMPOSITE FIBERS

Homogeneous dispersions of CNTs in viscoelastic polymer fluids can be spun into composite fibers. Melt spun, solution spun, and electro-spun CNT composite fibers have been produced. These methods have been shown to be very effective in aligning the CNTs with the direction of flow the fiber axis. This class of oriented CNT composites is of particular interest since direct axial loading of the CNTs is possible. It is in this fashion that high strength and light-weight CNT composites are envisioned. However, the ability to sustain stable CNT-polymer composite fiber spinning is largely dependent on the homogeneity of the dispersion. CNT agglomerates can disrupt the flow as fiber attenuation is applied leading to breakage during processing, or in the case of large agglomerates, complete blockage of the small diameter orifices through which the fibers are extruded. Achieving large attenuations or draw ratios, which serve to orient both the CNTs and the polymer chains with the fiber axis resulting in increased fiber properties, is primarily dependent on the quality of the initial CNT dispersion [22, 92–93, 98, 130–131, 135].

In a typical process, the CNTs are dispersed in the molten polymer using a high shear mixer followed by extrusion through a cylindrical die orifice or spinneret. Before the composite fiber cools and solidifies, attenuation is accomplished by continuous collection on a rotating drum. Mechanical testing of the fibers can then be performed using a variety of techniques including single filament tensile testing. CNT composite fibers spun from polymer solutions involve a coagulation process to solidify the fibers in which the solvent is replaced by a second miscible solvent, often water, which acts as a non-solvent to the polymer causing it to solidify. Complete removal of residual solvent in the fiber is done via a drying process. Electrospun fibers are produced by generating a high voltage between a negatively charged spinning solution and a conductive collector. The advantage of this technique is the production of ultrafine fibers (<100 nm) as very fine jets of polymer are accelerated toward the collector, but unlike melt or solution spinning, the production of continuous filament is difficult. Fibers have been electrospun from dispersions of CNTs in poly(vinly alcohol) [136], poly (acrylonitrile) [137], and poly(vinylidene fluoride) [138].

7.7 GEL-SPINNING OF CNT/POLYMER FIBERS

CNTs can act as a nucleation agent for polymer crystallization and as a template for polymer orientation. SWCNTs with their small diameter and long length can act as ideal nucleating agents. Furthermore, continuous carbon fibers with perfect structure, low density, and tensile strength close to the theoretical value may be feasible if processing conditions can be developed such that CNT ends, catalyst particles, voids, and entanglements are eliminated. Such a CNT fiber could have ten times the specific strength of the strongest commercial fiber available today [139]. In the current manufacture process of carbon fibers, polymer solution is extruded directly into a coagulation bath. However, higher strength and modulus PAN and PAN/SWCNT based fibers can be made through gel-spinning.

In gel spinning, the fiber coming out of the spinneret typically goes into a cold medium where it forms a gel (a process similar to "gello" formation). These gel fibers can be drawn to very high draw ratios. Gel-spun fibers in the gel bath are mostly unoriented and they are drawn anywhere from 10 to 50 times after gelation. Structure of these fibers is formed during this drawing process [140].

Small-diameter fibers possess high strength and gel-spinning results in high draw ratio and consequently high orientation and modulus.

7.8   ELECTROSPINNING OF CNT/POLYMER FIBERS

Electrospinning is an electrostatic induced self-assembly process, which has been developed for decades, and a variety of polymeric materials have been electrospun into ultra-fine filaments [135, 141]. Electrospinning of CNT/polymer fibrils is motivated by the idea to align the CNTs in a polymer matrix and produce CNT/polymer nanocomposites in a continuous manner. The alignment of CNTs enhances the axial mechanical and physical properties of the filaments. The concept of electrospinning process can be explained using Figure 7.4. Here, a high voltage is applied between an oppositely charged polymer fluid and a metallic collection screen. The fluid is contained in a glass syringe which has a capillary tip (spinneret). When the voltage reaches a critical value, the electric field overcomes the surface tension of the suspended polymer and a jet of ultra-fine fibers is produced. As the solvent evaporates, a mesh of nano to micro size fibers is accumulated on the collection screen. The fiber diameter and mesh thickness can be controlled through the variation of the electric field strength, polymer solution concentration and the duration of electrospinning. Aligned fiber collection can be made through suitable arrangement of parallel electrodes.

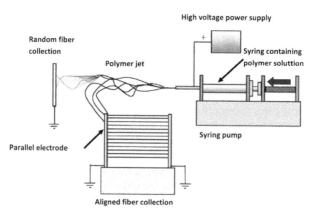

FIGURE 7.4   Schematic setup of the electrospinning of random and aligned fiber assemblies.

7.9   SOLUTION MIXING

In this approach, a dispersion of CNTs in a suitable solvent and polymers are mixed in solution. The CNT/polymer composite is formed by precipitation or by evaporation of the solvent. It is well known that it is very difficult to properly disperse pristine CNTs in a solvent by simple stirring. A high power ultrasonication process is more effective in forming a dispersion of CNTs. Ultrasonic irradiation has been extensively used in dispersion, emulsifying, crushing, and activating the particles. By taking advantage of the multi-effects of ultrasound, the aggregates and entanglements of CNTs can be effectively broken down. In this case, ultrasonic wave as well as mechanically stirring play important roles in the formation of the composites with a uniform particle size [142–143]. The chemical effects of ultrasound are associated with the rapid (microsecond time scale), violent collapse of cavitation bubbles created as the ultrasonic waves pass through a liquid medium [144]. Sonochemical theory and the corresponding studies suggested that ultrasonic cavitation can generate a high local temperature of 5000K and a local pressure of 500 atm, which is a very rigorous environment [132]. Uniform dispersions of CNTs in PS can be achieved by using sonication [145].

7.10   MELT MIXING

For solution mixing, the matrix polymer must be soluble in at least one solvent. This is problematic for many polymers. Melt mixing is a common and simple method particularly useful for thermoplastic polymers [146–147]. In melt processing, CNTs are mechanically dispersed into a polymer matrix using a high temperature and high shear force mixer or compounder [9, 148–150]. This approach is simple and compatible with current industrial practices. The shear forces help to break nanotube aggregates or prevent their formation. A homogeneous dispersion of MWNTs achieved through the matrix polymer, associated with significant enhancements in mechanical properties [13]. The disadvantage of this method is that the dispersion of CNTs in a polymer matrix is quite poor compared to the dispersion that may be achieved through solution mixing. In addition, the CNTs must be lower due to the high viscosities of the composites at higher loading of CNTs.

7.11   IN-SITU POLYMERIZATION

In this polymerization method, the CNTs are dispersed in monomer followed by polymerization. A higher percentage of CNTs may be easily dispersed in this method, and form a strong interaction with the polymer matrixes. This method is useful for the preparation of composites with polymers that cannot be processed by solution or melt mixing, e.g., insoluble and thermally unstable polymers. Conducting polymers are attached to CNTs surfaces by in-situ polymerization to improve the process ability, and electrical, magnetic and optical properties of CNTs [151–153]. A novel concept has been proposed to prepare PU-MWNTs composite via in-situ polymerization of a prepolymer in the presence of carboxylated MWNTs [154]. The in-situ polymerization of caprolactam in the presence of SWNTs allowed the continuous spinning of SWNTs-PA6 fibers [42].

7.12   APPLICATIONS

As developed in the preceding sections, because of their excellent mechanical, electrical, and magnetic properties, as well as nanometer scale diameter and high aspect ratio, CNTs can be very useful materials in composites to improve a particular property for specific applications (Table 17.3). The addition of CNTs to conjugated polymers can be improved the quantum efficiency of conjugated polymers because the interaction between the highly delocalized electrons of CNTs and the electrons correlated with the lattice of the polymer skeleton [46, 155]. Such composites are widely used in photovoltaic devices [61, 155–158] and light-emitting diodes [159]. CNT-conducting polymer composites have a potential application in supercapacitors [160–162].

TABLE 7.3   Application of CNT-polymer composites

Nanotube type	Polymer type	Applications	References
SWNT	Poly(3-octylthiophene)	Photovoltaic devices	[157]
MWNT; SWNT	Polyaniline, polypyrrole, poly-(3,4-ethylenedioxythiophene), poly(3-methyl-thiophene	Supercapacitors	[87]
SWNT	Nafion	Actuators	[163]
MWNT	poly(vinyl alcohol), poly(2-acrylamido-2-methyl-1-pro-panesulfonic acid)		[164]
MWNT	Nafion	Fuel cell	[165]
MWNT-functional	Sulfonated poly(arylene sulfone),		[163]
SWNT; MWNT	Polypyrrole	Biosensors	[166]
SWNT	Poly(methyl methacrylate)	Biocatalytic films	[167]
SWNT-functional; MWNT-functional	DNA(polynucleotide)	Gene delivery	[167]

CNTs are also widely used in actuators [168–171]. The addition of CNTs to PANI fibers increased the electromechanical actuation because the CNTs improved the mechanical, electronic and electrochemical properties of the PANI fibers [172–174]. Composites based on CNTs are studied for a variety of sensor applications [175–177]. For example, polypyrrole or PANI deposited on single-walled CNT networks that can be used as solid state pH sensors [178]. A DNA sensor was created from a composite of polypyrrole and CNTs functionalized with carboxylic groups to covalently immobilize DNA onto CNTs [179]. In general, the presence of CNTs tends to increase the overall and selectivity of biosensors. The thermal transport properties of polymer composites can be improved with the addition of CNTs due to the excellent thermal conductivity of CNTs. Such composite are quite attractive for usages as printed circuit boards, connectors, thermal interface materials, heat sinks, lids, housings, etc. [92, 180].

The superior properties of CNTs are not limited to electrical and thermal conductivities, but also include mechanical properties, such as stiffness, toughness, and strength. CNTs with their high aspect ratio and excellent mechanical properties have the potential to strengthen and toughen hydroxyapatite without offsetting its bioactivity, thus opening up a wider range of clinical applications[181–182] . Bianco et al. studied the application of CNTs as new vectors for the delivery of therapeutic molecules [181]. CNTs have been shown to cross cell membranes easily and to deliver peptides, proteins [183], and nucleic acids into cells [50, 181, 183–186].

CNTs were employed to reinforce the interfaces between ultrahigh molecular weight PE polymer particles, enhancing composite strength, stiffness, impact toughness as well as structural damping [187]. These composites are attractive for applications in aerospace and naval engineering. The high strength and toughness-to-weight characteristics of CNTs may also prove valuable as part of composite components in fuel cells that are deployed in transport applications, where durability is extremely important [188–189].

7.13   TEXTILE ASSEMBLIES OF CNTS

The challenges in making nanotube fibers/yarns with desirable properties are in achieving the maximum possible alignment of the nanotubes or their bundles within the yarn, increasing the nanotube packing density within the yarn and enhancing the internal bonding among the nanotubes [190].

It is interesting to note that a plied nanotube yarn has several levels of hierarchical structures. First, the CVD synthesized MWCNTs are 300 lm long and 10 nm in diameter and they formed about 20 nm diameter bundles in a nanotube forest. Simultaneous draw and twist of the bundles produced 10 lm diameter single yarn [191].

It is reported that no visible damage to the nanotube yarns is imparted by the braiding process and the 3-D braids are very fine, extremely flexible, hold sufficient load, and are well suited for the use in any other textile formation process, or directly as reinforcement for composites. The reported elastic and strength properties of carbon nanotube composites so far are rather low in comparison with conventional continuous carbon fiber composites. It is believed that the properties can be substantially improved if the processing methods and structures are optimized [191].

The electrical conductivity of CNT yarns, 3-D braids and their composites can be investigated too. It is noted that 3-D textile composites, including 3-D woven and 3-D braided materials, combine high in-plane mechanical properties with substantially improved transverse strength, damage tolerance and impact resistance. However, even relatively small volume content of the out-of-plane fibers results in considerable increase of interstitial resin pockets, which contributes to lower in-plane properties. Utilizing fine CNT yarns could dramatically reduce the through- the-thickness yarn size while still sufficiently improving the composite transverse properties. The electrical conductivity of 3-D hybrid composites are many times greater than that of commonly produced nanocomposites made from low volume fraction dispersion of relatively short CNTs in epoxy. Figure 7.5 gives schematics of 3-D woven and braided hybrids with very fine CNT yarns for multi- functional composite reinforcements [192].

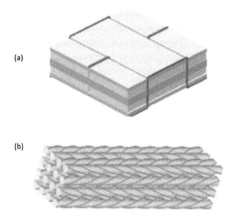

FIGURE 7.5 (a) Schematic of 3-D woven fabric incorporating very fine CNT yarn as through thickness (Z) reinforcement; (b) Schematic of square 3-D braided hybrid preform with large axial fiber bundles and very fine CNT off-axis braided yarns

7.14   CONCLUDING REMARKS

In conclusion, significant progress was achieved in the area of nanotube–poly-mer composites over last few years. A range of new composites have demon-strated astonishing mechanical parameters and conductivity values. Obviously, it is impossible to make a comprehensive overview of all aspects of this large subject in the framework of one topic, so we confine ourselves to discussing the most characteristic and important recent examples, where the homogeneous dis-persion of CNTs within polymer matrices plays a crucial role in the fabrication of multifunctional composites. More detailed information is available in topi-cal reviews devoted to particular issues. Progress in polymer/carbon nanotube composite research considered here will included studies on functionalization of CNTs, and applications of polymer/CNT composites. There are several methods for the dispersion of nanotubes in the polymer matrix such as solution mix-ing, melt mixing, electrospinning, in-situ polymerization and chemical func-tionalization of the carbon nanotubes, etc. In addition, we examine the fibers composed of pure CNTs or CNTs embedded in a polymeric matrix produced by various techniques and textile assemblies of CNTs. The critical challenge is the development of methods to improve the dispersion of CNTs in a polymer matrix because their enhanced dispersion in polymer matrices greatly improves the mechanical, electrical and optical properties of composites. Despite various methods, such as melt processing, solution processing, in situ polymerization, and chemical functionalization, there are still opportunities and challenges to be found in order to improve dispersion and modify interfacial properties. One of challenges is to achieve the optimal functionalization of CNTs, which can maxi-mize interfacial adhesion between CNTs and the polymer matrix. In conclusion, the CNT functionalization and matrix polymer design for dispersion of CNTs and interfacial adhesion between CNTs and a polymer matrix are the key chal-lenges for development of high performance CNT composites.

KEYWORDS

- **Carbon nanotube**
- **Carbon nanotube-polymer composites**
- **Functionalized Carbon nanotube-polymer composites**

REFERENCES

1. Li, Y., et al.;Preferential growth of semiconducting single-walled carbon nanotubes by a plasma enhanced CVD method.*Nano.Lett.***2004**,*4*(2), 317–321.

2. Dürkop, T.;Cobas, E.; and Fuhrer, M.;High-mobility Semiconducting Nanotubes. In AIP Conference Proceedings;**2003**.

3. Ajayan, P., et al;Aligned carbon nanotube arrays formed by cutting a polymer resin—nanotube composite.*Science*.**1994**,*265*(5176), 1212–1214.

4. Curran, S., et al.;Evolution and evaluation of the polymer/nanotube composite.*Synthet.Metal.*1999, *103*(1), 2559–2562.

5. Jeon, I.-Y.; et al.;Functionalization of Carbon Nanotubes. Carbon Nanotubes-Polymer Nanocomposites RijekaInTech Available from http://www. intechopen. com/books/carbon-nanotubespolymer-nanocomposites/functionalization-of-carbon-nanotubes (accessed 25 May 2012), 2011: p. 91-110.

6. Thess, A., et al.;Crystalline Ropes of Metallic Carbon Nanotubes. Science-AAAS-Weekly Paper Edition, **1996**, **273**(5274), 483–487.

7. Paiva, M., et al.;Mechanical and morphological characterization of polymer–carbon nanocomposites from functionalized carbon nanotubes.*Carbon*.**2004**, *42*(14), 2849–2854.

8. Sano, M., Kamino, A.;and Shinkai, S.;Construction of carbon nanotube "stars" with dendrimers.*AngewandteChemie*.**2001**, *113*(24), 4797–4799.

9. Spitalsky, Z., et al.;Carbon nanotube–polymer composites: chemistry, processing, mechanical and electrical properties.Progress in Polymer Science, **2010**.*35*(3), 357–401.

10. Saeed, K. and Park, S.Y.;Preparation of multiwalled carbon nanotube/nylon-6 nanocomposites by in situ polymerization.*J.Appl.Polym.Sci.***2007**, 106(6), 3729–3735.

11. Sahoo, N.G., et al.;Polymer nanocomposites based on functionalized carbon nanotubes.*Progr. Polym.Sci*, **2010**, **35**(7), 837–867.

12. Safadi, B., Andrews, R.;and Grulke, E.;Multiwalled carbon nanotube polymer composites: synthesis and characterization of thin films.*J.Appl.Polym.Sci.***2002**, *84*(14), 2660–2669.

13. Liu, T., et al.;Morphology and mechanical properties of multiwalled carbon nanotubes reinforced nylon-6 composites.*Macromolecules*.**2004**, *37*(19), 7214–7222.

14. Vigolo, B., et al.;Improved structure and properties of single-wall carbon nanotube spun fibers.*Appl.Phys.Lett.***2002**, *81*(7), 1210–1212.

15. Saeed, K., et al., Preparation of electrospun nanofibers of carbon nanotube/polycaprolactone nanocomposite.*Polymer*.**2006**, *47*(23), 8019–8025.

16. Zhang, X., et al.;Poly (vinyl alcohol)/SWNT composite film.*Nano Lett.***2003**, *3*(9), 1285–1288.

17. Holzinger, M., et al.;Sidewall functionalization of carbon nanotubes.*Angewandte Chemie International Edition*.**2001**, *40*(21), 4002–4005.

18. Hirsch, A.;Functionalization of single-walled carbon nanotubes.*Angewandte.Chemie.International.Edition*.**2002**, *41*(11), 1853–1859.

19. Schlittler, R., et al.;Single crystals of single-walled carbon nanotubes formed by self-assembly.*Science*.**2001**, *292*(5519), 1136-1139.

20. Lin, T., et al.;Chemistry of Carbon Nanotubes.*Aust. J.Chem*.**2003**, *56*(7), 635–651.

21. Kilbride, B.E., et al.;Experimental observation of scaling laws for alternating current and direct current conductivity in polymer-carbon nanotube composite thin films.*J.Appl.Phys*.**2002**. *92*(7), 4024–4030.

22. Byrne, M.T.; and Gun'ko, Y.K.;Recent Advances in Research on Carbon Nanotube–Polymer Composites.*Adv.Mater.***2010**, *22*(15), 1672–1688.

23. Hirsch, A.;Functionalization of Single-Walled Carbon Nanotubes.*Angewandte Chemie International Edition*.**2002**, *41*(11), 1853–1859.

24. Banerjee, S., Hemraj-Benny, T.;and Wong, S.S.;Covalent Surface Chemistry of Single-Walled Carbon Nanotubes.*Adv.Mater.***2005**, *17*(1), 17–29.

25. Nalwa, H.S.;Nanostructured materials and nanotechnology: concise edition. **2001:** Gulf Professional Publishing.

26. Ebbesen, T.; et al.;Purification of nanotubes.*Nature.***1994**. *367*(6463), 519–519.

27. Liu, J., et al.;Fullerene pipes.*Science.***1998**,*280*(5367), 1253–1256.

28. Niyogi, S., et al.;Chemistry of single-walled carbon nanotubes.*Acc.Chem.Res.***2002**, *35*(12), 1105–1113.

29. Georgakilas, V., et al.;*Multipurpose organically modified carbon nanotubes: from functionalization to nanotube composites.J.Am.Chem.Soc.***2008**, *130*(27), 8733–8740.

30. Park, H., Zhao, J.; and Lu, J.P.;Effects of sidewall functionalization on conducting properties of single wall carbon nanotubes.*Nano.Lett.***2006**, *6*(5), 916–919.

31. Zhang, X., et al., *Properties and structure of nitric acid oxidized single wall carbon nanotube films.* The Journal of Physical Chemistry B, 2004. *108*(42): p. 16435-16440.

32. Zeng, H.L., Gao, C.; and Yan, D.Y.;*Poly (ε-caprolactone)-Functionalized Carbon Nanotubes and Their Biodegradation Properties.Advan.Funct.Mater.***2006**. *16*(6), 812–818.

33. Chen, G.X., et al., Synthesis of Poly (L-lactide)-Functionalized multiwalledcarbon nanotubes by ring-opening polymerization.*Macromolecular.Chem.Phys*, **2007**, *208*(4), 389–398.

34. Qu, L., et al.;Soluble nylon-functionalized carbon nanotubes from anionic ring-opening polymerization from nanotube surface. Macromolecules, **2005**,*38*(24), 10328–10331.

35. Yang, M., et al.;Functionalization of multiwalled carbon nanotubes with polyamide 6 by anionic ring-opening polymerization.*Carbon.***2007**, *45*(12), 2327–2333.

36. Park, S.J., et al.;Synthesis and Dispersion Characteristics of Multi-Walled Carbon Nanotube Composites with Poly (methyl methacrylate) Prepared by In-Situ Bulk Polymerization.Macromol.Rap.Commun.**2003**, *24*(18), 1070–1073.

37. Baskaran, D., Mays, J.W.; and Bratcher, M.S.;Polymer-grafted multiwalled carbon nanotubes through surface-initiated polymerization.*Angewandte Chemie International Edition*, **2004**.*43*(16), 2138–2142.

38. Kong, H., Gao, C.;and Yan, D.;Functionalization of multiwalled carbon nanotubes by atom transfer radical polymerization and defunctionalization of the products.*Macromolecules.***2004**,*37*(11), 4022–4030.

39. Kong, H., et al.;Polyelectrolyte-functionalized multiwalled carbon nanotubes: preparation, characterization and layer-by-layer self-assembly.*Polymer*, **2005**, *46*(8), 2472–2485.

40. Kong, H., Gao, C.;and Yan, D.;Constructing amphiphilic polymer brushes on the convex surfaces of multi-walled carbon nanotubes by in situ atom transfer radical polymerization.*J.Mater.Chem.***2004**, *14*(9), 1401–1405.

41. Kong, H., et al.;Poly (N-isopropylacrylamide)-coated carbon nanotubes: Temperature-sensitive molecular nanohybrids in water.*Macromolecules.***2004**, *37*(18), 6683–6686.

42. Gao, J., et al.;Continuous spinning of a single-walled carbon nanotube-nylon composite fiber.*J.Am.Chem.Soc.***2005**, *127*(11), 3847–3854.

43. Sun, Y.-P., et al.;Soluble dendron-functionalized carbon nanotubes: preparation, characterization, and properties §.*Chem.Mater.***2001**, *13*(9), 2864–2869.

44. Qu, L., et al.;Polyimide-functionalized carbon nanotubes: synthesis and dispersion in nanocomposite films.*Macromolecules.***2004**, *37*(16), 6055–6060.

45. Jana, R. and Cho, J.W.;Thermal stability, crystallization behavior, and phase morphology of poly (ε-caprolactone) diol-grafted-multiwalled carbon nanotubes. J.Appl.Polym.Sci.**2008**, *110*(3), 1550–1558.

46. Lin, Y., et al.;Polymeric carbon nanocomposites from carbon nanotubes functionalized with matrix polymer.*Macromolecules.***2003**, *36*(19), 7199–7204.

47. Lou, X., et al.;Surface modification of multiwalled carbon nanotubes by poly (2-vinylpyri-dine): Dispersion, selective deposition, and decoration of the nanotubes.*Adv.Mater.***2004**. *16*(23–24), 2123–2127.

48. Xu, G., et al.;Covalent functionalization of multi-walled carbon nanotube surfaces by conjugated polyfluorenes.*Polymer.***2007**, *48*(26), 7510–7515.

49. Georgakilas, V., et al.;Amino acid functionalisation of water soluble carbon nanotubes.*Chem. Commun.* **2002**, 24, 3050–3051.

50. Bianco, A.;Kostarelos, K.;and Prato, M.;Applications of carbon nanotubes in drug delivery. *Curr.Opin.Chem.Biol.***2005**, *9*(6), 674–679.

51. Hayden, H., Gun'ko, Y.K.;and Perova, T.S.;Chemical modification of multi-walled carbon nanotubes using a tetrazine derivative.*Chem.Phys.Lett.***2007**, *435*(1), 84–89.

52. Ménard-Moyon, C., et al.;Unexpected outcome in the reaction of triazolinedione with carbon nanotubes.*Eur. J.Org.Chem.***2007**, *2007*(29), 4817–4819.

53. Ménard-Moyon, C., et al.;Functionalization of single-wall carbon nanotubes by tandem high-pressure/Cr (CO) 6 activation of diels-alder cycloaddition.*J.Am.Chem.Soc.***2006**, *128*(46), 14764–14765.

54. Cho, E., et al.;Ab initio study on the carbon nanotube with various degrees of functionalization.*Chem.Phys.Lett.***2006**, *419*(1), 134–138.

55. Holzinger, M., et al., 2+ 1 cycloaddition for cross-linking SWCNTs.*Carbon.* **2004**, *42*(5), 941–947.

56. Liu, Y.;Yao, Z.;and Adronov, A.;Functionalization of single-walled carbon nanotubes with well-defined polymers by radical coupling.*Macromolecules.***2005**, *38*(4), 1172–1179.

57. Liu, J., et al., Efficient microwave-assisted radical functionalization of single-wall carbon nanotubes.*Carbon.***2007**, *45*(4), 885–891.

58. Sui, G., et al.;Preparation and properties of natural rubber composites reinforced with pre-treated carbon nanotubes.*Polym.Adv.Technol.***2008**, *19*(11), 1543–1549.

59. Yuen, S.M., et al., Molecular motion, morphology, and thermal properties of multiwall carbon nanotube/polysilsesquioxane composite.*J.Polym.Sci.Part B: Polym. Phys.***2008**, *46*(5), 472–482.

60. Rasheed, A., et al.;Polymer nanotube nanocomposites: Correlating intermolecular interaction to ultimate properties.*Polymer.***2006**, *47*(13), 4734–4741.

61. Kymakis, E., et al., Carbon nanotube/PEDOT: PSS electrodes for organic photovoltaics.*The Eur.Phys. J. Appl.Phys.***2006**, *36*(3), 257–259.

62. Sahoo, N.G., et al.;Effect of functionalized carbon nanotubes on molecular interaction and properties of polyurethane composites.*Macromol.Chem.Physics.***2006**, *207*(19), 1773–1780.

63. Ruelle, B., et al.;*Functionalization of MWCNTs with atomic nitrogen.Micron.***2009**, *40*(1), 85–88.

64. Choi, J.-Y., et al.;In-situ grafting of hyperbranched poly (ether ketone) s onto multiwalled carbon nanotubes via the A3+ B2 approach.*Macromolecules.***2007**,*40*(13), 4474–4480.

65. Rao, A., et al.;Diameter-selective Raman scattering from vibrational modes in carbon nanotubes.*Science.***1997**, *275*(5297), 187–191.

66. Duesberg, G., et al.;Separation of carbon nanotubes by size exclusion chromatography.*Chem. Commun.***1998**(3), 435–436.

67. Bredeau, S., et al., From carbon nanotube coatings to high-performance polymer nanocomposites.*Polym.Int.***2008**. *57*(4), 547–553.

68. Das, D. and Das, P.K.;Superior activity of structurally deprived enzyme− carbon nanotube hybrids in cationic reverse micelles.*Langmuir.***2009**, *25*(8), 4421–4428.

69. Shin, J.Y.;Premkumar, T.;and Geckeler, K.E.;Dispersion of single-walled carbon nanotubes by using surfactants: are the type and concentration important?*Chem.Eur.J.***2008**, *14*(20),6044–6048.

70. Doe, C., et al.;Charged rod-like nanoparticles assisting single-walled carbon nanotube dispersion in water.*Adv.Funct.Mater.***2008**, *18*(18), 2685–2691.

71. Park, I., et al.;Multiwalled carbon nanotubes functionalized with PS via emulsion polymerization.*Macromol.Res.***2007**, 15(6), 498–505.

72. Park, I., et al.;Selective sequestering of multi-walled carbon nanotubes in self-assembled block copolymer.*Sens.Actuat B: Chem.***2007**, *126*(1), 301–305.

73. Paredes, J.; and Burghard, M.;Dispersions of individual single-walled carbon nanotubes of high length.*Langmuir.***2004**, *20*(12), 5149–5152.

74. Moore, V.C., et al., Individually suspended single-walled carbon nanotubes in various surfactants.*Nano.Lett.***2003**, *3*(10), 1379–1382.

75. Islam, M., et al.;High weight fraction surfactant solubilization of single-wall carbon nanotubes in water.*Nano Lett.***2003**, *3*(2), 269–273.

76. Kim, H.-S., et al.;Multiple light scattering measurement and stability analysis of aqueous carbon nanotube dispersions.*J.Phys.Chem.Solid*, **2008**, *69*(5), 1209–1212.

77. Wang, H., et al.;Dispersing single-walled carbon nanotubes with surfactants: a small angle neutron scattering study.*Nano Lett.***2004**, *4*(9), 1789–1793.

78. Lisunova, M.O., et al.;Stability of the aqueous suspensions of nanotubes in the presence of nonionic surfactant.*J.Coll.Interf.Sci.***2006**, *299*(2), 740–746.

79. Guo, Z.;Sadler, P.J.;and Tsang, S.C.;Immobilization and visualization of DNA and proteins on carbon nanotubes.*Adv.Mater.***1998**, *10*(9), 701–703.

80. Barone, P.W.; and Strano, M.S.;Reversible control of carbon nanotube aggregation for a glucose affinity sensor.*AngewandteChemie.***2006**, *118*(48), 8318–8321.

81. Numata, M., et al.;Creation of hierarchical carbon nanotube assemblies through alternative packing of complementary semi-artificial β-1, 3-Glucan/Carbon nanotube composites.*Chem.-A Eur. J.***2008**, *14*(8), 2398–2404.

82. Yan, L.Y., et al.;Individually dispersing single-walled carbon nanotubes with novel neutral pH water-soluble chitosan derivatives.*J.Phys.Chem C.***2008**, *112*(20), 7579–7587.

83. Rouse, J.H.;Polymer-assisted dispersion of single-walled carbon nanotubes in alcohols and applicability toward carbon nanotube/sol-gel composite formation.*Langmuir.***2005**, *21*(3), 1055–1061.

84. Yuan, W.Z., et al.;Wrapping carbon nanotubes in pyrene-containing poly (phenylacetylene) chains: solubility, stability, light emission, and surface photovoltaic properties.*Macromolecules.***2006**,*39*(23), 8011–8020.

85. Curran, S.A., et al., A composite from poly (m-phenylenevinylene-co-2, 5-dioctoxy-p-phenylenevinylene) and carbon nanotubes: A novel material for molecular optoelectronics.*Adv.Mater.***1998**, *10*(14), 1091–1093.

86. McCarthy, B., et al.;Microscopy studies of nanotube-conjugated polymer interactions.*Synthet Metals.***2001**, *121*(1–3), 1225–1226.

87. Nativ-Roth, E., et al.;Physical adsorption of block copolymers to SWNT and MWNT: a non-wrapping mechanism.*Macromolecules.***2007**, *40*(10), 3676–3685.

88. Liu, J., et al.;Stable non-covalent functionalisation of multi-walled carbon nanotubes by pyrene–polyethylene glycol through π–π stacking.*New. J.Chem.***2009**, *33*(5), 1017–1024.

89. Coleman, J.N., et al.;High performance nanotube-reinforced plastics: understanding the mechanism of strength increase.Adv.Funct.Mater.**2004**, *14*(8), 791–798.

90. Liao, S.-H., et al.;One-step functionalization of carbon nanotubes by free-radical modification for the preparation of nanocomposite bipolar plates in polymer electrolyte membrane fuel cells.*J.Mater.Chem.***2008**, *18*(33), 3993–4002.

91. Shi, D., et al.;Plasma coating of carbon nanofibers for enhanced dispersion and interfacial bonding in polymer composites.*Appl.Phys.Lett.***2003**,*83*(25), 5301–5303.

92. Coleman, J.N., et al.;Small but strong: a review of the mechanical properties of carbon nano-tube–polymer composites.*Carbon*.**2006**, *44*(9), 1624–1652.
93. Dalmas, F., et al.;Multiwalled carbon nanotube/polymer nanocomposites: processing and properties.*J.Polym.Sci. Part B: Polym.Phys*.**2005**, *43*(10), 1186–1197.
94. Pötschke, P., et al.;Orientation of multiwalled carbon nanotubes in composites with polycar-bonate by melt spinning.*Polymer*.**2005**, *46*(23), 10355–10363.
95. Blighe, F.M., et al.;On the factors controlling the mechanical properties of nanotube films. *Carbon*.**2008**, *46*(1), 41–47.
96. Zhu, J., et al.;Reinforcing epoxy polymer composites through covalent integration of func-tionalized nanotubes.*Adv.Funct.Mater.***2004**, *14*(7), 643–648.
97. Miaudet, P., et al.;Hot-drawing of single and multiwall carbon nanotube fibers for high tough-ness and alignment.*Nano.Lett.***2005**, *5*(11), 2212–2215.
98. Chou, T.-W., et al.;An assessment of the science and technology of carbon nanotube-based fibers and composites.*Compos.Sci.Technol*.**2010**, *70*(1), 1–19.
99. Mamedov, A.A., et al.;Molecular design of strong single-wall carbon nanotube/polyelectro-lyte multilayer composites.*Nat.Mater.***2002**, *1*(3), 190–194.
100. Kang, T.J., et al.;Ultra-thin and conductive nanomembranearrays for nanomechanicaltrans-ducers.*Adv.Mater*.**2008**, *20*(16), 3131–3137.
101. Park, H.J., et al., Preparation of transparent conductive multilayered films using active pen-tafluorophenyl ester modified multiwalled carbon nanotubes.*Langmuir*.**2008**, *24*(18), 10467–10473.
102. Shim, B.S. et al.;Smart electronic yarns and wearable fabrics for human biomonitoring made by carbon nanotube coating with polyelectrolytes.*Nano.Lett.***2008**, *8*(12), 4151–4157.
103. O'Connor, I., et al.;High-strength, high-toughness Composite Fibers by Swelling Kevlar in Nanotube Suspensions.*Small*.**2009**, *5*(4), 466–469.
104. O'Connor, I., et al.;Development of transparent, conducting composites by surface infiltration of nanotubes into commercial polymer films.*Carbon*.**2009**, *47*(8), 1983–1988.
105. Bergin, S.D., et al.;Towardssolutions of single-walled carbon nanotubes in common solvents. *Adv.Mater.***2008**, *20*(10), 1876–1881.
106. Giordani, S., et al.;Debundling of single-walled nanotubes by dilution: observation of large populations of individual nanotubes in amide solvent dispersions.*J.Phys.Chem. B*.**2006**. *110*(32): 15708–15718.
107. Kukovecz, Á., et al.;Multiwall carbon nanotube films surface-doped with electroceramics for sensor applications.*physica status solidi (b)*.**2008**,*245*(10), 2331–2334.
108. Cooper, S.M., et al.;Gas permeability of a buckypaper membrane.*Nano.Lett.***2003**, *3*(2), 189–192.
109. Gui, X., et al.;Carbon nanotube sponges.*Adv.Mater*.**2010**, *22*(5), 617–621.
110. Zhao, X., et al.;Strain monitoring in glass fiber reinforced composites embedded with carbon nanopaper sheet using Fiber Bragg Grating (FBG) sensors.*Compos. Part B: Eng*.**2009**, *40*(2), 134–140.
111. Abot, J., et al.;Novel carbon nanotube array-reinforced laminated composite materials with higher interlaminar elastic properties.*Compos.Sci.Technol*.**2008**, *68*(13), 2755–2760.
112. Zhang, D., et al.;Transparent, conductive, and flexible carbon nanotube films and their ap-plication in organic light-emitting diodes.*Nano.Lett.***2006**, *6*(9), 1880–1886.
113. Saito, R.;Physical properties of carbon nanotubes (paperback). 1998.
114. Lachman, N., et al.;Fracture behavior of carbon nanotube/carbon microfiber hybrid polymer composites.*J.Mater.Sci*.**2013**, *48*(16), 5590–5595.
115. Ruan, S., et al., Toughening high performance ultrahigh molecular weight polyethylene using multiwalled carbon nanotubes.*Polym*.**2003**, *44*(19), 5643–5654.

116. Hwang, G.L., Shieh, Y.T.;and Hwang, K.C.;Efficient Load Transfer to Polymer-Grafted Multiwalled Carbon Nanotubes in Polymer Composites.*Adv.Funct.Mater.***2004**, *14*(5), 487–491.

117. Fuentes, G., et al., Formation and electronic properties of BC_ {3} single-wall nanotubes upon boron substitution of carbon nanotubes.*Phys.Rev.B.***2004**, *69*(24), 245403.

118. Moniruzzaman, M.; and Winey, K.I.;Polymer nanocomposites containing carbon nanotubes. *Macromolecules*, **2006**, *39*(16), 5194–5205.

119. Assouline, E., et al., *Nucleation ability of multiwall carbon nanotubes in polypropylene composites.J.Polym.Sci. Part B: Polym.Phys.***2003**, *41*(5), 520–527.

120. Gojny, F.H., et al.;Influence of different carbon nanotubes on the mechanical properties of epoxy matrix composites–a comparative study.*Compos.Sci.Technol.***2005**, *65*(15), 2300–2313.

121. Cheng, Q., et al.;Fabrication and properties of aligned multiwalled carbon nanotube-reinforced epoxy composites.*J.Mater.Res.***2008**, *23*(11), 2975–2983.

122. Cheng, Q., et al.;Carbon nanotube/epoxy composites fabricated by resin transfer molding. *Carbon*, **2010**, *48*(1), 260–266.

123. Hughes, M.; and Spinks, G.M.;Multiwalled carbon nanotube actuators.*Adv.Mater.***2005**, *17*(4), 443–446.

124. Wang, Z., et al.;Processing and property investigation of single-walled carbon nanotube (SWNT) buckypaper/epoxy resin matrix nanocomposites.*Compos. Part A: Appl.Sci. Manuf.***2004**,*35*(10), 1225–1232.

125. Kim, Y.A., et al.;Fabrication of High-Purity, Double-Walled Carbon Nanotube Buckypaper. *Chem. Vapor.Deposit.***2006**, *12*(6), 327–330.

126. Hussain, F., et al.;Review article: polymer-matrix nanocomposites, processing, manufacturing, and application: an overview.*J.Compos.Mater.***2006**, *40*(17), 1511–1575.

127. Qian, D., et al.;Load transfer and deformation mechanisms in carbon nanotube-polystyrene composites.*Appl.Phys.Lett.***2000**, *76*, 2868.

128. Jin, L., Bower, C.;and Zhou, O.;Alignment of carbon nanotubes in a polymer matrix by mechanical stretching.*Appl.Phys.Lett.***1998**, *73*, 1197.

129. Xie, X.-L., Mai, Y.-W.;and Zhou, X.-P.;Dispersion and alignment of carbon nanotubes in polymer matrix: a review.*Mater.Sci.Eng. R. Reports*, **2005**, *49*(4), 89–112.

130. Cooper, C.A., et al.;Distribution and alignment of carbon nanotubes and nanofibrils in a polymer matrix.*Compos.Sci.Technol.***2002**, *62*(7), 1105–1112.

131. Alig, I., et al.;Destruction and formation of a conductive carbon nanotube network in polymer melts: in-line experiments.*Polymer.***2008**, *49*(7), 1902–1909.

132. Bellayer, S., et al.;Preparation of homogeneously dispersed multiwalled carbon nanotube/ polystyrene nanocomposites via melt extrusion using trialkyl imidazolium compatibilizer. *Adv.Funct.Mater.***2005**, *15*(6), 910–916.

133. Tibbetts, G.G., et al.;A review of the fabrication and properties of vapor-grown carbon nanofiber/polymer composites.*Compos.Sci.Technol.***2007**, *67*(7), 1709–1718.

134. Al-Saleh, M.H.; and Sundararaj, U.; A review of vapor grown carbon nanofiber/polymer conductive composites.*Carbon.***2009**,*47*(1), 2–22.

135. Fang, J., et al.;Applications of electrospun nanofibers.*Chinese Sci. Bull.* **2008**, *53*(15), 2265–2286.

136. Chen, W., et al.;Enhanced mechanical properties and morphological characterizations of poly (vinyl alcohol)–carbon nanotube composite films. *Appl.Surf.Sci.***2005**, *252*(5), 1404–1409.

137. Hou, H., et al.;Electrospun polyacrylonitrile nanofibers containing a high concentration of well-aligned multiwall carbon nanotubes.*Chem. Mater.***2005**, *17*(5), 967–973.

138. Weisenberger, M.C.; Andrews, R.;and Rantell, T.;Carbon nanotube polymer composites: recent developments in mechanical properties.In:Physical Properties of Polymers Handbook, J. Mark, Editor., Springer New York, **2007**; 585–598.

139. Chae, H.G., et al.; Carbon nanotube reinforced small diameter polyacrylonitrile based carbon fiber. *Compos.Sci. Technol.* **2009**, *69*(3), 406–413.

140. Chae, H.G., et al.; Stabilization and carbonization of gel spun polyacrylonitrile/single wall carbon nanotube composite fibers. *Polymer.* **2007**, *48*(13), 3781–3789.

141. Ko, F., et al.; Electrospinning of continuous carbon nanotube-filled nanofiber yarns. *Adv.Mater.* **2003**, *15*(14), 1161–1165.

142. Chen, L., et al.; Fabrication and characterization of polycarbonate/carbon nanotubes composites. *Compos. Part A Appl. Sci. Manuf.* **2006**, *37*(9), 1485–1489.

143. Higgins, B.A.; and Brittain, W.J.; Polycarbonate carbon nanofiber composites. *Eur. Polym. J.* **2005**, *41*(5), 889–893.

144. Xia, H., Wang, Q.; and Qiu, G.; Polymer-encapsulated carbon nanotubes prepared through ultrasonically initiated in situ emulsion polymerization. *Chem.Mater.* **2003**, *15*(20), 3879–3886.

145. Sahoo, N.G., et al.; Influence of carbon nanotubes and polypyrrole on the thermal, mechanical and electroactive shape-memory properties of polyurethane nanocomposites. *Compos.Sci. Technol.* **2007**, *67*(9), 1920–1929.

146. Paul, D.R.; *Polym.Blend.* Vol. 1. 1978: Elsevier.

147. Folkes, M. and Hope, P.; Polymer Blend and Alloy. **1993**, Springer.

148. Pötschke, P., Bhattacharyya, A.R.; and Janke, A.; Carbon nanotube-filled polycarbonate composites produced by melt mixing and their use in blends with polyethylene. *Carbon,* **2004**, *42*(5), 965–969.

149. Pötschke, P., et al.; Rheological and dielectric characterization of melt mixed polycarbonate-multiwalled carbon nanotube composites. *Polymer.* **2004**, *45*(26), 8863–8870.

150. Zhang, Z., et al.; Enhanced interactions between multi-walled carbon nanotubes and polystyrene induced by melt mixing. *Carbon,* **2006**, *44*(4), 692–698.

151. Xu, Y., et al.; Growing multihydroxyl hyperbranched polymers on the surfaces of carbon nanotubes by in situ ring-opening polymerization. *Macromolecules.* **2004**, *37*(24), 8846–8853.

152. Jiang, X., Bin, Y.; and Matsuo, M.; Electrical and mechanical properties of polyimide–carbon nanotubes composites fabricated by in situ polymerization. *Polymer.* **2005**, *46*(18), 7418–7424.

153. Zeng, H., et al.; In situ polymerization approach to multiwalled carbon nanotubes-reinforced nylon 1010 composites: mechanical properties and crystallization behavior. *Polymer.* **2006**. *47*(1), 113–122.

154. Jung, Y.C.; Sahoo, N.G.; and Cho, J.W.; Polymeric Nanocomposites of Polyurethane Block Copolymers and Functionalized Multi-Walled Carbon Nanotubes as Crosslinkers. *Macromol. Rap.Commun.* **2006**, *27*(2), 126–131.

155. Bhattacharyya, S., Kymakis, E.; and Amaratunga, G.; Photovoltaic properties of dye functionalized single-wall carbon nanotube/conjugated polymer devices. *Chem.Mater.* **2004**. *16*(23), 4819–4823.

156. Ago, H., et al.; Composites of carbon nanotubes and conjugated polymers for photovoltaic devices. *Adv.Mater.* **1999**, *11*(15), 1281–1285.

157. Kymakis, E. and Amaratunga, G.; Single-wall carbon nanotube/conjugated polymer photovoltaic devices. *Appl.Phys.Lett.* **2002**, *80*(1), 112–114.

158. Kymakis, E., Alexandrou, I.; and Amaratunga, G.; High open-circuit voltage photovoltaic devices from carbon-nanotube-polymer composites. *J. Appl.Phys.* **2003**, *93*(3), 1764–1768.

159. Woo, H., et al.; Organic light emitting diodes fabricated with single wall carbon nanotubes dispersed in a hole conducting buffer: the role of carbon nanotubes in a hole conducting polymer. *Synthetic. Metal.* **2001**, *116*(1), 369–372.

160. Peng, C., et al., Carbon nanotube and conducting polymer composites for supercapacitors. *Progr.Nat.Sci.* **2008**, *18*(7), 777–788.

161. Endo, M., M.S. Strano, and P.M. Ajayan, Potential applications of carbon nanotubes.In:Carbon nanotubes. **2008**, Springer.pp 13–61.

162. Jurewicz, K., et al.; *Supercapacitors from nanotubes/polypyrrole composites.Chem.Phys. Lett.***2001**, *347*(1), 36–40.

163. Landi, B.J., et al.; Single wall carbon nanotube-Nafion composite actuators.*Nano Lett.***2002**. *2*(11): p. 1329-1332.

164. Dong, B., et al., Preparation and electrochemical characterization of polyaniline/multi-walled carbon nanotubes composites for supercapacitor.*Mater.Sci.Eng. B.***2007**, *143*(1), 7–13.

165. Xiao, Q. and Zhou, X.;The study of multiwalled carbon nanotube deposited with conducting polymer for supercapacitor. *Electrochimica Acta*, **2003**.*48*(5), 575–580.

166. Koerner, H., et al., Remotely actuated polymer nanocomposites—stress-recovery of carbon-nanotube-filled thermoplastic elastomers.*Nat. Mater.***2004**, *3*(2), 115–120.

167. Mottaghitalab, V., et al., Polyaniline fibres containing single walled carbon nanotubes: enhanced performance artificial muscles.*Synthet.Metal.* **2006**,*156*(11), 796–803.

168. Fennimore, A., et al.;*Rotational actuators based on carbon nanotubes.*Nature.**2003**,*424*(6947): 408–410.

169. Li, C., Thostenson, E.T.;and Chou, T.-W.;Sensors and actuators based on carbon nanotubes and their composites: a review.*Compos.Sci.Technol.* **2008**, *68*(6), 1227–1249.

170. Baughman, R.H.; Zakhidov, A.A.; and de Heer, W.A.; Carbon nanotubes--the route toward applications. *Science.***2002**,*297*(5582), 787–792.

171. Baughman, R.H., et al.;Carbon nanotube actuators.*Science.***1999**, *284*(5418), 1340–1344.

172. Spinks, G.M., et al.;Carbon-nanotube-reinforced Polyaniline Fibers for High-Strength Artificial Muscles.*Adv. Mater,* **2006**, *18*(5), 637–640.

173. Tahhan, M., et al., *Carbon nanotube and polyaniline composite actuators.Smart Materials and Structures*, **2003**, *12*(4), 626.

174. Yun, S.; and Kim, J.; Multiwalled-carbon nanotubes and polyaniline coating on electro-active paper for bending actuator.*J.Phys D. App.Phys.***2006**, *39*(12), 2580.

175. Gooding, J.J.; Nanostructuring electrodes with carbon nanotubes: A review on electrochemistry and applications for sensing. *Electrochimica. Acta.***2005**, **50**(15), 3049–3060.

176. Zhao, Q.;Gan, Z.;and Zhuang, Q.; Electrochemical sensors based on carbon nanotubes. *Electroanalysis.***2002**, *14*(23), 1609–1613.

177. Sherigara, B.S.; Kutner, W.; and D'Souza, F.;Electrocatalytic properties and sensor applications of fullerenes and carbon nanotubes.*Electroanalysis*. **2003**. **15**(9): p. 753-772.

178. Ferrer-Anglada, N., Kaempgen, M.;and Roth, S.;Transparent and flexible carbon nanotube/polypyrrole and carbon nanotube/polyaniline pH sensors.*physica status solidi (b),*.**2006**, *243*(13), 3519–3523.

179. Cheng, G., et al., A sensitive DNA electrochemical biosensor based on magnetite with a glassy carbon electrode modified by muti-walled carbon nanotubes in polypyrrole.*Analytica.Chimica. Acta.***2005**, *533*(1), 11–16.

180. Choi, E., et al.;Enhancement of thermal and electrical properties of carbon nanotube polymer composites by magnetic field processing. *J.Appl.Phys.***2003**, *94*(9), 6034–6039.

181. Bianco, A., et al.;Biomedical applications of functionalised carbon nanotubes.*Chem.Commun.* **2005**, (5), 571–577.

182. Yang, W., et al.;Carbon nanotubes for biological and biomedical applications.*Nanotechnology.***2007**, *18*(41), 412001.

183. Klumpp, C., et al.;Functionalized carbon nanotubes as emerging nanovectors for the delivery of therapeutics.*Biochimica etBiophysicaActa (BBA)-Biomembranes*, **2006**, *1758*(3), 404–412.

184. Pantarotto, D., et al., Translocation of bioactive peptides across cell membranes by carbon nanotubes.*Chem.Commun.***2004**, *1*, 16–17.

185. Liu, Z., et al.;Preparation and characterization of platinum-based electrocatalysts on multi-walled carbon nanotubes for proton exchange membrane fuel cells.*Langmuir*, **2002**.*18*(10), 4054–4060.

186. Pastorin, G., et al., Double functionalisation of carbon nanotubes for multimodal drug delivery.*Chemical Commun.***2006**(11), 1182–1184.

187. Liao, K.; and Li, S.; Interfacial characteristics of a carbon nanotube–polystyrene composite system.Appl.Phys.Lett.**2001**, *79*(25), 4225–4227.

188. Ebbesen, T.W.;Carbon nanotubes: preparartion and properties. **1997**; CRC press.

189. Ebbesen, T.W.;Carbon nanotubes: preparation & properties.**1996**.

190. Li, Y.-L.; Kinloch, I.A.; and Windle, A.H.; Direct spinning of carbon nanotube fibers from chemical vapor deposition synthesis.*Science*.**2004**, *304*(5668), 276–278.

191. Bradford, P.D. and Bogdanovich, A.E.; Fabrication and properties of multifunctional, carbon nanotube yarn reinforced 3-D textile composites. In Proceedings of the 16th international conference on composite materials (ICCM-16). Kyoto, Japan, **2007.**

192. Bradford, P.D. and Bogdanovich, A.E.; Electrical conductivity study of carbon nanotube yarns, 3-D hybrid braids and their composites. *J.Compos. Mater.***2008**, *42*(15), 1533–1545.

CHAPTER 8

CNT/POLYMER COMPOSITES FROM CHEMISTRY, MECHANICS AND PHYSICS ASPECTS

A. K. HAGHI[1] and G. E. ZAIKOV[2]

[1]University of Guilan, Rasht, Iran

[2]Russian Academy of Sciences, Russia

CONTENTS

8.1   INTRODUCTION

Carbon nanotubes (CNTs) were first observed by Iijima, almost two decades ago,[1] and since then, extensive work has been carried out to characterize their properties [2–4]. A wide range of characteristic parameters has been reported for carbon nanotube nanocomposites. There are contradictory reports that show the influence of CNTs on a particular property (e.g., Young's modulus) to be improving, in different or even deteriorating [5]. However, from the experimental point of view, it is a great challenge to characterize the structure and to manipulate the fabrication of polymer nanocomposites. The development of such materials is still largely empirical and a finer degree of control of their properties cannot be achieved so far. Therefore, computer modeling and simulation will play an ever increasing role in predicting and designing material properties, and guiding such experimental work as synthesis and characterization. For polymer nanocomposites, computer modeling and simulation are especially useful in the hierarchical characteristics of the structure and dynamics of polymer nanocomposites ranging from molecular scale, microscale to mesoscale and macroscale, in particular, the molecular structures and dynamics at the interface between nanoparticles and polymer matrix. The purpose of this review is to discuss the application of modeling and simulation techniques to polymer nanocomposites. This includes a broad subject covering methodologies at various length and time scales and many aspects of polymer nanocomposites. We organize the review as follows. In Section 8.1 we will discuss about the properties of CNTs and nanocomposite. In Section 8.2, we introduce briefly the computational methods used so far for the systems of polymer nanocomposites which can be roughly divided into three types: molecular scale methods (e.g., molecular dynamics (MD), Monte Carlo (MC)), microscale methods (e.g., Brownian dynamics (BD), dissipative particle dynamics (DPD), lattice Boltzmann (LB), time-dependent Ginzburg–Lanau method, dynamic density functional theory (DFT) method), and mesoscale and macroscale methods (e.g., micromechanics, equivalent-continuum and self-similar approaches, finite element method (FEM)).[6] Many researchers used this method for determine the mechanical properties of nanocomposite that in Section 8.3 will be discussed. In Section 8.4 modeling of interfacial load transfer between CNT and polymer in nanocomposite will be introduced and finally we conclude the review by emphasizing the current challenges and future research directions.

8.2   CNTs AND NANOCOMPOSITE PROPERTIES

8.2.1   INTRODUCTION TO CNTs

CNTs are one dimensional carbon materials with aspect ratio greater than 1000. They are cylinders composed of rolled-up graphite planes with diameters in nanometer scale [7–10]. The cylindrical nanotube usually has at least one end capped with a hemisphere of fullerene structure. Depending on the process for CNT fabrication, there are two types of CNTs: single-walled CNTs (SWCNTs) and multiwalled CNTs (MWCNTs) [8–11]. SWCNTs consist of a single graphene layer rolled up into a seamless cylinder whereas MWCNTs consist of two or more concentric cylindrical shells of graphene sheets coaxially arranged around a central hollow core with van der Waals forces between adjacent layers. According to the rolling angle of the graphene sheet, CNTs have three chiralities: armchair, zigzag and chiral one. The tube chirality is defined by the chiral vector, $Ch = na1 + ma2$ (Figure 8.1), where the integers (n, m) are the number of steps along the unit vectors (a1 and a2) of the hexagonal lattice [9, 10]. Using this (n, m) naming scheme, the three types of orientation of the carbon atoms around the nanotube circumference are specified. If $n = m$, the nanotubes are called "armchair". If $m = 0$, the nanotubes are called "zigzag". Otherwise, they are called "chiral". The chirality of nanotubes has significant impact on their transport properties, particularly the electronic properties. For a given (n, m) nanotube, if $(2n + m)$ is a multiple of 3, then the nanotube is metallic, otherwise the nanotube is a semiconductor. Each MWCNT contains a multilayer of graphene, and each layer can have different chiralities, so the prediction of its physical properties is more complicated than that of SWCNT. Figure 8.1 shows the CNT with different chiralities.

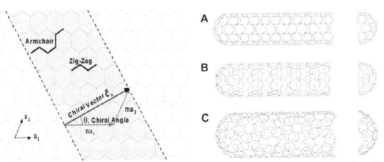

FIGURE 8.1   Schematic diagram showing how a hexagonal sheet of graphene is rolled to form a CNT with different chiralities (A: armchair; B: zigzag; C: chiral).

8.2.2 CLASSIFICATION OF CNT/POLYMER NANOCOMPOSITES

Polymer composites, consisting of additives and polymer matrices, including thermoplastics, thermosets and elastomers, are considered to be an important group of relatively inexpensive materials for many engineering applications. Two or more materials are combined to produce composites that possess properties that are unique and cannot be obtained each material acting alone. For example, high modulus carbon fibers or silica particles are added into a polymer to produce reinforced polymer composites that exhibit significantly enhanced mechanical properties including strength, modulus and fracture toughness. However, there are some bottlenecks in optimizing the properties of polymer composites by employing traditional micron-scale fillers. The conventional filler content in polymer composites is generally in the range of 10–70 wt. %, which in turn results in a composite with a high density and high material cost. In addition, the modulus and strength of composites are often traded for high fracture toughness.[12] Unlike traditional polymer composites containing micron-scale fillers, the incorporation of nanoscale CNTs into a polymer system results in very short distance between the fillers, thus the properties of composites can be largely modified even at an extremely low content of filler. For example, the electrical conductivity of CNT/epoxy nanocomposites can be enhanced several orders of magnitude with less than 0.5 wt.% of CNTs [13]. As described previously, CNTs are amongst the strongest and stiffest fibers ever known. These excellent mechanical properties combined with other physical properties of CNTs exemplify huge potential applications of CNT/polymer nanocomposites. Ongoing experimental works in this area have shown some exciting results, although the much-anticipated commercial success has yet to be realized in the years ahead. In addition, CNT/polymer nanocomposites are one of the most studied systems because of the fact that polymer matrix can be easily fabricated without damaging CNTs based on conventional manufacturing techniques, a potential advantage of reduced cost for mass production of nanocomposites in the future. Following the first report on the preparation of a CNT/polymer nanocomposite in 1994, [14] many research efforts have been made to understand their structure–property relationship and find useful applications in different fields, and these efforts have become more pronounced after the realization of CNT fabrication in industrial scale with lower costs in the beginning of the twenty-first century [15]. According to the specific application, CNT/polymer nanocomposites can be classified as structural or functional composites [16]. For the structural composites, the unique mechanical properties of CNTs, such as the high modulus, tensile strength and strain to fracture, are explored to obtain structural materials with much improved mechanical properties. As for CNT/polymer functional composites, many other unique properties of CNTs, such as electrical, thermal, optical and damping properties along with their excellent mechanical properties, are utilized to develop multifunctional composites for applications in the fields of heat resistance, chemical sensing, electrical and thermal management, photoemission, electromagnetic absorbing, and energy storage performances, and etc.

8.3 MODELING AND SIMULATION TECHNIQUES

8.3.1 MOLECULAR SCALE METHODS

The modeling and simulation methods at molecular level usually employ atoms, molecules or their clusters as the basic units considered. The most popular methods include molecular mechanics (MM), MD and MC simulation. Modeling of polymer nanocomposites at this scale is predominantly directed toward the thermodynamics and kinetics of the formation, molecular structure and interactions. The diagram in Figure 8.1 describes the equation of motion for each method and the typical properties predicted from each of them [17–22] We introduce here two widely used molecular scale methods: MD and MC.

8 3.1.1 MOLECULAR DYNAMICS (MD)

MD is a computer simulation technique that allows one to predict the time evolution of a system of interacting particles (e.g., atoms, molecules, granules, etc.) and estimate the relevant physical properties [23, 24]. Specifically, it generates such information as atomic positions, velocities and forces from which the macroscopic properties (e.g., pressure, energy, heat capacities) can be derived by means of statistical mechanics. MD simulation usually consists of three constituents: (i) a set of initial conditions (e.g., initial positions and velocities of all particles in the system); (ii) the interaction potentials to represent the forces among all the particles; (iii) the evolution of the system in time by solving a set of classical Newtonian equations of motion for all particles in the system. The equation of motion is generally given by

$$\vec{F}_1(t) = m_i \frac{d^2 \vec{r}_1}{dt^2} \qquad (8.1)$$

where $\vec{F}_{i(t)}$ is the force acting on the ith atom or particle at time t which is obtained as the negative gradient of the interaction potential U, mi is the atomic mass and $\vec{r}_i$ the atomic position. A physical simulation involves the proper selection of interaction potentials, numerical integration, periodic boundary conditions, and the controls of pressure and temperature to physically meaningful thermodynamic ensembles. The interaction potentials together with their parameters, i.e., the so-called force field, describe in detail how the particles in a system interact with each other, i.e., how the potential energy of a system depends on the particle coordinates. Such a force field may be obtained by quantum method (e.g., ab initio), empirical method (e.g., Lennard-Jones, Mores, Born-Mayer) or quantum-empirical method (e.g., embedded atom model, glue model, bond-order potential). The criteria for selecting a force field include the accuracy, transferability and computational speed. A typical interaction potential U may consist of a number of bonded and nonbonded interaction terms:

$$U(\vec{r_1}, \vec{r_2}, \vec{r_3}, ..., \vec{r_n}) = \sum_{i_{bond}}^{N_{bond}} U_{bond}(i_{bond}, \vec{r_a}, \vec{r_b}) + \sum_{i_{angle}}^{N_{angle}} U_{angle}(i_{angle}, \vec{r_a}, \vec{r_b}, \vec{r_c}) + \sum_{i_{torsion}}^{N_{torsion}} U_{torsion}(i_{torsion}, \vec{r_a}, \vec{r_b}, \vec{r_c}, \vec{r_d})$$

$$+ \sum_{i_{inversion}}^{N_{inversion}} U_{inversion}(i_{inversion}, \vec{r_a}, \vec{r_b}, \vec{r_c}, \vec{r_d}) + \sum_{i=1}^{N-1}\sum_{j>i}^{N} U_{vdw}(i, j, \vec{r_a}, \vec{r_b}) + \sum_{i=1}^{n-1}\sum_{j>i}^{N} U_{electrostatic}(i, j, \vec{r_a}, \vec{r_b}) \qquad (8.2)$$

The first four terms represent bonded interactions, i.e., bond stretching Ubond, bond-angle bend Uangle and dihedral angle torsion Utorsion and inversion interaction Uinversion, while the last two terms are nonbonded interactions, i.e., van der Waals energy Uvdw and electrostatic energy Uelectrostatic. In the equation, $\vec{r_a}, \vec{r_b}, \vec{r_c}, \vec{r_d}$ are the positions of the atoms or particles specifically involved in a given interaction; N_{bond}, N_{angle}, $N_{torsion}$ and $N_{inversion}$ stand for the total numbers of these respective interactions in the simulated system; i_{bond}, i_{angle}, $i_{torsion}$ and $i_{inversion}$ uniquely specify an individual interaction of each type; i and j in the van der Waals and electrostatic terms indicate the atoms involved in the interaction. There are many algorithms for integrating the equation of motion using finite difference methods. The algorithms of varlet, velocity varlet, leap-frog and Beeman, are commonly used in MD simulations [23]. All algorithms assume that the atomic position $\vec{r}$, velocities $\vec{v}$ and accelerations $\vec{a}$ can be approximated by a Taylor series expansion:

$$\vec{r}(t+\delta t) = \vec{r}(t) + \vec{v}(t)\delta t + \frac{1}{2}\vec{a}(t)\delta^2 t + ... \qquad (8.3)$$

$$\vec{v}(t+\delta t) = \vec{v}(t)\delta t + \frac{1}{2}\vec{b}(t)\delta^2 t + ... \qquad (8.4)$$

$$\vec{a}(t+\delta t) = \vec{a}(t) + \vec{b}(t)\delta t + ... \qquad (8.5)$$

Generally speaking, a good integration algorithm should conserve the total energy and momentum and be time-reversible. It should also be easy to implement and computationally efficient, and permit a relatively long time step. The Verlet algorithm is probably the most widely used method. It uses the positions

$\vec{r}(t)$ and accelerations $\vec{a}(t)$ at time t, and the positions $\vec{r}(t - \delta t)$ from the previous step (t-δ) to calculate the new positions $\vec{r}(t + \delta t)$ at (t+δt), we have:

$$\vec{r}(t + \delta t) = \vec{r}(t) + \vec{v}(t)\delta t + \frac{1}{2}\vec{a}(t)\delta t^2 + ... \tag{8.6}$$

$$\vec{r}(t - \delta t) = \vec{r}(t) - \vec{v}(t)\delta t + \frac{1}{2}\vec{a}(t)\delta t^2 + ... \tag{8.7}$$

$$\vec{r}(t + \delta t) = 2\vec{r}(t)\delta t - \vec{r}(t - \delta t) + \vec{a}(t)\delta t^2 + ... \tag{8.8}$$

The velocities at time t and $t + \frac{1}{2\delta t}$ can be respectively estimated

$$\vec{v}(t) = \left[\vec{r}(t + \delta t) - \vec{r}(t - \delta t)\right] / 2\delta t \tag{8.9}$$

$$\vec{v}(t + 1/2\delta t) = \left[\vec{r}(t + \delta t) - \vec{r}(t - \delta t)\right] / \delta t \tag{8.10}$$

MD simulations can be performed in many different ensembles, such as grand canonical (μVT), microcanonical (NVE), canonical (NVT) and isothermal–isobaric (NPT). The constant temperature and pressure can be controlled by adding an appropriate thermostat (e.g., Berendsen, Nose, Nose–Hoover and Nose–Poincare) and barostat (e.g., Andersen, Hoover and Berendsen), respectively. Applying MD into polymer composites allows us to investigate into the effects of fillers on polymer structure and dynamics in the vicinity of polymer–filler interface and also to probe the effects of polymer–filler interactions on the materials properties.

8.3.1.2 MONTE CARLO (MC)

MC technique, also called Metropolis method, [24] is a stochastic method that uses random numbers to generate a sample population of the system from which one can calculate the properties of interest. A MC simulation usually consists of three typical steps. In the first step, the physical problem under investigation is translated into an analogous probabilistic or statistical model. In the second step, the probabilistic model is solved by a numerical stochastic sampling experiment. In the third step, the obtained data are analyzed by using statistical methods. MC provides only the information on equilibrium properties (e.g., free energy, phase equilibrium), different from MD which gives nonequilibrium as well as equilibrium properties. In a NVT ensemble with N atoms, one hypothesizes a new configuration by arbitrarily or systematically moving one atom from position i→j. Due to such atomic movement, one can compute the change in the system Hamiltonian ΔH:

$$\Delta H = H(j)-H(i) \tag{8.11}$$

where H(i) and H(j) are the Hamiltonian associated with the original and new configuration, respectively.

This new configuration is then evaluated according to the following rules. If ΔH <0, then the atomic movement would bring the system to a state of lower energy. Hence, the movement is immediately accepted and the displaced atom remains in its new position. If ΔH≥0, the move is accepted only with a certain probability $Pi \to j$ which is given by

$$Pi \to j \propto \exp(-\frac{\Delta H}{K_BT}) \tag{12}$$

where K_B is the Boltzmann constant. According to Metropolis et al. [25] one can generate a random

number ζ between 0 and 1 and determine the new configuration according to the following rule:

$$\xi \leq \exp(-\frac{\Delta H}{K_BT}) \; ; \text{the move is accepted;} \tag{8.13}$$

$$\xi \rangle \exp(-\frac{\Delta H}{K_BT}) \; ; \text{the move is not accepted.} \tag{8.14}$$

If the new configuration is rejected, one counts the original position as a new one and repeats the process by using other arbitrarily chosen atoms. In a μVT ensemble, one hypothesizes a new configuration j by arbitrarily choosing one atom and proposing that it can be exchanged by an atom of a different kind. This procedure affects the chemical composition of the system. Also, the move is accepted with a certain probability. However, one computes the energy change ΔU associated with the change in composition. The new configuration is examined according to the following rules. If ΔU <0, the move of compositional change is accepted. However, if ΔU ≥0, the move is accepted with a certain probability which is given by

$$Pi \to j \propto \exp(-\frac{\Delta U}{K_BT}) \tag{8.15}$$

where ΔU is the change in the sum of the mixing energy and the chemical potential of the mixture. If the new configuration is rejected one counts the original configuration as a new one and repeats the process by using some other arbitrarily or systematically chosen atoms. In polymer nanocomposites, MC methods have been used to investigate the molecular structure at nanoparticle surface and evaluate the effects of various factors.

8.3.2 MICROSCALE METHODS

The modeling and simulation at microscale aim to bridge molecular methods and continuum methods and avoid their shortcomings. Specifically, in nanoparticle–polymer systems, the study of structural evolution (i.e., dynamics of phase separation) involves the description of bulk flow (i.e., hydrodynamic behavior) and the interactions between nanoparticle and polymer components. Note that hydrodynamic behavior is relatively straightforward to handle by continuum methods but is very difficult and expensive to treat by atomistic methods. In contrast, the interactions between components can be examined at an atomistic level but are usually not straightforward to incorporate at the continuum level. Therefore, various simulation methods have been evaluated and extended to study the microscopic structure and phase separation of these polymer nanocomposites, including BD, DPD, LB, time-dependent Ginsburg–Landau (TDGL) theory, and dynamic DFT. In these methods, a polymer system is usually treated with a field description or microscopic particles that incorporate molecular details implicitly. Therefore, they are able to simulate the phenomena on length and time scales currently inaccessible by the classical MD methods.

8.3.2.1 BROWNIAN DYNAMICS (BD)

BD simulation is similar to MD simulations [26]. However; it introduces a few new approximations that allow one to perform simulations on the microsecond timescale whereas MD simulation is known up to a few nanoseconds. In BD the explicit description of solvent molecules used in MD is replaced with an implicit continuum solvent description. Besides, the internal motions of molecules are typically ignored, allowing a much larger time step than that of MD. Therefore, BD is particularly useful for systems where there is a large gap of time scale governing the motion of different components. For example, in polymer–solvent mixture, a short time-step is required to resolve the fast motion of the solvent molecules, whereas the evolution of the slower modes of the system requires a larger time-step. However, if the detailed motion of the solvent molecules is concerned, they may be removed from the simulation and their effects on the polymer are represented by dissipative ($-\gamma P$) and random ($\sigma \zeta$ (t)) force terms. Thus, the forces in the governing Eq. (8.16) is replaced by a Langevin equation,

$$F_i(t) = \sum_{i \neq j} F_{ij}^{\ c} - \gamma P_i + \sigma \zeta_i(t)$$

(8.16)

where F_{ij}^c is the conservative force of particle j acting on particle i, γ and σ are constants depending on the system, Pi the momentum of particle i, and $\zeta(t)$ a Gaussian random noise term. One consequence of this approximation of the fast degrees of freedom by fluctuating forces is that the energy and momentum are no longer conserved, which implies that the macroscopic behavior of the system will not be hydrodynamic. In addition, the effect of one solute molecule on another through the flow of solvent molecules is neglected. Thus, BD can only reproduce the diffusion properties but not the hydrodynamic flow properties since the simulation does not obey the Navier–Stokes equations.

8.3.2.2 DISSIPATIVE PARTICLE DYNAMICS (DPD)

DPD was originally developed by Hoogerbrugge and Koelman [27]. It can simulate both Newtonian and non-Newtonian fluids, including polymer melts and blends, on microscopic length and time scales. Like MD and BD, DPD is a particle-based method. However, its basic unit is not a single atom or molecule but a molecular assembly (i.e., a particle).DPD particles are defined by their mass M_1, position r_i and momentum P_i. The interaction force between two DPD particles i and j can be described by a sum of conservative F_{ij}^c, dissipative F_{ij}^D and random forces F_{ij}^R :[28–30]

$$F_{ij} = F_{ij}^{\ C} + F_{ij}^{\ D} + F_{ij}^{R}$$

(8.17)

While the interaction potentials in MD are high-order polynomials of the distance r_{ij} between two particles, in DPD the potentials are softened so as to approximate the effective potential at microscopic length scales. The form of the conservative force in particular is chosen to decrease linearly with increasing r_{ij}. Beyond a certain cut-off separation r_c, the weight functions and thus the forces are all zero. Because the forces are pair wise and momentum is conserved, the macroscopic behavior directly incorporates Navier–Stokes hydrodynamics. However, energy is not conserved because of the presence of the dissipative and random force terms which are similar to those of BD, but incorporate the effects of Brownian motion on larger length scales. DPD has several advantages over MD, for example, the hydrodynamic behavior is observed with far fewer particles than required in a MD simulation because of its larger particle size. Besides, its force forms allow larger time steps to be taken than those in MD.

8.3.2.3   LATTICE BOLTZMANN (LB)

LB [31] is another microscale method that is suited for the efficient treatment of polymer solution dynamics. It has recently been used to investigate the phase separation of binary fluids in the presence of solid particles. The LB method is originated from lattice gas automaton which is constructed as a simplified, fictitious molecular dynamic in which space, time and particle velocities are all discrete. A typical lattice gas automaton consists of a regular lattice with particles residing on the nodes. The main feature of the LB method is to replace the particle occupation variables (Boolean variables), by single-particle distribution functions (real variables) and neglect individual particle motion and particle–particle correlations in the kinetic equation. There are several ways to obtain the LB equation from either the discrete velocity model or the Boltzmann kinetic equation, and to derive the macroscopic Navier–Stokes equations from the LB equation. An important advantage of the LB method is that microscopic physical interactions of the fluid particles can be conveniently incorporated into the numerical model. Compared with the Navier– Stokes equations, the LB method can handle the interactions among fluid particles and reproduce the microscale mechanism of hydrodynamic behavior. Therefore it belongs to the MD in nature and bridges the gap between the molecular level and macroscopic level. However, its main disadvantage is that it is typically not guaranteed to be numerically stable and may lead to physically unreasonable results, for instance, in the case of high forcing rate or high interparticle interaction strength.

8.3.2.4   TIME-DEPENDENT GINZBURG-LANDAU METHOD

TDGL is a microscale method for simulating the structural evolution of phase separation in polymer blends and block copolymers. It is based on the Cahn–Hilliard–Cook (CHC) nonlinear diffusion equation for a binary blend and falls under the more general phase-field and reaction-diffusion models [32–34]. In the TDGL method, a free-energy function is minimized to simulate a temperature quench from the miscible region of the phase diagram to the immiscible region. Thus, the resulting time-dependent structural evolution of the polymer blend can be investigated by solving the TDGL/CHC equation for the time dependence of the local blend concentration. Glotzer and coworkers have discussed and applied this method to polymer blends and particle-filled polymer systems [35]. This model reproduces the growth kinetics of the TDGL model, demonstrating that such quantities are insensitive to the precise form of the double-well potential of the bulk free-energy term. The TDGL and CDM methods have recently been used to investigate the phase separation of polymer nanocomposites and polymer blends in the presence of nanoparticles [36–40].

8.3.2.5   DYNAMIC DFT METHOD

Dynamic DFT method is usually used to model the dynamic behavior of polymer systems and has been implemented in the software package Mesodyn TM from Accelrys [41]. The DFT models the behavior of polymer fluids by combining Gaussian mean-field statistics with a TDGL model for the time evolution of conserved order parameters. However, in contrast to traditional phenomenological free-energy expansion methods employed in the TDGL approach, the free energy is not truncated at a certain level, and instead retains the full polymer path integral numerically. At the expense of a more challenging computation, this allows detailed information about a specific polymer system beyond simply the Flory–Huggins parameter and mobilities to be included in the simulation. In addition, viscoelasticity, which is not included in TDGL approaches, is included at the level of the Gaussian chains. A similar DFT approach has been developed by Doi and coworkers [42, 43] and forms the basis for their new software tool Simulation Utilities for Soft and Hard Interfaces (SUSHI), one of a suite of molecular and mesoscale modeling tools (called OCTA) developed for the simulation of polymer materials.[44] The essence of dynamic DFT method is that the instantaneous unique conformation distribution can be obtained from the off-equilibrium density profile by coupling a fictitious external potential to the Hamiltonian. Once such distribution is known, the free energy is then calculated by standard statistical thermodynamics. The driving force for diffusion is obtained from the spatial gradient of the first functional derivative of the free energy with respect to the density. Here, we describe briefly the equations for both polymer and particle in the diblock polymer–particle composites [38].

8.3.3   MESOSCALE AND MACROSCALE METHODS

Despite the importance of understanding the molecular structure and nature of materials, their behavior can be homogenized with respect to different aspects which can be at different scales. Typically, the observed macroscopic behavior is usually explained by ignoring the discrete atomic and molecular structure and assuming that the material is continuously distributed throughout its volume. The continuum material is thus assumed to have an average density and can be subjected to body forces such as gravity and surface forces. Generally speaking, the macroscale methods (or called continuum methods hereafter) obey the fundamental laws of: (i) continuity, derived from the conservation of mass; (ii) equilibrium, derived from momentum considerations and Newton's second law; (iii) the moment of momentum principle, based on the model that the time rate of change of angular momentum with respect to an arbitrary point is equal to the resultant moment; (iv) conservation of energy, based on the first law of thermodynamics; and (v) conservation of entropy, based on the second law of thermodynamics. These laws provide the basis for the continuum model and must be coupled with the appropriate constitutive equations and the equations of state to provide all the equations necessary for solving a continuum problem. The continuum method relates the deformation of a continuous medium to the external forces acting on the medium and the resulting internal stress and strain. Computational approaches range from simple closed-form analytical expressions to micromechanics and complex structural mechanics calculations based on beam and shell theory. In this section, we introduce some continuum methods that have been used in polymer nanocomposites, including micromechanics models (e.g., Halpin–Tsai model, Mori–Tanaka model), equivalent-continuum model, self-consistent model and finite element analysis.

8.3.4 MICROMECHANICS

Since the assumption of uniformity in continuum mechanics may not hold at the microscale level, micromechanics methods are used to express the continuum quantities associated with an infinitesimal material element in terms of structure and properties of the micro constituents. Thus, a central theme of micromechanics models is the development of a representative volume element (RVE) to statistically represent the local continuum properties. The RVE is constructed to ensure that the length scale is consistent with the smallest constituent that has a first-order effect on the macroscopic behavior. The RVE is then used in a repeating or periodic nature in the full-scale model. The micromechanics method can account for interfaces between constituents, discontinuities, and coupled mechanical and nonmechanical properties. Our purpose is to review the micromechanics methods used for polymer nanocomposites. Thus, we only discuss here some important concepts of micromechanics as well as the Halpin–Tsai model and Mori–Tanaka model.

8.3.4.1 BASIC CONCEPTS

When applied to particle reinforced polymer composites, micromechanics models usually follow such basic assumptions as (i) linear elasticity of fillers and polymer matrix; (ii) the fillers are axisymmetric, identical in shape and size, and can be characterized by parameters such as aspect ratio; (iii) well-bonded filler–polymer interface and the ignorance of interfacial slip, filler–polymer debonding or matrix cracking. The first concept is the linear elasticity, i.e., the linear relationship between the total stress and infinitesimal strain tensors for the filler and matrix as expressed by the following constitutive equations:

For filler $$\sigma^f = C^f \varepsilon^f \tag{8.18}$$

For matrix $$\sigma^m = C^m \varepsilon^m \tag{8.19}$$

where C is the stiffness tensor. The second concept is the average stress and strain. Since the pointwise stress field $\sigma(x)$ and the corresponding strain field $\varepsilon(x)$ are usually nonuniform in polymer composites, the volume–average stress $\bar{\sigma}$ and strain $\bar{\varepsilon}$ are then defined over the representative averaging volume V, respectively,

$$\bar{\sigma} = \frac{1}{V} \int \sigma(x) dv \tag{8.20}$$

$$\bar{\varepsilon} = \frac{1}{V} \int \varepsilon(x) dv \tag{8.21}$$

Therefore, the average filler and matrix stresses are the averages over the corresponding volumes v_f and v_m, respectively,

$$\bar{\sigma_f} = \frac{1}{V_f} \int \sigma(x) dv \tag{8.22}$$

$$\bar{\sigma_m} = \frac{1}{V_m} \int \sigma(x) dv \tag{8.23}$$

The average strains for the fillers and matrix are defined, respectively, as

$$\bar{\varepsilon_f} = \frac{1}{V_f} \int \varepsilon(x) dv \tag{8.24}$$

$$\bar{\varepsilon_m} = \frac{1}{V_m} \int \varepsilon(x) dv \tag{8.25}$$

Based on the above definitions, the relationships between the filler and matrix averages and the overall averages can be derived as follows:

$$\overline{\sigma} = \overline{\sigma_f} v_f + \overline{\sigma_m} v_m \tag{8.26}$$

$$\overline{\varepsilon} = \overline{\varepsilon_f} v_f + \overline{\varepsilon_m} v_m \tag{8.27}$$

where v_f, v_m are the volume fractions of the fillers and matrix, respectively.

The third concept is the average properties of composites which are actually the main goal of a micromechanics model. The average stiffness of the composite is the tensor C that maps the uniform strain to the average stress

$$\overline{\sigma} = \overline{\varepsilon} C \tag{8.28}$$

The average compliance S is defined in the same way:

$$\overline{\varepsilon} = \overline{\sigma} S \tag{8.29}$$

Another important concept is the strain–concentration and stress–concentration tensors A and B which are basically the ratios between the average filler strain (or stress) and the corresponding average of the composites.

$$\overline{\varepsilon_f} = \overline{\varepsilon} A \tag{8.30}$$

$$\overline{\sigma_f} = \overline{\sigma} B \tag{8.31}$$

Using the above concepts and equations, the average composite stiffness can be obtained from the strain concentration tensor A and the filler and matrix properties:

$$C = C_m + v_f (C_f - C_m) A \tag{8.32}$$

8.3.4.2 HALPIN– TSAI MODEL

The Halpin–Tsai model is a well-known composite theory to predict the stiffness of unidirectional composites as a functional of aspect ratio. In this model, the longitudinal E11 and transverse E22 engineering moduli are expressed in the following general form:

$$\frac{E}{E_m} = \frac{1+\zeta\eta v_f}{1-\eta v_f} \tag{8.33}$$

where E and E_m represent the Young's modulus of the composite and matrix, respectively, v_f is the volume fraction of filler, and η is given by:

$$\eta = \frac{\dfrac{E}{E_m}-1}{\dfrac{E_f}{E_m}+\zeta_f} \tag{8.34}$$

where E_f represents the Young's modulus of the filler and ζ_f the shape parameter depending on the filler geometry and loading direction. When calculating longitudinal modulus E_{11}, ζ_f is equal to l/t, and when calculating transverse modulus E_{22}, ζ_f is equal to w/t. Here, the parameters of l, w and t are the length, width and thickness of the dispersed fillers, respectively. If $\zeta_f \to 0$, the Halpin–Tsai theory converges to the inverse rule of mixture (lower bound):

$$\frac{1}{E} = \frac{v_f}{E_f} + \frac{1-v_f}{E_m} \tag{8.35}$$

Conversely, if $\zeta_f \to \infty$, the theory reduces to the rule of mixtures (upper bound),

$$E = E_f v_f + E_m(1-v_f) \tag{8.36}$$

8.3.4.3   MORI-TANAKA MODEL

The Mori–Tanaka model is derived based on the principles of Eshelby's inclusion model for predicting an elastic stress field in and around an ellipsoidal filler in an infinite matrix. The complete analytical solutions for longitudinal E_{11} and transverse E_{22} elastic moduli of an isotropic matrix filled with aligned spherical inclusion are:[45, 46].

$$\frac{E_{11}}{E_m} = \frac{A_0}{A_0 + v_f(A_1 + 2v_0 A_2)} \tag{37}$$

$$\frac{E_{22}}{E_m} = \frac{2A_0}{2A_0 + v_f(-2A_3 + (1-v_0 A_4) + (1+v_0)A_5 A_0)} \tag{8.38}$$

where *Em* represents the Young's modulus of the matrix, v_f the volume fraction of filler, v_0 the Poisson's ratio of the matrix, parameters, A0, A1,...,A5 are functions of the Eshelby's tensor and the properties of the filler and the matrix, including Young's modulus, Poisson's ratio, filler concentration and filler aspect ratio [45].

8.3.4.5 EQUIVALENT-CONTINUUM AND SELF-SIMILAR APPROACHES

Numerous micromechanical models have been successfully used to predict the macroscopic behavior of fiber-reinforced composites. However, the direct use of these models for nanotube-reinforced composites is doubtful due to the significant scale difference between nanotube and typical carbon fiber. Recently, two methods have been proposed for modeling the mechanical behavior of single-walled carbon nanotube (SWCN) composites:

Equivalent-continuum approach and self-similar approach [47]. The equivalent-continuum approach was proposed by Odegard et al. [48]. In this approach, MD was used to model the molecular interactions between SWCN/polymer and a homogeneous equivalent-continuum reinforcing element (e.g., a SWCN surrounded b polymer) was constructed as shown in Figure 8.2. Then, micromechanics are used to determine the effective bulk properties of the equivalent-continuum reinforcing element embedded in a continuous polymer. The equivalent-continuum approach consists of four major steps, as briefly described below. Step 1: MD simulation is used to generate the equilibrium structure of a SWCN–polymer composite and then to establish the RVE of the molecular model and the equivalent-continuum model. Step 2: The potential energies of deformation for the molecular model and effective fiber are derived and equated for identical loading conditions. The bonded and nonbonded interactions within a polymer molecule are quantitatively described by MM. For the SWCN/polymer system, the total potential energy U_m of the molecular model is

$$U^m = \sum U^r(K_r) + \sum U^\theta(K_\theta) + \sum U^{vdw}(K_{vdw}) \qquad (8.39)$$

where U^r, U^q and U_{vdw} are the energies associated with covalent bond stretching, bond-angle bending, and van der Waals interactions, respectively. An equivalent-truss model of the RVE is used as an intermediate step to link the molecular and equivalent-continuum models. Each atom in the molecular model is represented by a pin-joint, and each truss element represents an atomic bonded or nonbonded interaction. The potential energy of the truss model is

$$U^t = \sum U^a (E^a) + \sum U^b (E^b) + \sum U^c (E^c) \qquad (8.40)$$

where U^a, U^b and U^c are the energies associated with truss elements that represent covalent bond stretching, bond-angle bending, and van der Waals interactions, respectively. The energies of each truss element are a function of the Young's modulus, E.

Step 3: A constitutive equation for the effective fiber is established. Since the values of the elastic stiffness tensor components are not known a priori, a set of loading conditions are chosen such that each component is uniquely determined from

$$U^f = U^t = U^m \qquad (8.41)$$

Step 4: Overall constitutive properties of the dilute and unidirectional SWCN/polymer composite are determined with Mori–Tanaka model with the mechanical properties of the effective fiber and the bulk polymer. The layer of polymer molecules that are near the polymer/nanotube interface (Figure 8.2) is included in the effective fiber, and it is assumed that the matrix polymer surrounding the effective fiber has mechanical properties equal to those of the bulk polymer. The self-similar approach was proposed by Pipes and Hubert [49] which consists of three major steps:

First, a helical array of SWCNs is assembled. This array is termed as the SWCN nanoarray where 91 SWCNs make up the cross-section of the helical nanoarray. Then, the SWCN nanoarrays is surrounded by a polymer matrix and assembled into a second twisted array, termed as the SWCN nanowire. Finally, the SWCN nanowires are further impregnated with a polymer matrix and assembled into the final helical array-the SWCN microfiber. The self-similar geometries described in the nanoarray, nanowire and microfiber (Figure 8.3) allow the use of the same mathematical and geometric model for all three geometries [49]

8.3.4.6   FINITE ELEMENT METHOD

FEM is a general numerical method for obtaining approximate solutions in space to initial-value and boundary-value problems including time-dependent processes. It employs preprocessed mesh generation, which enables the model to fully capture the spatial discontinuities of highly inhomogeneous materials. It also allows complex, nonlinear tensile relationships to be incorporated into the analysis. Thus, it has been widely used in mechanical, biological and geological systems. In FEM, the entire domain of interest is spatially discretized into an assembly of simply shaped subdomains (e.g., hexahedra or tetrahedral in three dimensions, and rectangles or triangles in two dimensions) without gaps and without overlaps. The subdomains are interconnected at joints (i.e., nodes). The implementation of FEM includes the important steps shown in Figure 8.4. The energy in FEM is taken from the theory of linear elasticity and thus the input parameters are simply the elastic moduli and the density of the material. Since these parameters are in agreement with the values computed by MD, the simulation is consistent across the scales. More specifically, the total elastic energy in the absence of tractions and body forces within the continuum model is given by:[50]

$$U = U_v + U_k \qquad (8.42)$$

$$U_k = 1/2 \int dr p(r) \left| \dot{U}_r \right|^2 \qquad (8.43)$$

$$U_v = \frac{1}{2} \int dr \sum_{\mu,\nu,\lambda,\sigma=1}^{3} \varepsilon_{\mu\nu}(r) C_{\mu\nu\lambda\sigma} \lambda\sigma(r) \qquad (8.44)$$

where U_v is the Hookian potential energy term which is quadratic in the symmetric strain tensor e, contracted with the elastic constant tensor C. The Greek indices (i.e., m, n, l, s) denote Cartesian directions. The kinetic energy U_k involves the time rate of change of the displacement field $\dot{U}$, and the mass density ρ.

These are fields defined throughout space in the continuum theory. Thus, the total energy of the system is an integral of these quantities over the volume of the sample dυ. The FEM has been incorporated in some commercial software packages and open source codes (e.g., ABAQUS, ANSYS, Palmyra, and OOF) and widely used to evaluate the mechanical properties of polymer composites. Some attempts have recently been made to apply the FEM to nanoparticle-reinforced polymer nanocomposites. In order to capture the multiscale material behaviors, efforts are also underway to combine the multiscale models spanning from molecular to macroscopic levels [51, 52].

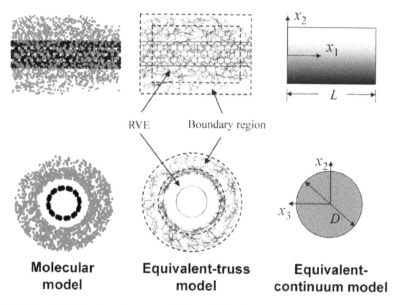

FIGURE 8.2   Equivalent-continuum modeling of effective fiber [48].

8.4   MULTI SCALE MODELING OF MECHANICAL PROPERTIES

In Odegard's study, [48] a method has been presented for linking atomistic simulations of nanostructured materials to continuum models of the corresponding
bulk material. For a polymer composite system reinforced with single-walled
carbon nanotubes (SWNT), the method provides the steps whereby the nanotube, the local polymer near the nanotube, and the nanotube/polymer interface
can be modeled as an effective continuum fiber by using an equivalent-continuum model. The effective fiber retains the local molecular structure and bonding information, as defined by molecular dynamics, and serves as a means for
linking the equivalent-continuum and micromechanics models. The micromechanics method is then available for the prediction of bulk mechanical properties of SWNT/polymer composites as a function of nanotube size, orientation,
and volume fraction. The utility of this method was examined by modeling tow
composites that both having a interface. The elastic stiffness constants of the
composites were determined for both aligned and three-dimensional randomly
oriented nanotubes, as a function of nanotube length and volume fraction. They
used Mori–Tanaka model [53] for random and oriented fibers position and compare their model with mechanical properties, the interface between fiber and
matrix was assumed perfect. Motivated by micrographs showing that embedded

nanotubes often exhibit significant curvature within the polymer, Fisher et al. [54] have developed a model combining finite element results and micromechanical methods (Mori-Tanaka)to determine the effective reinforcing modulus of a wavy embedded nanotube with perfect bonding and random fiber orientation assumption. This effective reinforcing modulus (ERM) is then used within a multiphase micromechanics model to predict the effective modulus of a polymer reinforced with a distribution of wavy nanotubes. We found that even slight nanotube curvature significantly reduces the effective reinforcement when compared to straight nanotubes. These results suggest that nanotube waviness may be an additional mechanism limiting the modulus enhancement of nanotube-reinforced polymers. Bradshaw et al. [55] investigated the degree to which the characteristic waviness of nanotubes embedded in polymers can impact the effective stiffness of these materials. A 3D finite element model of a single infinitely long sinusoidal fiber within an infinite matrix is used to numerically compute the dilute strain concentration tensor. A Mori–Tanaka model utilizes this tensor to predict the effective modulus of the material with aligned or randomly oriented inclusions. This hybrid finite element micromechanical modeling technique is a powerful extension of general micromechanics modeling and can be applied to any composite microstructure containing nonellipsoidal inclusions. The results demonstrate that nanotube waviness results in a reduction of the effective modulus of the composite relative to straight nanotube reinforcement. The degree of reduction is dependent on the ratio of the sinusoidal wavelength to the nanotube diameter. As this wavelength ratio increases, the effective stiffness of a composite with randomly oriented wavy nanotubes converges to the result obtained with straight nanotube inclusions.

The effective mechanical properties of carbon nanotube-based composites are evaluated by Liu and Chen [56] using a 3D nanoscale RVE based on 3D elasticity theory and solved by the finite element method. Formulas to extract the material constants from solutions for the RVE under three loading cases are established using the elasticity. An extended rule of mixtures, which can be used to estimate the Young's modulus in the axial direction of the RVE and to validate the numerical solutions for short CNTs, is also derived using the strength of materials theory. Numerical examples using the FEM to evaluate the effective material constants of a CNT-based composites are presented, which demonstrate that the reinforcing capabilities of the CNTs in a matrix are significant. With only about 2 and 5 percent volume fractions of the CNTs in a matrix, the stiffness of the composite in the CNT axial direction can increase as many as 0.7 and 9.7 times for the cases of short and long CNT fibers, respectively. These simulation results, which are believed to be the first of its kind for CNT-based composites, are consistent with the experimental results reported in the literature Schadler et al., [57] Wagner et al., [58] Qian et al. [59]. The developed extended rule of mixtures is also found to be quite effective in evaluating the stiffness of the CNT-based composites in the CNT axial direction. Many research issues need to be addressed in the modeling and simulations of CNTs in a matrix material for the development of nanocomposites. Analytical methods and simulation models to extract the mechanical properties of the CNT-based nanocomposites need to be further developed and verified with experimental results. The analytical method and simulation approach developed in this paper are only a preliminary study. Different type of RVEs, load cases and different solution methods should be investigated. Different interface conditions, other than perfect bonding, need to be investigated using different models to more accurately account for the interactions of the CNTs in a matrix material at the nanoscale. Nanoscale interface cracks can be analyzed using simulations to investigate the failure mechanism in nanomaterials. Interactions among a large number of CNTs in a matrix can be simulated if the computing power is available. Single-walled and multiwalled CNTs as reinforcing fibers in a matrix can be studied by simulations to find out their advantages and disadvantages. Finally, large multiscale simulation models for CNT-based composites, which can link the models at the nano, micro and macro scales, need to be developed, with the help of analytical and experimental work [56]. The three RVEs proposed in [60] and shown in Figure 8.3 are relatively simple regarding the models and scales and pictures in Figure 8.4 are three loading cases for the cylindrical RVE. However, this is only the first step toward more sophisticated and large scale simulations of CNT-based composites. As the computing power and confidence in simulations of CNT-based composites increase, large scale 3D models containing hundreds or even more CNTs, behaving linearly or nonlinearly, with coatings or of different sizes, distributed evenly or randomly, can be employed to investigate the interactions among the CNTs in a matrix and to evaluate the effective material properties. Other numerical methods can also be attempted for the modeling and simulations of CNT-based composites, which may offer some advantages over the FEM approach. For example, the boundary element method, Liu et al.; [60] Chen and Liu, [61] accelerated with the fast multipole techniques, Fu et al., [62] Nishimura et al., [63] and the mesh free methods Qian et al. [64] may enable one to model an RVE with thousands of CNTs in a matrix on a desktop computer. Analysis of the CNT-based composites using the boundary element method is already underway and will be reported subsequently.

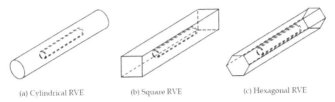

(a) Cylindrical RVE   (b) Square RVE   (c) Hexagonal RVE

FIGURE 8.3 Three nanoscale representative volume elements for the analysis of CNT-based nanocomposites [56]

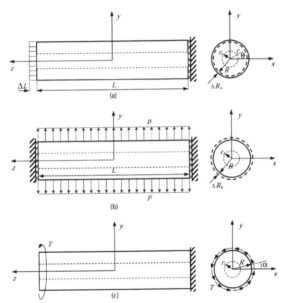

FIGURE 8.4 Three loading cases for the cylindrical RVE used to evaluate the effective material properties of the CNT-based composites. (a) Under axial stretch DL; (b) under lateral uniform load P; (c) under torsional load T. [56].

The effective mechanical properties of CNT based composites are evaluated using square RVEs based on 3D elasticity theory and solved by the FEM. Formulas to extract the effective material constants from solutions for the square RVEs under two loading cases are established based on elasticity. Square RVEs with multiple CNTs are also investigated in evaluating the Young's modulus and Poisson's ratios in the transverse plane. Numerical examples using the FEM are presented, which demonstrate that the load-carrying capabilities of the CNTs in a matrix are significant. With the addition of only about 3.6 percent volume fraction of the CNTs in a matrix, the stiffness of the composite in the CNT axial direction can increase as much as 33 percent for the case of long CNT fibers [65]. These simulation results are consistent with both the experimental ones reported in the literature [56–59, 66]. It is also found that cylindrical RVEs tend to overestimate the effective Young's moduli due to the fact that they overestimate the volume fractions of the CNTs in a matrix. The square RVEs, although more demanding in modeling and computing, may be the preferred model in future simulations for estimating the effective material constants, especially when multiple CNTs need to be considered. Finally, the rules of mixtures, for both long and short CNT cases, are found to be quite accurate in estimating the effective Young's moduli in the CNT axial direction. This may suggest that 3D FEM modeling may not be necessary in obtaining the effective material constants in the CNT direction, as in the studies of the conventional fiber reinforced composites. Efforts in comparing the results presented in this paper using the continuum approach directly with the MD simulations are underway. This is feasible now only for a smaller RVE of one CNT embedded in a matrix. In future research, the MD and continuum approach should be integrated in a multiscale modeling and simulation environment for analyzing the CNT-based composites. More efficient models of the CNTs in a matrix also need to be developed, so that a large number of CNTs, in different shapes and forms (curved or twisted), or randomly distributed in a matrix, can be modeled. The ultimate validation of the simulation results should be done with the nanoscale or microscale experiments on the CNT-reinforced composites [64].

Griebel and Hamaekers [67] reviewed the basic tools utilized in computational nanomechanics and materials, including the relevant underlying principles and concepts. These tools range from subatomic and initio methods to classical molecular dynamics and multiple-scale approaches. The energetic link between the quantum mechanical and classical systems has been discussed, and limitations of the standing alone molecular dynamics simulations have been shown on a series of illustrative examples. The need for multiscale simulation methods to take nanoscale aspects of material behavior was therefore emphasized; that was followed by a review and classification of the mainstream and emerging multiscale methods. These simulation methods include the broad areas of quantum mechanics, molecular dynamics and multiple-scale approaches, based on coupling the atomistic and continuum models. They summarize the strengths and limitations of currently available multiple-scale techniques, where the emphasis is made on the latest perspective approaches, such as the bridging scale method, multiscale boundary conditions, and multiscale fluidics. Example problems, in which multiple-scale simulation methods yield equivalent results to full atomistic simulations at fractions of the computational cost, were shown.

They Compare their results with Odegard, et al., [48] the micromechanic method was BEM Halpin-Tsai Eq. [68] with aligned fiber by perfect bonding.

The solutions of the strain-energy-changes due to a SWNT embedded in an infinite matrix with imperfect fiber bonding are obtained through numerical method by Wan *et al. [69]*. A "critical" SWNT fiber length is defined for full load transfer between the SWNT and the matrix, through the evaluation of the strain-energy-changes for different fiber lengths The strain-energy-change is also used to derive the effective longitudinal Young's modulus and effective bulk modulus of the composite, using a dilute solution. The main goal of their research was investigation of strain-energy-change due to inclusion of SWNT using FEM. To achieve full load transfer between the SWNT and the matrix, the length of SWNT fibers should be longer than a 'critical' length if no weak interphase exists between the SWNT and the matrix [69].

A hybrid atomistic/continuum mechanics method is established in the Feng *et al. [70]* study the deformation and fracture behaviors of CNTs in composites. The unit cell containing a CNT embedded in a matrix is divided in three regions, which are simulated by the atomic-potential method, the continuum method based on the modified Cauchy–Born rule, and the classical continuum mechanics, respectively. The effect of CNT interaction is taken into account via the Mori–Tanaka effective field method of micromechanics. This method not only can predict the formation of Stone-Wales (5-7-7-5) defects, but also simulate the subsequent deformation and fracture process of CNTs. It is found that the critical strain of defect nucleation in a CNT is sensitive to its chiral angle but not to its diameter. The critical strain of Stone–Wales defect formation of zigzag CNTs is nearly twice that of armchair CNTs. Due to the constraint effect of matrix, the CNTs embedded in a composite are easier to fracture in comparison with those not embedded. With the increase in the Young's modulus of the matrix, the critical breaking strain of CNTs decreases.

Estimation of effective elastic moduli of nanocomposites was performed by the version of effective field method developed in the framework of quasi-crystalline approximation when the spatial correlations of inclusion location take particular ellipsoidal forms [71] The independent justified choice of shapes of inclusions and correlation holes provide the formulae of effective moduli which are symmetric, completely explicit and easily to use. The parametric numerical analyses revealed the most sensitive parameters influencing the effective moduli which are defined by the axial elastic moduli of nanofibers rather than their transversal moduli as well as by the justified choice of correlation holes, concentration and prescribed random orientation of nanofibers [72].

Li and Chou [73, 74] have reported a multiscale modeling of the compressive behavior of carbon nanotube/polymer composites. The nanotube is modeled at the atomistic scale, and the matrix deformation is analyzed by the continuum finite element method. The nanotube and polymer matrix are assumed to be bonded by van der Waals interactions at the interface. The stress distributions at the nanotube/polymer interface under isostrain and isostress loading conditions have been examined. They have used beam elements for SWCNT using molecular structural mechanics, truss rod for vdW links and cubic elements for matrix .the rule of mixture was used as for comparison in this research. The buckling forces of nanotube/polymer composites for different nanotube lengths and diameters are computed. The results indicate that continuous nanotubes can most effectively enhance the composite buckling resistance.

Anumandla and Gibson [75] describe an approximate, yet comprehensive, closed-form micromechanics model for estimating the effective elastic modulus of carbon nanotube-reinforced composites. The model incorporates the typically observed nanotube curvature, the nanotube length, and both 1D and 3D random arrangement of the nanotubes. The analytical results obtained from the closed-form micromechanics model for nanoscale representative volume elements and results from an equivalent finite element model for effective reinforcing modulus of the nanotube reveal that the reinforcing modulus is strongly dependent on the waviness, wherein, even a slight change in the nanotube curvature can induce a prominent change in the effective reinforcement provided. The micromechanics model is also seen to produce reasonable agreement with experimental data for the effective tensile modulus of composites reinforced with multiwalled nanotubes (MWNTs) and having different MWNT volume fractions.

Effective elastic properties for carbon nanotube-reinforced composites are obtained through a variety of micromechanics techniques [76]. Using the in-plane elastic properties of graphene, the effective properties of CNTs are calculated utilizing a composite cylinders micromechanics technique as a first step in a two-step process. These effective properties are then used in the self-consistent and Mori–Tanaka methods to obtain effective elastic properties of composites consisting of aligned single or multiwalled CNTs embedded in a polymer matrix. Effective composite properties from these averaging methods are compared to a direct composite cylinders approach extended from the work of Hashin and Rosen, [77] and Christensen and Lo, [78] Comparisons with finite element simulations are also performed. The effects of an interphase layer between the nanotubes and the polymer matrix as result of functionalization is also investigated using a multilayer composite cylinders approach. Finally, the modeling of the clustering of nanotubes into bundles due to interatomic forces is accomplished herein using a tessellation method in conjunction with a multiphase Mori–Tanaka technique. In addition to aligned nanotube composites, modeling of the effective elastic properties of randomly dispersed nanotubes into a matrix is performed using the Mori–Tanaka method, and comparisons with experimental data are made.

Selmi et al. [79] deal with the prediction of the elastic properties of polymer composites reinforced with SWNTs. Their contribution is the investigation of several micromechanical models, while most of the papers on the subject deal with only one approach. They implemented four homogenization schemes, a sequential one and three others based on various extensions of the Mori–Tanaka (M–T) mean-field homogenization model: two-level (M–T/M–T), two-step (M–T/M–T) and two-step (M–T/Voigt). Several composite systems are studied, with various properties of the matrix and the graphene, short or long nanotubes, fully aligned or randomly oriented in 3D or 2D. Validation targets are experimental data or finite element results, either based on a 2D periodic unit cell or a 3D representative volume element. The comparative study showed that there are cases where all micromechanical models give adequate predictions, while for some composite materials and some properties, certain models fail in a rather spectacular fashion. It was found that the two-level (M–T/M–T) homogenization model gives the best predictions in most cases. After the characterization of the discrete nanotube structure using a homogenization method based on energy equivalence, the sequential, the two-step (M–T/M–T), the two-step (M–T/Voigt), the two-level (M–T/M–T) and finite element models were used to predict the elastic properties of SWNT/polymer composites. The data delivered by the micromechanical models are compared against those obtained by finite element analyzes or experiments. For fully aligned, long nanotube polymer composite, it is the sequential and the two-level (M–T/M–T) models which delivered good predictions. For all composite morphologies (fully aligned, two-dimensional in-plane random orientation, and three-dimensional random orientation), it is the two-level (M–T/M–T) model which gave good predictions compared to finite element and experimental results in most situations. There are cases where other micromechanical models failed in a spectacular way.

Luo et al. [80] have used multiscale homogenization (MH) and FEM for wavy and straight SWCNTs, they have compare their results with Mori-Tanaka, Cox, Halpin-Tsai, Fu, et al., [81] Lauke [82]. Trespass et al. [83] used 3D elastic beam for C-C bond and 3D space frame for CNT and progressive fracture model for prediction of elastic modulus, they used rule of mixture for compression of their results. Their assumption was embedded a single SWCNT in polymer with perfect bonding. The multiscale modeling, MC, FEM and equivalent-continuum method was used by Spanos and Kontsos [84] and compared with the results of Zhu et al. [85] and Paiva et al. [86]

Md. A. Bhuiyan et al. [87] studied the effective modulus of CNT/PP composites is evaluated using FEA of a 3D RVE which includes the PP matrix, multiple CNTs and CNT/PP interphase and accounts for poor dispersion and nonhomogeneous distribution of CNTs within the polymer matrix, weak CNT/ polymer interactions, CNT agglomerates of various sizes and CNTs orientation and waviness. Currently, there is no other model, theoretical or numerical, that accounts for all these experimentally observed phenomena and captures their individual and combined effect on the effective modulus of nanocomposites. The model is developed using input obtained from experiments and validated against experimental data. CNT-reinforced PP composites manufactured by extrusion and injection molding are characterized in terms of tensile modulus, thickness and stiffness of CNT/PP interphase, size of CNT agglomerates and CNT distribution using tensile testing, AFM and SEM, respectively. It is concluded that CNT agglomeration and waviness are the two dominant factors that hinder the great potential of CNTs as polymer reinforcement. The proposed model provides the upper and lower limit of the modulus of the CNT/PP composites and can be used to guide the manufacturing of composites with engineered properties for targeted applications. CNT agglomeration can be avoided by employing processing techniques such as sonication of CNTs, stirring, calendaring, etc., whereas CNT waviness can be eliminated by increasing the injection pressure during molding and mainly by using CNTs with smaller aspect ratio. Increased pressure during molding can also promote the alignment of CNTs along the applied load direction. The 3D modeling capability presented in this study gives an insight on the upper and lower bound of the CNT/PP composites modulus quantitatively by accurately capturing the effect of various processing parameters. It is observed that when all the experimentally observed factors are considered together in the FEA the modulus prediction is in good agreement with the modulus obtained from the experiment. Therefore, it can be concluded that the FEM models proposed in this study by systematically incorporating experimentally observed characteristics can be effectively used for the determination of mechanical properties of nanocomposite materials. Their result is in agreement with the results reported in, [88]. The theoretical micromechanical models, shown in Figure 8.5, are used to confirm that our FEM model predictions follow the same trend with the one predicted by the models as expected.

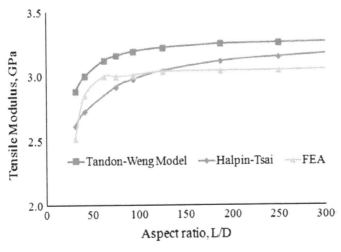

FIGURE 8.5   Effective modulus of 5 wt.% CNT/PP composites: theoretical models vs. FEA.

For reasons of simplicity and in order to minimize the mesh dependency on the results the hollow CNTs are considered as solid cylinders of circular cross-sectional area with an equivalent average diameter, shown in Figure 8.6, calculated by equating the volume of the hollow CNT to the solid one [87–88].

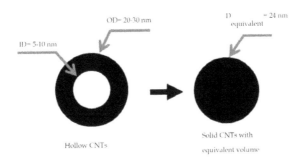

FIGURE 8.6   Schematic of the CNTs considered for the FEA.

The micromechanical models used for the comparison was Halpin–Tsai (H–T) [89] and Tandon–Weng (T–W) [90] model and the comparison was performed for 5 wt.% CNT/PP. It was noted that the H–T model results to lower modulus compared to FEA because H–T equation does not account for maximum packing fraction and the arrangement of the reinforcement in the composite. A modified H–T model that account for this has been proposed in the literature [91]. The effect of maximum packing fraction and the arrangement of the reinforcement within the composite becomes less significant at higher aspect ratios [92].

A finite element model of carbon nanotube, interphase and its surrounding polymer is constructed to study the tensile behavior of embedded short CNTs in polymer matrix in presence of vdW interactions in interphase region by shokrieh and rafiee [93]. The interphase is modeled using nonlinear spring elements capturing the force-distance curve of vdW interactions. The constructed model is subjected to tensile loading to extract longitudinal Young's modulus. The obtained results of this work have been compared with the results of previous research of the same authors [94] on long embedded carbon nanotube in polymer matrix. It shows that the capped short CNTs reinforce polymer matrix less efficient than long CNTs.

Despite the fact that researches have succeeded to grow the length of CNTs up to 4 cm as a world record in US Department of Energy Los Alamos National Laboratory [95] and also there are some evidences on producing CNTs with lengths up to millimeters, [96, 97] CNTs are commercially available in different lengths ranging from 100 nm to approximately 30 lm in the market based on employed process of growth [98–101]. Chemists at Rice University have identified a chemical process to cut CNTs into short segments [102] As a consequent, it can be concluded that the SWCNTs with lengths smaller than 1000 nm do not contribute significantly in reinforcing polymer matrix. On the other hand, the efficient length of reinforcement for a CNT with (10, 10) index is about 1.2 lm and short CNT with length of 10.8 lm can play the same role as long CNT reflecting the uppermost value reported in our previous research [94]. Finally, it is shown that the direct use of Halpin–Tsai equation to predict the modulus of SWCNT/ composites overestimates the results. It is also observed that application of previously developed long equivalent fiber stiffness [94] is a good candidate to be used in Halpin–Tsai equations instead of Young's modulus of CNT. Halpin–Tsai equation is not an appropriate model for smaller lengths, since there is not any reinforcement at all for very small lengths.

Earlier, a nanomechanical model has been developed by Chowdhury et al. [103] to calculate the tensile modulus and the tensile strength of randomly oriented short CNTs reinforced nanocomposites, considering the statistical variations of diameter and length of the CNTs. According to this model, the entire composite is divided into several composite segments which contain CNTs of almost the same diameter and length. The tensile modulus and tensile strength of the composite are then calculated by the weighted sum of the corresponding modulus and strength of each composite segment. The existing micromechanical approach for modeling the short fiber composites is modified to account for the structure of the CNTs, to calculate the modulus and the strength of each segmented CNT-reinforced composites. Multi-walled CNTs with and without intertube bridging (see Figure 8.7) have been considered. Statistical variations of the diameter and length of the CNTs are modeled by a normal distribution. Simulation results show that CNTs intertube bridging, length and diameter affect the nanocomposites modulus and strength. Simulation results have been compared with the available experimental results and the comparison concludes that the developed model can be effectively used to predict tensile modulus and tensile strength of CNTs reinforced composites.

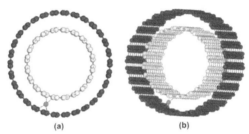

(a)   (b)

FIGOUR 8.7   Schematic of MWNT with intertube bridging. (a) Top view and (b) oblique view [103].

The effective elastic properties of carbon nanotube-reinforced polymers have been evaluated by Tserpes and Chanteli [104] as functions of material and geometrical parameters using a homogenized RVE. The RVE consists of the polymer matrix, a multiwalled carbon nanotube (MWCNT) embedded into the matrix and the interface between them. The parameters considered are the nanotube aspect ratio, the nanotube volume fraction as well as the interface stiffness and thickness. For the MWCNT, both isotropic and orthotropic material properties have been considered. Analyses have been performed by means of a 3D FE model of the RVE. The results indicate a significant effect of nanotube volume fraction. The effect of nanotube aspect ratio appears mainly at low values and diminishes after the value of 20. The interface mostly affects the effective elastic properties at the transverse direction. Having evaluated the effective elastic properties of the MWCNT/polymer at the microscale, the RVE has been used to predict the tensile modulus of a polystyrene specimen reinforced by randomly aligned MWCNTs for which experimental data exist in the literature. A very good agreement is obtained between the predicted and experimental tensile moduli of the specimen. The effect of nanotube alignment on the specimen's tensile modulus has been also examined and found to be significant since as misalignment increases the effective tensile modulus decreases radically. The proposed model can be used for the virtual design and optimization of CNT/polymer composites since it has proven capable of assessing the effects of different material and geometrical parameters on the elastic properties of the composite and predicting the tensile modulus of CNT-reinforced polymer specimens.

8.5   MODELING OF THE INTERFACE

8.5.1   INTRODUCTION

The superior mechanical properties of the nanotubes alone do not ensure mechanically superior composites because the composite properties are strongly influenced by the mechanics that govern the nanotube–polymer interface. Typically in composites, the constituents do not dissolve or merge completely and therefore, normally, exhibit an interface between one another, which can be considered as a different material with different mechanical properties. The structural strength characteristics of composites greatly depend on the nature of bonding at the interface, the mechanical load transfer from the matrix (polymer) to the nanotube and the yielding of the interface. As an example, if the composite is subjected to tensile loading and there exists perfect bonding between the nanotube and polymer and/or a strong interface then the load (stress) is transferred to the nanotube; since the tensile strength of the nanotube (or the interface) is very high the composite can withstand high loads. However, if the interface is weak or the bonding is poor, on application of high loading either the interface fails or the load is not transferred to the nanotube and the polymer fails due to their lower tensile strengths. Consider another example of transverse crack propagation. When the crack reaches the interface, it will tend to propagate along the interface, since the interface is relatively weaker (generally) than the nanotube (with respect to resistance to crack propagation). If the interface is weak, the crack will cause the interface to fracture and result in failure of the composite. In this aspect, CNTs are better than traditional fibers (glass, carbon) due to their ability to inhibit nano and micro cracks. Hence, the knowledge and understanding of the nature and mechanics of load (stress) transfer between the nanotube and polymer and properties of the interface is critical for manufacturing of mechanically enhanced CNT/polymer composites and will enable in tailoring of the interface for specific applications or superior mechanical properties. Broadly, the interfacial mechanics of CNT/polymer composites is appealing from three aspects: mechanics, chemistry, and physics. From a mechanics point of view, the important questions are:

1. The relationship between the mechanical properties of individual constituents, i.e., nanotube and polymer, and the properties of the interface and the composite overall.
2. The effect of the unique length scale and structure of the nanotube on the property and behavior of the interface.
3. Ability of the mechanics modeling to estimate the properties of the composites for the design process for structural applications.

From a chemistry point of view, the interesting issues are

1. The chemistry of the bonding between polymer and nanotubes, especially the nature of bonding (e.g., covalent or noncovalent and electrostatic).
2. The relationship between the composite processing and fabrication conditions and the resulting chemistry of the interface.
3. The effect of functionalization (treatment of the polymer with special molecular groups like hydroxyl or halogens) on the nature and strength of the bonding at the interface.

From the physics point of view, researchers are interested in

1. The CNT–polymer interface serves as a model nanomechanical or a lower dimensional system (1D) and physicists are interested in the nature of forces dominating at the nanoscale and the effect of surface forces (which are expected to be significant due to the large surface to volume ratio).
2. The length scale effects on the interface and the differences between the phenomena of mechanics at the macro (or meso) and the nanoscale.

8.5.2 SOME MODELING METHOD IN INTERFACE MODELING

Computational techniques have extensively been used to study the interfacial mechanics and nature of bonding in CNT–polymer composites. The computational studies can be broadly classified as atomistic simulations and continuum methods. The atomistic simulations are primarily based on molecular dynamic simulations (MD) and density functional theory (DFT) [105], [106–110] (some references). The main focus of these techniques was to understand and study the effect of bonding between the polymer and nanotube (covalent, electrostatic or van der Waals forces) and the effect of friction on the interface. The continuum methods extend the continuum theories of micromechanics modeling and fiber-reinforced composites (elaborated in the next section) to CNT/polymer composites [111–114] and explain the behavior of the composite from a mechanics point of view.

On the experimental side, the main types of studies that can be found in literature are as follows:

1. Researchers have performed experiments on CNT/polymer bulk composites at the macroscale and observed the enhancements in mechanical properties (like elastic modulus and tensile strength) and tried to correlate the experimental results and phenomena with continuum theories like micromechanics of composites or Kelly Tyson shear lag model [105, 115–120]

2. Raman spectroscopy has been used to study the reinforcement provided by CNTs to the polymer, by straining the CNT–polymer composite and observing the shifts in Raman peaks [121–125].

3. In situ TEM straining has also been used to understand the mechanics, fracture and failure processes of the interface. In these techniques, the CNT–polymer composite (an electron transparent thin specimen) is strained inside a TEM and simultaneously imaged to get real-time and spatially resolved (1 nm) information [110, 126]

8.5.3 NUMERICAL APPROACH

A MD model may serve as a useful guide, but its relevance for a covalent-bonded system of only a few atoms in diameter is far from obvious. Because of this, the phenomenological multiple column models that considers the interlayer radial displacements coupled through the van der Waals forces is used. It should also be mentioned the special features of load transfer, in tension and in compression, in MWNT/epoxy composites studied by Schadler et al.[57] who detected that load transfer in tension was poor in comparison to load transfer in compression, implying that during load transfer to MWNTs, only the outer layers are stressed in tension due to the telescopic inner wall sliding (reaching at the shear stress 0.5 MPa), [127] whereas all the layers respond in compression. It should be mentioned that NTCMs usually contain not individual, separated SW-CNTs, but rather bundles of closest-packed SWCNTs, [128] where the twisting of the CNTs produces the radial force component giving the rope structure more stable than wires in parallel. Without strong chemically bonding, load transfer between the CNTs and the polymer matrix mainly comes from weak electrostatic and van der Waals interactions, as well as stress/deformation arising from mismatch in the coefficients of thermal expansion [129] Numerous researchers [130] have attributed lower than- predicted CNT/polymer composite properties to the availability of only a weak interfacial bonding. So, Frankland et al. [106] demonstrated by MD simulation that the shear strength of a polymer/nanotube interface with only van der Waals interactions could be increased by over an order of magnitude at the occurrence of covalent bonding for only 1 percent of the nanotube's carbon atoms to the polymer matrix. The recent force-field-based molecular mechanics calculations [131] demonstrated that the binding energies and frictional forces play only a minor role in determining the strength of the interface. The key factor in forming a strong bond at the interface is having a helical conformation of the polymer around the nanotube; polymer wrapping around nanotube improves the polymer-nanotube interfacial strength, although configurationally thermodynamic considerations do not necessarily support these architectures for all polymer chains [132]. Thus, the strength of the interface may result from molecular-level entanglement of the two phases and forced long-range ordering of the polymer. To ensure the robustness of data reduction schemes that are based on continuum mechanics, a careful analysis of continuum approximations used in macromolecular models and possible limitations of these approaches at the nanoscale are additionally required that can be done by the fitting of the results obtained by the use of the proposed phenomenological interface model with the experimental data of measurement of the stress distribution in the vicinity of a nanotube.

Meguid et al. [133] investigated the interfacial properties of carbon nanotube (CNT) reinforced polymer composites by simulating a nanotube pull-out experiment. An atomistic description of the problem was achieved by implementing constitutive relations that are derived solely from interatomic potentials. Specifically, they adopt the Lennard-Jones (LJ) interatomic potential to simulate a nonbonded interface, where only the van der Waals (vdW) interactions between the CNT and surrounding polymer matrix was assumed to exist. The effects of such parameters as the CNT embedded length, the number of vdW interactions, the thickness of the interface, the CNT diameter and the cut-off distance of the LJ potential on the interfacial shear strength (ISS) are investigated and discussed. The problem is formulated for both a generic thermoset polymer and a specific two-component epoxy based on a diglycidyl ether of bisphenol A (DGEBA) and triethylene tetramine (TETA) formulation. The study further illustrated that by accounting for different CNT capping scenarios and polymer morphologies around the embedded end of the CNT, the qualitative correlation between simulation and experimental pull-out profiles can be improved. Only vdW interactions were considered between the atoms in the CNT and the polymer implying a nonbonded system. The vdW interactions were simulated using the LJ potential, while the CNT was described using the Modified Morse potential. The results reveal that the ISS shows a linear dependence on the vdW interaction density and decays significantly with increasing nanotube embedded length. The thickness of the interface was also varied and our results reveal that lower interfacial thicknesses favor higher ISS. When incorporating a 2.5Ψ cut-off distance to the LJ potential, the predicted ISS shows an error of approximately 25.7 relative to a solution incorporating an infinite cut-off distance. Increasing the diameter of the CNT was found to increase the peak pull-out force approximately linearly. Finally, an examination of polymeric and CNT capping conditions showed that incorporating an end cap in the simulation yielded high initial pull-out peaks that better correlate with experimental findings. These findings have a direct bearing on the design and fabrication of carbon nanotube-reinforced epoxy composites.

Fiber pull-out tests have been well recognized as the standard method for evaluating the interfacial bonding properties of composite materials. The output of these tests is the force required to pullout the nanotube from the surrounding polymer matrix and the corresponding interfacial shear stresses involved. The problem is formulated using a representative volume element (RVE) which consists of the reinforcing CNT, the surrounding polymer matrix, and the CNT/ polymer interface as depicted in Figure 8.8a, b shows a schematic of the pull-out process, where x is the pullout distance and L is the embedded length of the nanotube. The atomistic-based continuum (ABC) multiscale modeling technique is used to model the RVE. The approach adopted here extends the earlier work of Wernik and Meguid [134]

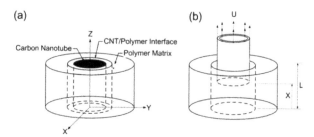

FIGURE 8.8   Schematic depictions of (a) the representative volume element and (b) the pull-out process [133]

The new features of the current work relate to the approach adopted in the modeling of the polymer matrix and the investigation of the CNT polymer interfacial properties as appose to the effective mechanical properties of the RVE. The idea behind the ABC technique is to incorporate atomistic interatomic potentials into a continuum framework. In this way, the interatomic potentials introduced in the model capture the underlying atomistic behavior of the different phases considered. Thus, the influence of the nanophase is taken into account via appropriate atomistic constitutive formulations. Consequently, these measures are fundamentally different from those in the classical continuum theory. For the sake of completeness, Wernik and Meguid provided a brief outline of the method detailed in their earlier work [133–134]

The cumulative effect of the vdW interactions acting on each CNT atom is applied as a resultant force on the respective node which is then resolved into its three Cartesian components. This process is depicted in Figure 8.9. During each iteration of the pull-out process, the above expression is reevaluated for each vdW interaction and the cumulative resultant force and its three Cartesian components are updated to correspond to the latest pull-out configuration. Figure 8.10 shows a segment of the CNT with the cumulative resultant vdW force vectors as they are applied to the CNT atoms.

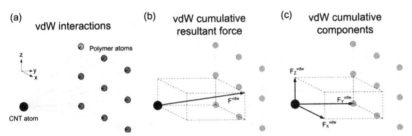

FIGURE 8.9 The process of nodal vdW force application. (a) vdW interactions on an individual CNT atom, (b) the cumulative resultant vdW force, and (c) the cumulative vdW Cartesian components.

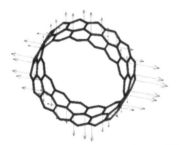

FIGURE 8.10 Segment of CNT with cumulative resultant vdW force vectors.

Yang et al. [135] investigated the CNT size effect and weakened bonding effect between an embedded CNT and surrounding matrix were characterized using MD simulations. Assuming that the equivalent-continuum model of the CNT atomistic structure is a solid cylinder, the transversely isotropic elastic constants of the CNT decreased as the CNT radius increased. Regarding the elastic stiffness of the nanocomposite unit cell, the same CNT size dependency was observed in all independent components, and only the longitudinal Young's modulus showed a positive reinforcing effect whereas other elastic moduli demonstrated negative reinforcing effects as a result of poor load transfer at the interface. To describe the size effect and weakened bonding effect at the interface, a modified multi-inclusion model was derived using the concepts of an effective CNT and effective matrix. During the scale bridging process incorporating the MD simulation results and modified multi-inclusion model, we found that both the elastic modulus of the CNT and the adsorption layer near the CNT contributed to the size-dependent elastic modulus of the nanocomposites. Using the proposed multiscale bridging model, the elastic modulus for nanocomposites at various volume fractions and CNT sizes could be estimated. Among three major factors (CNT waviness, the dispersion state, and adhesion between the CNT and matrix), the proposed model considered only the weakened bonding effect. However, the present multiscale framework can be easily applied in considering the aforementioned factors and describing the real nanocomposite microstructures. In addition, by considering chemically grafted molecules (covalent or noncovalent bonds) to enhance the interfacial load transfer mechanism in MD simulations, the proposed multiscale approach can offer a deeper understanding of the reinforcing mechanism, and a more practical analytical tool with which to analyze and design functional nanocomposites. The analytical estimation reproduced from the proposed multiscale model can also provide useful information in modeling finite element-based representative volume elements of nanocomposite microstructures for use in multifunctional design.

The effects of the interphase and RVE configuration on the tensile, bending and torsional properties of the suggested nanocomposite were investigated by Ayatollahi et al. [136]. It was found that the stiffness of the nanocomposite could be affected by a strong interphase much more than by a weaker interphase. In addition, the stiffness of the interphase had the maximum effect on the stiffness of the nanocomposite in the bending loading conditions. Furthermore, it was revealed that the ratio of Le/Ln in RVE can dramatically affect the stiffness of the nanocomposite especially in the axial loading conditions.

For CNTs not well bonded to polymers, Jiang et al. [137] established a cohesive law for carbon nanotube/polymer interfaces. The cohesive law and its properties (e.g., cohesive strength, cohesive energy) are obtained directly from the Lennard–Jones potential from the van der Waals interactions. Such a cohesive law is incorporated in the micromechanics model to study the mechanical behavior of carbon nanotube-reinforced composite materials. CNTs indeed improve the mechanical behavior of composite at the small strain. However, such improvement disappears at relatively large strain because the completely debonded nanotubes behave like voids in the matrix and may even weaken the composite. The increase of interface adhesion between CNTs and polymer matrix may significantly improve the composite behavior at the large strain [138].

Zalamea et al. [139] employed the shear transfer model as well as the shear lag model to explore the stress transfer from the outermost layer to the interior layers in MWCNTs. Basically, the interlayer properties between graphene layers were designated by scaling the parameter of shear transfer efficiency with respect to the perfect bonding. The authors pointed out that as the number of layers in MWCNTs increases, the stress transfer efficiency decreases correspondingly. Shen et al. [140] examined load transfer between adjacent walls of DWCNTs using MD simulation, indicating that the tensile loading on the outermost wall of MWCNTs cannot be effectively transferred into the inner walls. However, when chemical bonding between the walls is established, the effectiveness can be dramatically enhanced. It is noted that in the above investigations, the loadings were applied directly on the outermost layers of MWCNTs; the stresses in the inner layers were then calculated either from the continuum mechanics approach [139] or MD simulation [140]. Shokrieh and Rafiee [93, 94] examined the mechanical properties of nanocomposites with capped single-walled carbon nanotubes (SWCNTs) embedded in a polymer matrix. The load transfer efficiency in terms of different CNTs' lengths was the main concern in their examination. By introducing an interphase to represent the vdW interactions between SWCNTs and the surrounding matrix, Shokrieh and Rafiee [93, 94] converted the atomistic SWCNTs into an equivalent-continuum fiber in finite element analysis. The idea of an equivalent solid fiber was also proposed by Gao and Li [141] to replace the atomistic structure of capped SWCNTs in the nanocomposites' cylindrical unit cell. The modulus of the equivalent solid was determined based on the atomistic structure of SWCNTs through molecular structure mechanics [142]. Subsequently, the continuum-based shear lag analysis was carried out to evaluate the axial stress distribution in CNTs. In addition, the influence of end caps in SWCNTs on the stress distribution of nanocomposites was also taken into account in their analysis. Tsai and Lu [143] characterized the effects of the layer number, intergraphic layers interaction, and aspect ratio of MWCNTs on the load transfer efficiency using the conventional shear lag model and finite element analysis. However, in their analysis, the interatomistic characteristics of the adjacent graphene layers associated with different degrees of interactions were simplified by a thin interphase with different moduli. The atomistic interaction between the grapheme layers was not taken into account in their modeling of MWCNTs. In light of the forgoing investigations, the equivalent solid of SWCNTs was developed by several researchers and then implemented as reinforcement in continuum-based nanocomposite models. Nevertheless, for MWCNTs, the subjects concerning the development of equivalent-continuum solid are seldom explored in the literature. In fact, how to introduce the atomistic characteristics, i.e., the interfacial properties of neighboring graphene layers in MWCNTs, into the equivalent-continuum solid is a challenging task as the length scales used to describe the physical phenomenon are distinct. Thus, a multiscale based simulation is required to account for the atomistic attribute of MWCNTs into an equivalent-continuum solid. In Lu and Tsai study, [144] the multiscale approach was utilized to investigate the load transfer efficiency from surrounding matrix to DWCNTs. The analysis consisted of two stages. First, a cylindrical DWCNTs equivalent continuum was proposed based on MD simulation where the pullout extension on the outer layer was performed in an attempt to characterize the atomistic behaviors between neighboring graphite layers. Subsequently, the cylindrical continuum (denoting the DWCNTs) was embedded in a unit cell of nanocomposites, and the axial stress distribution as well as the load transfer efficiency of the DWCNTs was evaluated from finite element analysis. Both single-walled carbon nanotubes (SWCNTs) and DWCNTs were considered in the simulation and the results were compared with each other.

An equivalent cylindrical solid to represent the atomistic attributes of DW-CNTs was proposed in this study. The atomistic interaction of adjacent graphite layers in DWCNTs was characterized using MD simulation based on which a spring element was introduced in the continuum equivalent solid to demonstrate the interfacial properties of DWCNTs. Subsequently, the proposed continuum solid (denotes DWCNTs) was embedded in the matrix to form DWCNTs nanocomposites (continuum model), and the load transfer efficiency within the DWCNTs was determined from FEM analysis. For the demonstration purpose, the DWCNTs with four different lengths were considered in the investigation. Analysis results illustrate that the increment of CNTs' length can effectively improve the load transfer efficiency in the outermost layers, nevertheless, for the inner layers, the enhancement is miniature. On the other hand, when the covalent bonds between the adjacent graphene layers are crafted, the load-carrying capacity in the inner layer increases as so does the load transfer efficiency of DWCNTs. As compared to SWCNTs, the DWCNTs still possess the less capacity of load transfer efficiency even though there are covalent bonds generated in the DWCNTs.

8.6 CONCLUDING REMARKS

Many traditional simulation techniques (e.g., MC, MD, BD, LB, Ginzburg–Landau theory, micromechanics and FEM) have been employed, and some novel simulation techniques (e.g., DPD, equivalent-continuum and self-similar approaches) have been developed to study polymer nanocomposites. These techniques indeed represent approaches at various time and length scales from molecular scale (e.g., atoms), to microscale (e.g., coarse-grains, particles, monomers) and then to macroscale (e.g., domains), and have shown success to various degrees in addressing many aspects of polymer nanocomposites. The simulation techniques developed thus far have different strengths and weaknesses, depending on the need of research. For example, molecular simulations can be used to investigate molecular interactions and structure on the scale of 0.1–10 nm. The resulting information is very useful to understanding the interaction strength at nanoparticle–polymer interfaces and the molecular origin of mechanical improvement. However, molecular simulations are computationally very demanding, thus not so applicable to the prediction of mesoscopic structure and properties defined on the scale of 0.1–10 mm, for example, the dispersion of nanoparticles in polymer matrix and the morphology of polymer nanocomposites. To explore the morphology on these scales, mesoscopic simulations such as coarse-grained methods, DPD and dynamic mean-field theory are more effective. On the other hand, the macroscopic properties of materials are usually studied by the use of mesoscale or macroscale techniques such as micromechanics and FEM. But these techniques may have limitations when applied to polymer nanocomposites because of the difficulty to deal with the interfacial nanoparticle–polymer interaction and the morphology, which are considered crucial to the mechanical improvement of nanoparticle-filled polymer nanocomposites. Therefore, despite the progress over the past years, there are a number of challenges in computer modeling and simulation. In general, these challenges represent the work in two directions. First, there is a need to develop new and improved simulation techniques at individual time and length scales. Secondly, it is important to integrate the developed methods at wider range of time and length scales, spanning from quantum mechanical domain (a few atoms) to molecular domain (many atoms), to mesoscopic domain (many monomers or chains), and finally to macroscopic domain (many domains or structures), to form a useful tool for exploring the structural, dynamic, and mechanical properties, as well as optimizing design and processing control of polymer nanocomposites. The need for the second development is obvious. For example, the morphology is usually determined from the mesoscale techniques whose implementation requires information about the interactions between various components (e.g., nanoparticle–nanoparticle and nanoparticle–polymer) that should be derived from molecular simulations. Developing such a multiscale method is very challenging but indeed represents the future of computer simulation and modeling, not only in polymer nanocomposites but also other fields. New concepts, theories and computational tools should be developed in the future to make truly seamless multiscale modeling a reality. Such development is crucial in order to achieve the longstanding goal of predicting particle–structure property relationships in material design and optimization.

The strength of the interface and the nature of interaction between the polymer and carbon nanotube are the most important factors governing the ability of nanotubes to improve the performance of the composite. Extensive research has been performed on studying and understanding CNT/polymer composites from chemistry, mechanics and physics aspects. However, there exist various issues like processing of composites and experimental challenges, which need to be addressed to gain further insights into the interfacial processes.

KEYWORDS

- **Carbon nanotubes (CNT)**
- **Modeling**
- **Nanocomposites**
- **Polymer**

REFERENCES

1. Iijima, S. *Nature*. **1991**, *354*, 56.
2. Dresselhaus, M. S.; Dresselhaus, G.; and Eklund, P. C.; Science of Fullerenes and Carbon Nanotubes, Academic Press; **1996**.
3. Saito, R.; Dresselhaus, G.; and Dresselhaus, G.; Physical properties of carbon nanotubes, London: Imperial College Press; **1998**.
4. Harris, P. J. F.; Carbon nanotubes and related structures: new materials for the twenty-first century, Cambridge University, Cambridge; **1999**.
5. Wagner, H. D.; and Vaia, R. A.; *Mater. Today. 7*, 38; 2004.
6. Zeng, Q. H.; Yu, A. B.; and Lu, G. Q.; *Prog. Polym. Sci.* **2008**, *33*, 191.
7. Dai, H.; *Surface. Sci.* **2002**, *500*, 218.
8. Bethune, D. S.; Klang, C. H.; De Vries, M. S.; Gorman, G.; Savoy, R.; Vazquez, J.; and Beyers, R.; *Nature*. **1993**, *363*, 605.
9. Dresselhaus, M. S.; Dresselhaus, G.; and Saito, R.; *Carbon*, **1995**, *33*, 883.
10. Thostenson, E. T.; Ren, Z. F.; and Chou, T. W.; *Compos. Sci. Technol. 61*, **2001**, 1899.
11. Yakobson, B. I.; and Avouris, P.; *Top. Appl. Phys.* **2001**, *80*, 287.
12. Ajayan, P. M.; Schadler, L. S.; and Braun, P. V.; Nanocomposite Science and Technology, John Wiley & Sons; **2006**.
13. Li, J.; Ma, P. C.; Chow, W. S.; To, B. Z. Tang, J.; and Kim, K.; *Adv. Funct. Mater. 17*, **2007**, 3207.
14. Ajayan, P. M.; Stephan, O.; Colliex, C.; and Trauth, D.; *Science*, **1994**, *265*, 1212.
15. Summary of Searching Results. http://www.scopus.com. Accessed January 2010.
16. Du, J. H.; Bai, J.; and Cheng, H. M.; *Express. Polym. Lett.* **2007**, *1*, 253.
17. Lee, J. Y.; Baljon, A. R. C.; Loring, R. F.; and Panagiotopoulos, A. Z.; *J. Chem. Phys.* **1998**, *109*, 10321.
18. Smith, G. D.; Bedrov, D.; Li, L. W.; and Byutner, O.; *J. Chem. Phys.* **2002**, *117*, 9478.
19. Smith, J. S.; Bedrov, D.; and Smith, G. D.; *Compos. Sci. Technol.*, **2003**, *63*, 1599.
20. Zeng, Q. H.; Yu, A. B.; Lu, G. Q.; Standish, R. K.; *Chem. Mater.* **2003**, *15*, 4732.

21. Vacatello, M.; *Macromol. Theor. Simul.* **2003**, *12*, 86.
22. Zeng, Q. H.; Yu, A. B.; and Lu, G. Q.; *Nanotechnology*. **2005**, *16*, 2757.
23. Allen, M. P.; and Tildesley, D. J.; Computer Simulation of Liquids. Oxford University Press.; **1989**.
24. Frenkel, D.; and Smit, B.; Understanding molecular simulation: from algorithms to applications, 2nd ed. Academic Press; **2002**.
25. Metropolis, N.; Rosenbluth, A. W.; Marshall, N.; Rosenbluth, M. N.; Teller, A.T.; *J. Chem. Phys.* **1953**, *21*, 1087.
26. Carmesin, I.; and Kremer, K.; *Macromolecules*. **1988**, *21*, 2819.
27. Hoogerbrugge, P. J.; and Koelman, J. M. V. A.; *Europhys. Lett.* **1992**, *19*, 155.
28. Gibson, J. B.; Chen, K.; and Chynoweth, S.; *J. Colloid. Interface. Sci.* **1998**, *206*, 464.
29. Dzwinel, V.; and Yuen, D. A.; *Int. J. Mod. Phys. C.*, **2000**, *11*, 1037.
30. Dzwinel, W.; and Yuen, D. A.; *J. Colloid. Interface. Sci.* **2000**, *225*, 179.
31. Chen, S.; and Doolen, G. D.; *Annu. Rev. Fluid. Mech.* **1998**, *30*, 329.
32. Cahn, J. W.; *Acta Metall.* **1961**, *9*, 795.
33. Cahn, J. W.; and Hilliard, J. E.; *Acta Metall.* **1971**, *19*, 151.
34. Cahn, J. W.; *J. Chem. Phys.* **1959**, *30*, 1121.
35. Lee, B. P.; Douglas, J. F.; and Glotzer, S. C.; *Phys. Rev. E.*, **1999**, *60*, 5812.
36. Ginzburg, V.V.; Qiu, F.; Paniconi, M.; Peng, G. W.; Jasnow, D.; and Balazs, A. C.; *Phys. Rev. Lett.* **1999**, *82*, 4026.
37. Qiu, F.; Ginzburg, V. V.; Paniconi, M.; Peng, G. W.; Jasnow, D.; and Balazs, A.C.; *Langmuir.* **1999**, *15*, 4952.
38. Ginzburg, V. V.; Gibbons, C.; Qiu, F.; Peng, G. W.; and Balazs, A. C.; *Macromolecules*. **2000**, *33*, 6140.
39. Ginzburg, V. V.; Qiu, F.; and Balazs, A. C.; *Polymer*. **2002**, *43*, 461.
40. G. He, A.C. Balazs, *J. Comput. Theor. Nanosci.*, **2**, 99 (2005).
41. Altevogt, P.; Ever, O. A.; Fraaije, J. G. E. M.; Maurits, N. M.; and van Vlimmeren, B.A.C.; *J. Mol. Struct.* **1999**, *463*, 139.
42. Kawakatsu, T.; Doi, M.; and Hasegawa, A.; *Int. J. Mod. Phys. C.* **1999**, *10*, 1531.
43. Morita, H.; Kawakatsu, T.; and Doi, M.; *Macromolecules*. **2001**, *34*, 8777.
44. Doi, M.; OCTA-a free and open platform and softwares of multiscale simulation for soft materials /http://octa.jp/S. 2002.
45. Tandon, G. P.; and Weng, G. J.; *Polym. Compos.* **1984**, *5*, 327.
46. Fornes, T. D.; and Paul, D. R.; *Polymer*. **2003**, *44*, 4993.
47. Odegard, G. M.; Pipes, R. B.; and Hubert, P.; *Compos. Sci. Technol.* **2004**, *64*, 1011.
48. Odegard, G. M.; Gates, T. S.; Wise, K. E.; Park, C.; and Siochi, E. J.; *Compos. Sci. Technol.* **2003**, *63* 1671.
49. Pipes, R. B.; and Hubert, P.; *Compos. Sci. Technol.*, **2002**, *62*, 419.
50. Rudd, R. E.; and Broughton, J. Q.; *Phys. Stat. Sol. B.* **2000**, *217*, 251.
51. Starr, F.W.; and Glotzer, S. C.; Simulations of filled polymers on multiple length scales, In MRS Proceedings,. KK4-1. Cambridge University Press, **2000**; 661, p
52. Glotzer, S. C.; and Starr, F. W.; Towards multiscale simulations of filled and nanofilled polymers, In AIChE Symposium Series, 44–53. New York; American Institute of Chemical Engineers; **1998**, **2001**.
53. Mori, T.; and Tanaka, K.; *Acta Metallurgica*. **1973**, *21*, 571.
54. Fisher, F.T.; Bradshaw, R. D.; and Brinson, L. C.; *Comp. Sci. Tech.* **2003**, *63*, 1689.
55. Fisher, F.T.; Bradshaw, R. D.; and Brinson, L. C.; *Comp. Sci. Tech.* **2003**, *63*, 1705.
56. Liu, Y.J.; and Chen, X. L.; *Mech. Matter.* **2003**, *35*, 69.
57. Schadler, L. S.; Giannaris, S. C.; and Ajayan, P. M.; *Appl. Phys. Lett.* **1998**, *73*, 3842.
58. Wagner, H. D.; Lourie, O.; Feldman, Y.; and Tenne, R.; *Appl. Phys. Lett.* **1998**, *72*, 188.

59. Qian, D.; Dickey, E. C.; Andrews, R.; and Rantell, T.; *Appl. Phys. Lett.* **2000**, *76*, 2868.
60. Liu, Y. J.; and Chen, X. L.; Modeling and analysis of carbon nanotube-based composites using the FEM and BEM., Submitted to CMES: Computer Modeling in Engineering and Science, **2002**.
61. Liu, Y. J.; Xu, N.; and Luo, J. F.; *J. Appl. Mech.* **2000**, *67*, 41.
62. Fu, Y.; Klimkowski, K. J.; Rodin, G. J.; Berger, E.; Browne, J. C.; Singer, J. K.; Van De Geijn, R. A.; and Vemaganti, K. S.; *Int. J. Num. Meth. Eng.* **1998**, *42*, 1215.
63. Nishimura, N.; Yoshida, K.; and Kobayashi, S.; *Eng. Anal. Bound. Elem.* **1999**, *23*, 97.
64. Qian, D.; Liu, W. K.; and Ruoff, R. S.; *J. Phys. Chem. B.* **2001**, *105*, 10753.
66. Chen, X. L.; and Liu, Y. J.; *Comput. Mater. Sci.*, **29**, 1 **2004**.
67. Bower, C.; Rosen, R.; Jin, L.; Han, J.; and Zhou, O.; *Appl. Phys. Lett.* **1999**, *74*, 3317.
68. Gibson, R. F.; Principles of Composite Material Mechanics, CRC Press.
69. Wan, H.; Delale, F.; and Shen, L.; *Mech. Res. Commun.* **2005**, *32*, 481.
70. Shi, D.; Feng, X.; Jiang, H.; Huang, Y.Y.; and Hwang, K.; *Int. J. Fract.* **2005**, *134*, 369.
71. Buryachenko, V. A.; and Roy, A.; *Compos.: Part B.* **2005**, *36*, 405.
72. Buryachenko, V. A.; Roy, A.; Lafdi, K.; Andeson, K. L.; and Chellapilla, S.; *Comp. Sci. Tech.* **2005**, *65*, 2435.
73. Li, C.; and Chou, T. W.; *J. Nanosci. Nanotechnol.* **2003**, *3*, 423.
74. Li, C.; and Chou, T. W.; *Comp. Sci. Tech.* **2006**, *66*, 2409.
75. Anumandla, V.; and Gibson, R. F.; *Compos.: Part A.* **2006**, *37*, 2178.
76. Seidel, G. D.; and Lagoudas, D. C.; *Mech. Mater.* **2006**, *38*, 884.
77. Hashin, Z.; and Rosen, B.; *J. Appl. Mech.*, 31, 223 (1964).
78. Christensen, R.; and Lo, K.; *J. Mech. Phys. Solid.* **1979**, *27*, 315.
79. Selmi, A.; Friebel, C.; Doghri, I.; and Hassis, H.; *Comp. Sci. Tech.* **2007**, *67*, 2071.
80. Luo, D.; Wang, W. X.; and Takao, Y.; *Comp. Sci. Tech.* **2007**, *67*, 2947.
81. Fu, S. Y.; Yue, C. Y.; Hu, X.; and Mai, Y. W.; *Compos. Sci. Technol.* **2000**, *60*, 3001.
82. Lauke, B.; *J. Polym. Eng.* **1992**, *11*, 103.
83. Tserpes, K. I.; Panikos, P.; Labeas, G.; and Panterlakis, Sp.G.; *Theor. Appl. Fract. Mech.* **2008**, *49*, 51.
84. Spanos, P. D.; and Kontsos, A.; *Prob. Eng. Mech.* **2008**, *23*, 456.
85. Paiva, M. C.; Zhou, B.; Fernando, K. A. S.; Lin, Y.; Kennedy, J. M.; and Sun, Y-P. *Carbon.* **2004**, *42*, 2849.
86. J. Zhu, H. Peng, F. Rodriguez-Macias, J. Margrave, V. Khabashesku, A. Imam, K. Lozano, E. Barrera, *Adv. Func. Mater.*, 14, 643 (2004).
87. Bhuiyan, M. A.; Pucha, R.V.; Worthy, J.; Karevan, M.; and Kalaitzidou, K.; *Compos. Struct.* **2013**, *95*, 80.
88. Papanikos, P.; Nikolopoulos, D. D.; Tserpes, K. I.; *Comput. Mater. Sci.*, **2008**, *43*, 345.
89. Affdl, J. C. H.; Kardos, J. L.; *Polym. Eng. Sci.* **1976**, *16*, 344.
90. Tandon, G. P.; and Weng, G. J.; *Polym. Compos.* **1984**, *5*, 327.
91. Nielsen, L. E.; Mechanical Properties of Polymers and Composites, vol. 2. **1974**, New York: Marcel Dekker.
92. Tucker III, C. L.; and Liang, E.; *Compos. Sci. Technol.* **1999**, *59*, 655.
93. Shokrieh, M. M.; and Rafiee, R.; *Compos. Struct.* **2010**, *92*, 2415.
94. Shokrieh, M. M.; and Rafiee, R.; *J. Compos. Struct.* **2009**, *22*, 23.
95. Press Release, US Consulate, World-record-length carbon nanotube grown at US Laboratory, Mumbai-India, September 15 (2004).
96. Evans, J.; Chemistry World News 2004. <http:// www/rsc.org/chemistryworld/news>.
97. Pan, Z.; Xie, S. S.; Chang, B.; and Wang, C.; *Nature.* **1998**, *394*, 631.
98. http://www.carbonsolution.com.
99. http://www.fibermax.eu/shop/.

100. http://www.nanoamor.com.
101. www.thomas-swan.co.uk.
102. Rice University's chemical 'Scissors' yield short carbon nanotubes. New process yields nanotubes small enough to migrate through cells. Science Daily July 2003]. <http://www.sciencedaily.com/releases/2003/07/030723083644.htm>.
103. Chowdhury, S. C.; Haque, B. Z.; Okabe, T.; and Gillespie Jr., J. W.; *Compos. Part B, 43,* **2012,** 1756.
104. Tserpes, K. I.; and Chanteli, A.; *Compos. Struct.* **2013,** *99,* 366.
105. Gou, J.; Minaie, B.; Wang, B.; Liang, Z.; and Zhang, C.; *Comput. Mater. Sci.,* **2004,** *31,* 225.
106. Frankland, S. J. V.; and Harik, V. M.; *Surf. Sci.* **2003,** *525,* 103.
107. Natarajan, U.; Misra, S.; Mattice, and W. L.; *Comput. Theor. Polym. Sci.* **1998,** *8,* 323.
108. Lordi, V.; and Yao, N.; *J. Mater. Res.* **2000,** *15,* 2770.
109. Wong, M.; Paramsothy, M.; Xu, X. J.; Ren, Y.; Li, S.; and Liao, K.; *Polymer.* **2003** *44,* 7757.
110. Qian, D.; Liu, W. K.; and Ruoff, R. S.; *Compos. Sci. Technol.* **2003,** *63,* 1561.
111. Liu, Y. J.; and Chen, X. L.; *J. Boundary. Elem..* **2003,** *1,* 316.
112. Chen, X. L.; and Liu, Y. J.; *Comput. Mater. Sci.* **2004,** *29,* 1.
113. Chen, X. L.; and Liu, Y. J.; *Mech. Mater.* **2003,** *35,* 69.
114. Qian, D.; Dickey, E. C.; Andrews, R.; and Rantell, T.; *Appl. Phys. Lett.* **2000,** *76,* 2868.
115. Thostenson, E. T.; and Chou, T-W.; *J. Phys. D: Appl. Phys.* **2002,** *35,* 77.
116. Bower, C.; Rosen, R.; Jin, L.; Han, J.; and Zhou, O.; *Appl. Phys. Lett.* **1999,** *74,* 3317.
117. Cooper, C. A.; Cohen, S. R.; Barber, A. H.; and Wagner, H. D.; *Appl. Phys. Lett.* **2002,** *81,* 3873.
118. Qian, D.; and Dickey, E. C.; *J. Microsc.* **2001,** *204,* 39.
119. Schadler, L. S.; Giannaris, S. C.; and Ajayan, P. M.; *Appl. Phys. Lett.* **1998,** *73,* 3842.
120. Wagner, H. D.; *Chem. Phys. Lett.* **2002,** *361,* 57.
121. Ajayan, P. M.; Schadler, L. S.; Giannaris, C.; and Rubio, A.; *Adv. Mater.* **2000,** *12,* 750.
122. Cooper, C. A.; Young, R. J.; and Halsall, M.; *Compos. Part A: Appl. Sci. Manuf.* **2001,** *32,* 401.
123. Hadjiev, V. G.; Iliev, M. N.; Arepalli, S.; Nikolaev, P.; and Files, B. S.; *Appl. Phys. Lett.* **2001,** *78,* 3193.
124. Paipetis, A.; Galiotis, C.; Liu, Y. C.; and Nairn, J. A.; *J. Compos. Mater.* **1999,** *33,* 377.
125. Valentini, L.; Biagiotti, J.; Kenny, J. M.; and Lopez Manchado, M. A.; *J. Appl. Polym. Sci.* **2003,** *89,* 2657.
126. D. Qian, G.J. Wagner, W.K. Liu, M-F, Yu, R.S. Ruoff, *Appl. Mech. Rev.,* **55,** 495(2002).
127. Yu, M-F.; Yakobson, B. I.; and Ruo, R. S.; *J. Phys. Chem. B.* **2000,** *104,* 8764.
128. Qian, D.; Liu, W. K.; and Ruoff, R. S.; *Compos. Sci. Technol.* **2003,** *63,* 1561.
129. Liao, K.; and Li, S.; *Appl. Phys. Lett.* **2001.** *79,* 4225.
130. Andrews, R.; and Weisenberger, M. C.; *Curr. Opin. Solid. State. Mater. Sci.* **2004,** *8,* 31.
131. Lordi, V.; and Yao, N.; *J. Mater. Res.* **2000,** *5,* 2770.
132. Wagner, H. D.; and Vaia, R. A.; *Mater Today.* **2004,** *7,* 38.
133. Wernik, J. M.; Cornwell-Mott, B. J.; and Meguid, S. A.; *Int. J. Solids. Struct.* **2012,** *49,* 1852.
134. Wernik, J. M.; Meguid, S. A.; *Acta. Mech.* **2011,** *217,* 1.
135. Yang, S.; Yu, S.; Kyoung, W.; Han, D. S.; and Cho, M.; *Polymer.* **2012,** *53,* 623.
136. Ayatollahi, M. R.; Shadlou, S.; and Shokrieh, M. M.; *Compos. Struct.* **2011,** *93,* 2250.
137. Jiang, L. Y.; Huang, Y.; Jiang, H.; Ravichandran, G.; Gao, H.; Hwang, K. C.; and Liu, B.; *J. Mech. Phys. Solids.* **2006,** *54,* 2436.
138. Tan, H.; Jiang, L. Y.; Huang, Y.; Liu, B.; and Hwang, K. C.; *Compos. Sci. Technol.* **2007,** *67,* 2941.
139. Zalamea, L.; Kim, H.; and Pipes, R. B.; *Compos. Sci. Technol.* **2007,** *67,* 3425.
140. Shen, G. A.; Namilae, S.; and Chandra, N.; *Mater. Sci. Eng. A.,* **2006,** *429,* 66.

141. Gao, X. L.; and Li, K.; *Int. J. Solids. Struct.* **2005,** *42,* 1649.
142. Li, C.; and Chou, T. W.; *Int. J. Solids. Struct.* **2003,** *40,* 2487.
143. Tsai, J. L.; and Lu, T. C.; *Compos. Struct.* **2009,** *90,* 172.
144. Lu, T. C.; and Tsai, J. L.; *Composites: Part B.* **2013,** *44,* 394.

CHAPTER 9

TRENDS IN NANOCHEMISTRY FOR METAL/CARBON NANOCOMPOSITES

V. I. KODOLOV[1,2] and V. V. TRINEEVA[1,3]

[1]Basic Research – High Educational Centre of Chemical Physics & Mesoscopy, Udmurt Scientific Center, Ural Division, Russian Academy of Sciences, Izhevsk, Russia

[2]M.T. Kalashnikov Izhevsk State Technical University, Izhevsk, Russia

[3]Institute of Mechanics, Russian Academy of Sciences, Izhevsk, Russia

CONTENTS

9.1   INTRODUCTION

At last time the option exists that the basis of nanotechnology is self-organization of systems [1]. System self-organization refers to synergetics [2]. Quite often, especially recently, the papers are published, for example, by Malinetsky [3], in which it is considered that nanotechnology is based on self-organization of metastable systems. As assumed [4], self-organization can proceed by dissipative (synergetic) and continual (conservative) mechanisms. At the same time, the system can be arranged due to the formation of new stable ("strengthening") phases or due to the growth provision of the existing basic phase. This phenomenon underlies the arising nanochemistry.

Assuming that nanoparticle oscillation energies correlate with their dimensions and comparing this energy with the corresponding region of electromagnetic waves, we can assert that energy action of nanostructures is within the energy region of chemical reactions. Therefore one of the possible definitions of nanochemistry can be the following:

Nanochemistry is a science investigating nanostructures and nanosystems in metastable ("transition") states and processes flowing with them in near-"transition" state or in "transition" state with low activation energies.

To carry out the processes based on the notions of nanochemistry, the directed energy action on the system is required, with the help of chemical particle field as well, for the transition from the prepared near-"transition" state into the process product state (in our case—into nanostructures or nanocomposites). The perspective area of nanochemistry is the chemistry in nanoreactors. Nanoreactors can be compared with specific nanostructures representing limited space regions in which chemical particles orientate creating "transition state" prior to the formation of the desired nanoproduct. Nanorectors have a definite activity which predetermines the creation of the corresponding product. When nanosized particles are formed in nanoreactors, their shape and dimensions can be the reflection of shape and dimensions of the nanoreactor [5].

9.2   THEORETICAL PREMISES FOR OBTAINING AND APPLICATION OF THE NANOCOMPOSITES

Previously we [5] proposed the parameter called the nanosized interval (B), in which the nanostructures demonstrate their activity. Depending on the structure and composition of nanoreactor internal walls, distance between them, shape and size of nanoreactor, the nanostructures differing in activity are formed. The correlation between surface energy, taking into account the thickness of surface layer, and volume energy was proposed as a measure of the activity of nano-structures, nanoreactors, and nanosystems [1].

In this case, we obtain the absolute dimensionless characteristic (a) of nano-structure or nanoreactor activity

$$a = \varepsilon_s^0 \, d/\varepsilon_v^0 \, N/r(h) = \varepsilon_s^0 \, d/\varepsilon_v^0 \cdot 1/B, \tag{1}$$

where B equals r(h)/N, r—radius of rotating bodies, including the hollow ones, h—film thickness, and N—number changing depending on the nanostructure shape. Parameter d characterizes the nanostructure surface layer thickness, and corresponding energies of surface unit and volume unit are defined by the nano-structure composition.

The proposed scheme of obtaining carbon/metal containing nanostructures in nanoreactors of polymeric matrixes includes the selection of polymeric ma-trixes containing functional groups. 3d metals (iron, cobalt, nickel, and copper) are selected as the elements coordinating functional groups. The elements in-dicated easily coordinate with functional groups containing oxygen, nitrogen, and halogens. Depending on metal coordinating ability and conditions for nano-structure obtaining (in liquid or solid medium with minimal content of liquid) we obtain "embryos" of future nanostructures of different shapes, dimensions and composition. It is advisable to model coordination processes and further redox processes with the help of quantum chemistry apparatus, following step-by-step consideration in accordance with the planned scheme.

At the same time, the metal orientation proceeds in interface regions and nanopores of polymeric phase which conditions further direction of the pro-cess to the formation of metal/carbon nanocomposite. In other words, the birth and growth of nanosize structures occur during the process in the same way as known from the macromolecule physics [6], in which Avrami equations are successfully used. The application of Avrami equations to the processes of nanostructure formation was previously discussed in the papers dedicated to the formation of ordered shapes of macromolecules [6], formation of carbon nano-structures by electric arc method [7], obtaining of fiber materials [8].

As follows from Avrami equation:

$$1-\upsilon = \exp[-k\tau^n], \tag{2}$$

where υ—crystallinity degree, τ—duration, k—value corresponding to specific process rate, n—number of degrees of freedom changing from 1 to 6, the factor under the exponential is connected with the process rate with the duration (time) of the process. Under the conditions of the isothermal growth of the ordered system "embryo," it can be accepted that the nanoreactor activity will be pro-portional to the process rate in relation to the flowing process. Then the share of the product being formed (W) in nanoreactor will be expressed by the following equation:

$$W = 1 - \exp(-a\tau^n) = 1 - \exp[-(\varepsilon_s/\varepsilon_v)\tau^n] = 1 - \exp\{-[(\varepsilon^0_s d/\varepsilon^0_v)S/V]\,\tau^n\}, \qquad (3)$$

where a—nanoreactor activity, ε_s—surface energy reflecting the energy of interaction of reagents with nanoreactor walls, ε_v—nanoreactor volume energy, $\varepsilon^0_s d$—multiplication of surface layer energy by its thickness, ε^0_v—energy of nanoreactor volume unit, S—surface of nanoreactor walls, V—nanoreactor volume.

When the metal ion moves inside the nanoreactor with redox interaction of ion (mol) with nanoreactor walls, the balance setting in the pair "metal containing—polymeric phase" can apparently be described with the following equation:

$$zF\Delta\varphi = RT\ln K = RT\ln(N_p/N_r) = RT\ln(1 - W), \qquad (4)$$

where z—number of electrons participating in the process; $\Delta\varphi$—difference of potentials at the boundary "nanoreactor wall—reactive mixture"; F—Faraday number; R—universal gas constant; T—process temperature; K—process balance constant; N_p—number of moles of the product produced in nanoreactor; N_r—number of moles of reagents or atoms (ions) participating in the process which filled the nanoreactor; and W—share of nanoproduct obtained in nanoreactor.

In turn, the share of the transformed components participating in phase interaction can be expressed with the equation which can be considered as a modified Avrami equation:

$$W = 1 - \exp[-\tau^n \exp(zF\Delta\varphi/RT)], \qquad (5)$$

where τ—duration of the process in nanoreactor; n—number of degrees of freedom changing from 1 to 6.

During the redox process connected with the coordination process, the character of chemical bonds changes. Therefore correlations of wave numbers of the changing chemical bonds can be applied as the characteristic of the nanostructure formation process in nanoreactor:

$$W = 1 - \exp[-\tau^n(v_{HC}/v_{KC})], \qquad (6)$$

where v_{HC}—corresponds to wave numbers of initial state of chemical bonds, and v_{KC}—wave numbers of chemical bonds changing during the process.

Modified Avraami equations were tested to prognosticate the duration of the processes of obtaining metal/carbon nanofilms in the system "Cu—PVA" at

200°C [10]. The calculated time (2.5 hr) correspond to the experimental duration of obtaining carbon nanofilms on copper clusters.

The nanostructures formed in nanoreactors of polymeric matrixes can be presented as oscillators with rather high oscillation frequency. It should be pointed out that according to references [1] for nanostructures (fullerenes and nanotubes) the absorption in the range of wave numbers 1,300–1,450 cm^{-1} is indicative. These values of wave numbers correspond to the frequencies in the range 3.9–4.35·10^{13} Hz, that is, in the range of ultrasound frequencies.

If the medium into which the nanostructure is placed blocks its translational or rotational motion giving the possibility only for the oscillatory motion, the nanostructure surface energy can be identified with the oscillatory energy:

$$\varepsilon_s \approx \varepsilon_\kappa = m\upsilon_\kappa^2/2, \tag{7}$$

where m—nanostructure mass, a υ_κ—velocity of nanostructure oscillations. Knowing the nanostructure mass, its specific surface and having identified the surface energy, it is easy to find the velocity of nanostructure oscillations:

$$\upsilon_\kappa = \sqrt{2\varepsilon_\kappa/m} \tag{8}$$

If only the nanostructure oscillations are preserved, it can be logically assumed that the amplitude of nanostructure oscillations should not exceed its linear nanosize, that is, $\lambda < r$. Then the frequency of nanostructure oscillations can be found as follows:

$$\nu_\kappa = \upsilon_\kappa/\lambda \tag{9}$$

Therefore the wave number can be calculated and compared with the experimental results obtained from IR spectra.

The influence of nanostructures on the media and compositions was discussed based on quantum-chemical modeling [9]. After comparing the energies of interaction of fullerene derivatives with water clusters, it was found that the increase in the interactions in water medium under the nanostructure influence is achieved only with the participation of hydroxyfullerene in the interaction. The energy changes reflect the oscillatory process with periodic boosts and attenuations of interactions. The modeling results can identify that the transfer of nanostructure influence onto the molecules in water medium is possible with the proximity or concordance of oscillations of chemical bonds in nanostructure and medium. The process of nanostructure influence onto media has an oscillatory character and is connected with a definite orientation of particles in the medium in the same way as reagents orientate in nanoreactors of polymeric matrixes. To describe this process, it is advisable to introduce such critical parameters as critical content of nanoparticles, critical time and critical temperature [10]. The growth of the number of nanoparticles (n) usually leads to the increase in the number of interaction (N). Also such situation is possible when with the increase of n critical value, N value gets much greater than the number of active nanoparticles. If the temperature exceeds the critical value, this results in the distortion of self-organization processes in the composition being modified and decrease in nanostructure influence onto media.

9.3   MATERIALS AND METHODS

9.3.1   MATERIALS AND METHODS OF METAL/CARBON NANOCOMPOSITES SYNTHESIS

Based on theoretical notions the synthesis scheme of metal/carbon nanocomposites comprises two stages. At the first stage, nanoreactors are prepared in selected polymeric matrixes and filled with metal containing phase. As polymeric matrixes it is proposed to use polyvinyl alcohol, polyvinyl acetate and polyvinyl chloride differing by crystallinity degree, correlation of functional groups, swelling degree and dimensions of interlayer spaces. Metal containing phase represents chlorides or oxides of such 3d metals as Fe, Co, Ni, and Cu; it is also possible to use metallurgical dust. When using metal chlorides as metal containing phase, at the first stage water solutions of salts and polymers (PVA or PVC) are mixed. When mixing, the color changes in accordance with the complexes formed and when the water is removed xerogels in the form of colored films are formed. When at the first stage we apply metal oxides, the mechanic-chemical process is used in the presence of the active medium (water or water solution of acids or bases). Finally we also obtain the colored xerogels.

To investigate the processes at the initial stage, optical transmission microscopy, spectral photometry, IR and Raman spectroscopies , atomic force microscopy (AFM) are applied. For the corresponding correlations "polymer—metal containing phase" the dimensions, shape and energy characteristics of nanoreactors are found with the help of AFM [11, 12]. Depending on a metal participating in coordination, the structure and relief of xerogel surface change. The comparison of phase contrast pictures on the corresponding films indicates greater concentration of the extended polar structures in the films containing copper, in comparison with the films containing nickel and cobalt (Figure 9.1). The processing of the pictures of phase contrast to reveal the regions of energy interaction of cantilever with the surface in comparison with the background produces practically similar result with optical transmission microscopy.

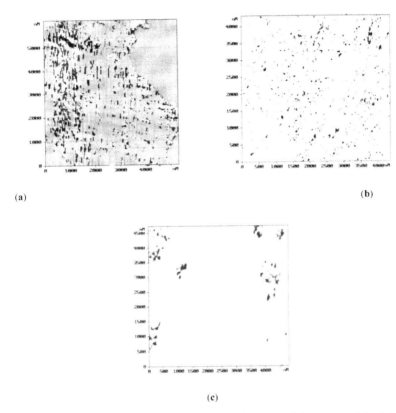

(a)   (b)

(c)

FIGURE 9.1   Pictures of phase contrast of PVA surfaces containing copper (**a**), nickel (**b**) and cobalt (**c**).

The results of AFM investigations of xerogels films (Figure 9.2) obtained from metal oxides and PVA [15] is distinguished in comparison with previous data, that testify to difference in reactivity of metal chlorides and metal oxides.

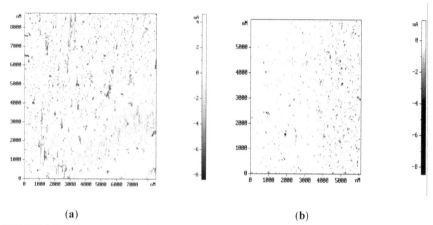

(a)   (b)

FIGURE 9.2 Phase contrast pictures of xerogels films (a) PVA—Ni and (b) PVA—Cu.

Below (Figure 9.3) the fields of energetic interaction of cantilever with surface of xerogels PVA—Ni (a) and PVA—Cu (b) are given.

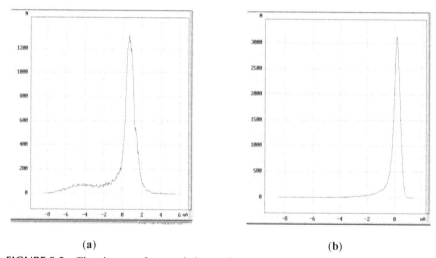

(a)   (b)

FIGURE 9.3 The pictures of energetic interaction of cantilever with surface of xerogels (a)
PVA—Ni and (b) PVA—Cu.

According to AFM results investigation the addition of Ni/C nanocomposite
in PVA leads to more strong coordination in comparison with analogous addition Cu/C nanocomposite.

The mechanism of formation of nanoreactors filled with metals was found with the help of IR spectroscopy [11].

Thus, at the first stage the coordination of metal containing phase and corresponding orientation in nanoreactor take place.

At the second stage it is required to give the corresponding energy impulse to transfer the "transition state" formed into carbon/metal nanocomposite of definite size and shape. To define the temperature ranges in which the structuring takes place, DTA-TG investigation is applied. It is found that in the temperature range under 200°C nanofilms, from carbon fibers associated with metal phase as well, are formed on metal or metal oxide clusters. When the temperature elevates up to 400°C, 3-D nanostructures are formed with different shapes depending on coordinating ability of the metal. In this case nanofilms are scrolled as seen in Figure 9.4.

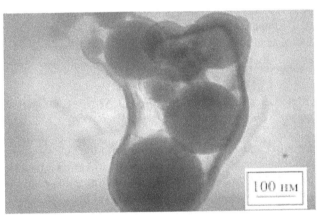

FIGURE 9.4 Microphotograph (transmission electron microscopy) demonstarating the moment of nanolfilm scrolling on metal nanopartciles.

To investigate the processes at the second stage of obtaining metal/carbon nanocomposites X-ray photoelectron spectroscopy, transmission electron microscopy and IR spectroscopy are applied.

The sample for IR spectroscopy was prepared when mixing metal/carbon nanocomposite powder with one drop of vaselene oil in agate mortar to obtain a homogeneous paste with further investigation of the paste obtained on the appropriate instrument. As the vaselene oil was applied when the spectra were taken, we can expect strong bands in the range 2,750–2,950 cm^{-1}. Two types of nanocomposites rather widely applied during the modification of various polymeric materials were investigated. These were: copper/carbon nanocomposite and nickel/carbon nanocomposite specified below.

In turn, the nanopowders obtained were tested with the help of high-resolution transmission electron microscopy, electron microdiffraction, laser analyzer, X-ray photoelectron spectroscopy, and IR spectroscopy.

The method of metal/carbon nanocomposite synthesis applied has the following advantages:

(i) Originality of stage-by-stage obtaining of metal/carbon nanocomposites with intermediary evaluation of the influence of initial mixture composition on their properties.

(ii) Wide application of independent modern experimental and theoretical analysis methods to control the technological process.

(iii) Technology developed allows synthesizing a wide range of metal/carbon nanocomposites depending on the process conditions.

(iv) Process does not require the use of inert or reduction atmospheres and specially prepared catalysts.

(v) Method of obtaining metal/carbon nanocomposites allows applying secondary raw material.

9.3.2 METHODS OF FINE DISPERSED SUSPENSIONS PREPARATION

To select the components of fine suspensions with the help of quantum-chemical modeling by the scheme described before [5], first, the interaction possibility of the material component being modified (or its solvent or surface-active substance) with metal/carbon nanocomposite is defined. The suspensions are prepared by the dispersion of the nanopowder in ultrasound station. The stability of fine suspension is controlled with the help of laser analyzer. The action on the corresponding regions participating in the formation of fine dispersed suspension or sol is determined with the help of IR spectroscopy. As an example, below you can see brief technique for obtaining fine suspension based on polyethylene polyamine.

Fine suspensions of metal/carbon nanocomposite were obtained mixing the nanopowder with polyethylene polyamine (PEPA) with further ultrasound processing on the stations Saphir UZV and UZTA-0.2/22-OM. The suspensions obtained were studied with the help of IR spectroscopy (IR-Fourier spectrophotometer FSM 1201).

9.4 RESULTS AND DISCUSSION

9.4.1 *CHARACTERISTICS OF NANOCOMPOSITES OBTAINED*

Under metal/carbon nanocomposite we understand the nanostructure containing metal clusters stabilized in carbon nanofilm structures. The carbon phase can be in the form of film structures or fibers. The metal particles are associated with carbon phase. The metal nanoparticles in the composite basically have the shapes close to spherical or cylindrical ones. Due to the stabilization and association of metal nanoparticles with carbon phase, chemically active metal particles are stable in air and during heating as the strong complex of metal nanoparticles with the matrix of carbon material is formed. The test results of nanocomposites obtained are given in Table 9.1.

TABLE 9.1 Characteristic of metal/carbon nanocomposites (Met/C HK)

Type Met/C HK	Cu/C	Ni/C	Co/C	Fe/C
Composition, (%): Carbon, (%)	50/50	60/40	65/35	70/30
Density (g/cm³)	1.71	2.17	1.61	2.1
Average Dimension (nm)	20(25)	11	15	17
Specific Surface (m²/g)	160 (average)	251	209	168
Metal Nanoparticle Shape	Close to spherical, there are dodecahedrons	There are spheres and rods	Nanocrystals	Close to spherical
Caron Phase Shape (shell)	Nanofibers associated with metal phase forming nanocoatings	Nanofilms scrolled in nanotubes	Nanofilms associated with nanocrystals of metal containing phase	Nanofilms forming nanobeads with metal containing phase
Atomic Magnetic Moment (Reference) (µB)	0.0	0.6	1.7	2.2
Atomic Magnetic Moment (Nanocomposite) (µB)	0.6	1.8	2.5	2.5

The nanocomposites described above were investigated with the help of IR spectroscopy by the technique indicated above. In this paper the IR spectra of Cu/C and Ni/C nanocomposites are discussed (Figure 9.5), which find a wider application as the material modifiers.

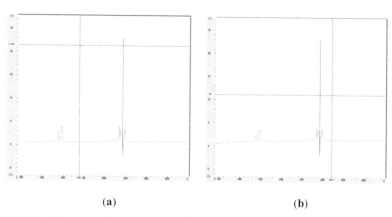

(a)   (b)

FIGURE 9.5   IR spectra of copper/carbon (**a**) and nickel/carbon (**b**) nanocomposite powder.

On IR spectra (Figure 9.5) of two nanocomposites the common regions of IR radiation absorption are registered. Further, the bands appearing in the spectra and having the largest relative area were evaluated. We can see the difference in the intensity and number of absorption bands in the range 1,300–1,460 cm^{-1}, which confirms the different structures of composites. In the range 600–800 cm^{-1} the bands with a very weak intensity are seen, which can be referred to the oscillations of double bonds (π-electrons) coordinated with metals. In case of Cu/C nanocomposite a weak absorption is found at 720 cm^{-1}. In case of Ni/C nanocomposite, except for this absorption, the absorption at 620 cm^{-1} is also observed.

In IR spectrum of copper/carbon nanocomposite two bands with a high relative area are found:

at 1,323 cm^{-1} (relative area—9.28)

at 1,406 cm^{-1} (relative area—25.18).

These bands can be referred to skeleton oscillations of polyarylene rings.

In IR spectrum of nickel/carbon nanocomposite the band mostly appears at 1,406 cm^{-1} (relative area—14.47).

According to the investigations with transmission electron microscopy the formation of carbon nanofilm structures consisting of carbon threads is characteristic for copper/carbon nanocomposite. In contrast, carbon fiber structures, including nanotubes, are formed in nickel/carbon nanocomposite. There are several absorption bands in the range 2,800–3,050 cm^{-1}, which are attributed to valence oscillations of C-H bonds in aromatic and aliphatic compounds. These absorption bonds are connected with the presence of vaselene oil in the sample. It is difficult to find the presence of metal in the composite as the metal is stabilized in carbon nanostructure. At the same time, it should be pointed out that apparently nanocomposites influence the structure of vaselene oil in different ways. The intensities and number of bands for Cu/C and Ni/C nanocomposites are different:

(i) for copper/carbon nanocomposite in the indicated range—5 bands, and total intensity corresponds by the relative area—64.63.

(ii) for nickel/carbon nanocomposite in the same range—4 bands with total intensity (relative area)—85.6.

The investigations of Carbon films in Metal/Carbon Nanocomposites and their peculiarities are carried out by Raman spectroscopy with the using of Laser Spectrometer Horiba LabRam HR 800. Below the Raman spectra and PEM microphotograph of Copper/Carbon (Cu/C) Nanocomposite are represented (Figures 9.6 and 9.7).

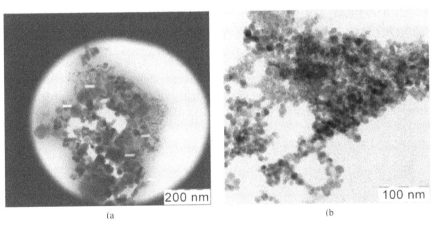

200 nm

100 nm

(a   (b

FIGURE 9.6   PEM microphotographs of Cu/C Nanocomposites (NC): (a)—Cu/C NC (28 nm); (b)—Cu/C NC (25 nm).

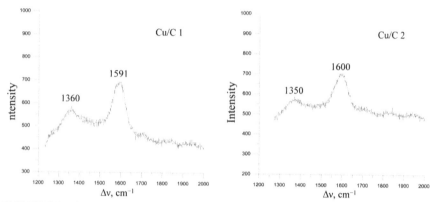

FIGURE 9.7   Raman spectra of Cu/C Nanocomposites: Cu/C 1 (28 nm); Cu/C 2 (25 nm).

Wave numbers and intensity relation testify to presence of nanoparticles containing the Copper atoms coordinated. At the same time the comparison of IR and Raman spectra shows their closeness on wave numbers and intensities relation.

9.4.2   CHARACTERISTICS OF METAL/CARBON NANOCOMPOSITES FINE DISPERSED SUSPENSIONS

The distribution of nanoparticles in water, alcohol, and water-alcohol suspensions prepared based on the above technique are determined with the help of laser analyzer. In Figure 9.8 you can see distributions of copper/carbon nanocomposite in the media different polarity and dielectric penetration.

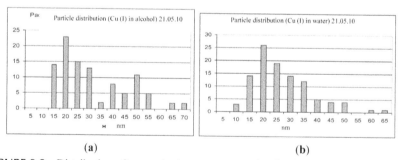

(a)   (b)

FIGURE 9.8   Distribution of copper/carbon nanocomposites in alcohol (a) in water (b).

When comparing the figures we can see that ultrasound dispersion of one and the same nanocomposite in media different by polarity results in the changes of distribution of its particles. In water solution the average size of Cu/C nanocomposite equals 20 nm, and in alcohol medium—greater by 5 nm.

Assuming that the nanocomposites obtained can be considered as oscillators transferring their oscillations onto the medium molecules, we can determine to what extent the IR spectrum of liquid medium will change, for example, PEPA applied as a hardener in some polymeric compositions, when we introduce small and supersmall quantities of nanocomposite into it.

IR spectra demonstrate the change in the intensity at the introduction of metal/carbon nanocomposite in comparison with the pure medium (IR spectra are given in Figure 9.9). The intensities of IR absorption bands are directly connected with the polarization of chemical bonds at the change in their length, valence angles at the deformation oscillations, that is, at the change in molecule normal coordinates.

When nanocomposites are introduced, the changes in the area and intensity of absorption bands, which indicates the coordination interactions and influence of nanostructure onto the medium are observed (Figures 9.9, 9.10, and 9.11).

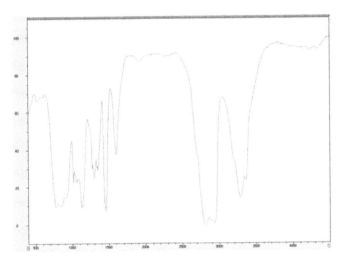

FIGURE 9.9   IR spectrum of polyethylene polyamine.

Special attention in PEPA spectrum should be paid to the peak at 1,598 cm⁻¹ attributed to deformation oscillations of N-H bond, where hydrogen can participate in different coordination and exchange reactions.

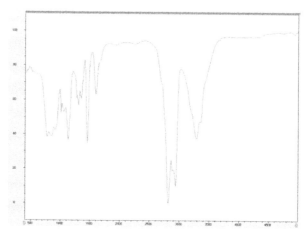

FIGURE 9.10   IR spectrum of copper/carbon nanocomposite fine suspension in polyethylene polyamine medium (ω (NC) = 1%).

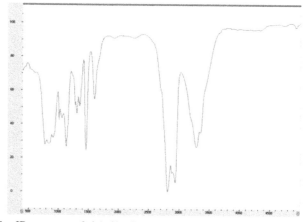

FIGURE 9.11   IR spectrum of nickel/carbon nanocomposite fine suspension in polyethylene polyamine (ω (NC) = 1%).

In the spectra wave numbers characteristic for symmetric $v_s(NH_2)$ 3,352 cm^{-1} and asymmetric $v_{as}(NH_2)$ 3,280 cm^{-1} oscillations of amine groups are present. There is a number of wave numbers attributed to symmetric $v_s(CH_2)$ 2,933 cm^{-1} and asymmetric valence $v_{as}(CH_2)$ 2,803 cm^{-1}, deformation wagging oscillations $v_D(CH_2)$ 1,349 cm^{-1} of methylene groups, deformation oscillations of NH v_D (NH) 1,596 cm^{-1} and NH$_2$ $v_D(NH_2)$ 1,456 cm^{-1} amine groups. The oscillations of skeleton bonds at $v(CN)$ 1,059–1,273 cm^{-1} and $v(CC)$ 837 cm^{-1} are the most vivid. The analysis of intensities of IR spectra of PEPA and fine suspensions of metal/carbon nanocomposites based on it revealed a significant change in the intensities of amine groups of dispersion medium (for $v_s(NH_2)$ in 1.26 times, and for $v_{as}(NH_2)$ in approximately 50 times).

Such demonstrations are presumably connected with the distribution of the influence of nanoparticle oscillations onto the medium with further structuring and stabilizing of the system. Under the influence of nanoparticle the medium changes which is confirmed by the results of IR spectroscopy (change in the intensity of absorption bands in IR region). Density, dielectric penetration, viscosity of the medium are the determining parameters for obtaining fine suspension with uniform distribution of particles in the volume. At the same time, the structuring rate and consequently the stabilization of the system directly depend on the distribution by particle sizes in suspension. At the wide range of particle distribution by sizes, the oscillation frequency of particles different by size can significantly differ, in this connection, the distortion in the influence transfer of nanoparticle system onto the medium is possible (change in the medium from the part of some particles can be balanced by the other). At the narrow range of nanoparticle distribution by sizes the system structuring and stabilization are possible. With further adjustment of the components such processes will positively influence the processes of structuring and self-organization of final composite system determining physical-mechanical characteristics of hardened or hard composite system.

The effects of the influence of nanostructures at their interaction into liquid medium depend on the type of nanostructures, their content in the medium and medium nature. Depending on the material modified, fine suspensions of nanostructures based on different media are used. Water and water solutions of surface-active substances, plasticizers, and foaming agents (when modifying foam concretes) are applied as such media to modify silicate, gypsum, cement and concrete compositions. To modify epoxy compounds and glues based on epoxy resins the media based on PEPA, isomethyltetrahydrophthalic anhydride, toluene and alcohol-acetone solutions are applied. To modify polycarbonates and derivatives of polymethyl methacrylate dichloroethane and dichloromethane media are used. To modify polyvinyl chloride compositions and compositions based on phenolformaldehyde and phenolrubber polymers alcohol or acetone-based media are applied. Fine suspensions of metal/carbon nanocomposites are produced using the above media for specific compositions. In IR spectra of all studied suspensions the significant change in the absorption intensity, especially in the regions of wave numbers close to the corresponding nanocomposite oscillations, is observed. At the same time, it is found that the effects of nanocomposite influence on liquid media (fine suspensions) decreases with time and the activity of the corresponding suspensions drops. The time period in which the appropriate activity of nanocomposites is kept changes in the interval 24 hr—1 month depending on the nanocomposite type and nature of the basic medium (liquid phase in which nanocomposites dispergate). For instance, IR spectroscopic investigation of fine suspension based on isomethyltetrahydrophthalic anhydride containing 0.001 percent of Cu/C nanocomposite indicates the decrease in the peak intensity, which sharply increased on the third day when nanocomposite was introduced (Figure 9.12).

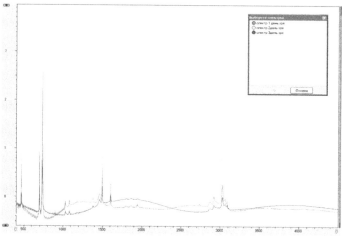

FIGURE 9.12 Changes in IR spectrum of copper/carbon nanocomposite fine suspension based on isomethyltetrahydrophthalic anhydride with time (a) IR spectrum on the first day after the nanocomposite was introduced, (b) IR spectrum on the second day, and (c) IR spectrum on the third day.

Similar changes in IR spectra take place in water suspensions of metal/carbon nanocomposites based on water solutions of surface-active nanocomposites. In Figure 9.13 you can see IR spectrum of iron/carbon nanocomposite based on water solution of sodium lignosulfonate in comparison with IR spectrum of water solution of surface-active substance.

As it is seen, when nanocomposite is introduced and undergoes ultrasound dispergation, the band intensity in the spectrum increases significantly. Also the shift of the bands in the regions 1,100–1,300 cm^{-1}; 2,100–2,200 cm^{-1} is observed, which can indicate the interaction between sodium lignosulfonate and nanocomposite. However, after two weeks the decrease in band intensity is seen. As the suspension stability evaluated by the optic density is 30 days, the nanocomposite activity is quite high in the period when IR spectra are taken. It can be expected that the effect of foam concrete modification with such suspension will be revealed if only 0.001 percent of nanocomposite is introduced.

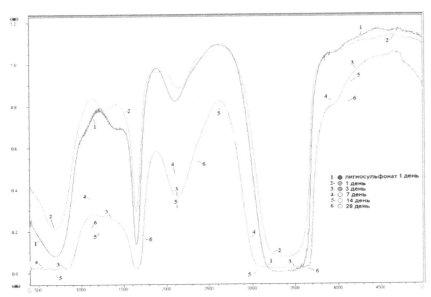

FIGURE 9.13   Comparison of IR spectra of water solution of sodium lignosulfonate (1) and fine suspension of iron/carbon nanocomposite (0.001%) based on this solution on the first day after nanocomposite introduction (2), on the third day (3), on the seventh day (4), 14th day (5), and 28th day (6).

The ultimate breaking stresses were compared in the process of compression of foam concretes modified with copper/carbon nanocomposites obtained in different nanoreactors of polyvinyl alcohol [13, 14]. The sizes of nanoreactors change depending on the crystallinity and correlation of acetate and hydroxyl groups in PVA which results in the change of sizes and activity of nanocomposites obtained in nanoreactors. It is observed that the sizes of nanocomposites obtained in nanoreactors of PVA matrixes 16/1(ros) (NC2), PVA 16/1 (imp) (NC1), PVA 98/10 (NC3), correlate as NC3 > NC2 > NC1. The smaller the nanoparticle size and greater its activity, the less amount of nanostructures is required for self-organization effect. At the same time, the oscillatory nature of the influence of these nanocomposites on the compositions of foam concretes is seen in the fact that if the amount of nanocomposite is 0.0018 percent from the cement mass, the significant decrease in the strength of NC1 and NC2 is observed. The increase in foam concrete strength after the modification with iron/carbon nanocomposite is a little smaller in comparison with the effects after the application of NC1 and NC2 as modifiers. The corresponding effects after the modification of cement, silicate, gypsum and concrete compositions with nanostructures is defined by the features of components and technologies applied. These features often explain the instability of the results after the modification of the foregoing compositions with nanostructures. Besides during the modification the changes in the activity of fine suspensions of nanostructures depending on the duration and storage conditions should be taken into account.

In this regard, it is advisable to use metal/carbon nanocomposites when modifying polymeric materials whose technology was checked on strictly controlled components.

At present a wide range of polymeric substances and materials: compounds, glues, binders for glass-, basalt and carbonplastics based on epoxy resins, phenol-rubber compositions, polyimide and polyimide compositions, materials on polycarbonate and polyvinyl chloride basis, as well as special materials, such as current-conducting glues and pastes, fireproof intumescent glues and coatings are being modified.

Below you can find the results of some working-outs [15]:

(i) The introduction of metal/carbon nanostructures (0.005%) in the form of fine suspension into PEPA or the mixture of amines into epoxy compositions allows increasing the thermal stability of the compositions by 75–100 degrees and consequently increase the application range of the existing products. This modification contributes to the increase in adhesive and cohesive characteristics of glues, lacquers and binders.

(ii) Hot vulcanization glue was modified with copper/carbon and nickel/carbon nanostructures using toluol-based fine suspensions. On the test results of samples of four different schemes the tear strength σ_t increased up to 50 percent and shear strength τ_s—up to 80 percent, concentration of metal/carbon nanocomposite introduced was 0.0001–0.0003 percent.

(iii) Fine suspension of nanostructures was produced in polycarbonate and dichloroethane solutions to modify polycarbonate-based compositions. The introduction of 0.01 percent of copper/carbon nanostructures leads to the significant decrease in temperature conductivity of the material (in 1.5 times). The increase in the transmission of visible light in the range 400–500 nm and decrease in the transmission in the range 560–760 nm were observed.

(iv) When modifying polyvinyl chloride film with fine suspension containing iron/carbon nanocomposite, the increase of the crystalline phase in the material was observed. The PVC film modified containing 0.0008 percent of NC does not accumulate the electrostatic charge on its surface.

The material obtained completely satisfies the requirements applied to PVC films for stretch ceilings.

(v) The introduction of nickel/carbon nanocomposite (0.01% of the mass of polymer filled on 65% of silver microparticles) into the epoxy polymer hardened with PEPA leads to the decrease in electric resistance to 10^{-5} Ohm·cm (10^{-4} Ohm·cm without nanocomposite).

9.5 CONCLUSION

In the paper the possibilities of developing new ideas about self-organization processes and about nanostructures and nanosystems are discussed on the example of metal/carbon nanocomposites. It is proposed to consider the obtaining of metal/carbon nanocomposites in nanoreactors of polymeric matrixes as self-organization process similar to the formation of ordered phases which can be described with Avrami equation. The application of Avrami equations during the synthesis of nanofilm structures containing copper clusters has been tested. The influence of nanostructures on active media is given as the transfer of oscillation energy of the corresponding nanostructures onto the medium molecules.

IR spectra of metal/carbon and their fine suspensions in different (water and organic) media have been studied for the first time. It has been found that the introduction of supersmall quantities of prepared nanocomposites leads to the significant change in band intensity in IR spectra of the media. The attenuation of oscillations generated by the introduction of nanocomposites after the time interval specific for the pair "nanocomposite—medium" has been registered.

Thus to modify compositions with fine suspensions it is necessary for the latter to be active enough that should be controlled with IR spectroscopy.

A number of results of material modification with fine suspensions of metal/carbon nanocomposites are given, as well as the examples of changes in the properties of modified materials based on concrete compositions, epoxy and phenol resins, polyvinyl chloride, polycarbonate and current-conducting polymeric materials.

KEYWORDS

- Metal/carbon nanocomposites
- Modification
- Nanochemistry methods
- Super small quantities of nanocomposites
- Synthesis in nanoreactors

REFERENCES

1. Kodolov, V. I.; and Khokhriakov, N. V.; Chemical Physics of the Processes of Formation and Transformation of Nanostructures and Nanosystems. Izhevsk: Izhevsk State Agricultural Academy; **2009**, *2 vol., 1,* 360, *2,* 415.
2. Melikhov, I. V.; and Bozhevolnov, V. E.; Variability and self-organization in nanosystems. *J. Nanopart. Res.* **2003**, *5,* 465–472.
3. Malinetsky, G. G.; Designing of the future and modernization in Russia. Preprint of M. V. Keldysh Institute of Applied Mechanics; **2010**, *41,* 32.
4. Tretyakov, Yu. D.; Self-organization processes. *Uspekhi Chimii.* **2003**, *72(8),* 731–764.
5. Kodolov, V. I.; Khokhriakov, N. V.; Trineeva, V. V.; and Blagodatskikh, I. I.; Activity of nanostructures and its representation in nanoreactors of polymeric matrixes and active media. *Chem. Phys. Mesoscopy.* **2008**, *10(4),* 448–460.
6. Wunderlich, B.; Physics of macromolecules. 3 vol. M.: Mir, **1979**, *2,* 574.
7. Fedorov, V. B.; Khakimova, D. K.; Shipkov, N. N.; and Avdeenko, M. A.; To thermodynamics of carbon materials. Doklady AS USSR, **1974**, *219(3),* 596–599; Fedorov, V. B.; Khakimova, D. K.; and Sharshorov, M. H.; et al. To kinetics of graphitation. Doklady AS USSR, **1975**, *222(2),* 399–402.
8. Theory of Chemical Fibers Formation. ed. by Serkov A. T.; M.: Chemistry, **1975**, 548.
9. Kodolov, V. I.; Khokhriakov, N. V.; and Kuznetsov, A. P.; To the issue of the mechanism of the influence of nanostructures on structurally changing media at the formation of "intellectual" composites. *Nanotechnics.* **2006**, *3(7),* 27–35.
10. Kodolov, V. I.; Khokhriakov, N. V.; Trineeva, V. V.; and Blagodatskikh, I. I.; Problems of Nanostructure Activity Estimation, Nanostructures Directed Production and Application. Nanomaterials Yearbook—2009. From Nanostructures, Nanomaterials and Nanotechnologies to Nanoindustry. New York: Nova Science Publishers, Inc., **2010**, 1–18.
11. Kodolov, V. I.; Blagodatskikh, I. I.; and Lyakhovich, A. M.; et al. Investigation of the formation processes of metal containing carbon nanostructures in nanoreactors of polyvinyl alcohol at early stages. *Chem. Phys. Mesoscopy.* **2007**, *9(4),* 422–429.
12. Trineeva, V. V.; Lyakhovich, A. M.; and Kodolov, V. I.; Forecasting of the formation processes of carbon metal containing nanostructures using the method of atomic force microscopy. *Nanotechnics.* **2009**, *4(20),* 87–90.
13. Akhmetshina, L. F.; Kodolov, V. I.; Tereshkin, I. P.; and Korotin, A. I.; Influence of carbon metal containing nanostructures on strength of concrete composites. *Nanotechnol. Construct.* **2010**, *6,* 35–46.
14. Kodolov, V. I.; Trineeva, V. V.; Kovyazina, O. A.; and Vasil'chenko, Yu. M.; Production and Application of Metal/Carbon Nanocomposites. Nanotechnol. Construct. 2010, 6, 46–55.
15. Kovyazina, O. A.; Trineeva, V. V.; Akhmetshina, L. F.; and Vasilchenko, Yu. M.; et al. Experience of the application of metal/carbon nanocomposites to modify materials. Abstracts of VII International Scientific-Practical Conference "Nanotechnologies for Production-2010." Fryazino; 53–54.

CHAPTER 10

A RESEARCH NOTE ON PRODUCTION TECHNOLOGY OF CARBON-METAL CONTAINING NANOPRODUCTS IN NANOREACTORS OF POLYMERIC MATRIXES

V. I. KODOLOV, V. V. TRINEEVA, O. A. KOVYAZINA, and A. YU. BONDAR

Basic Research-High Educational Center of Chemical Physics and Mesoscopy, Udmurt Scientific Center, Russian Academy of Sciences; Izhevsk, Udmurt Republic, Russia; E-mail: kodol@istu.udm.ru

CONTENTS

10.1 INTRODUCTION

Chemistry in nanoreactors has been rapidly developed during the last few years. At present, the term "nanoreactor" is regarded in a wider implication. As for the modern definition, the term "nanoreactor" may include the defective regions of metal salt polycrystals, interface regions in substances with lamellar structures, extended cavities formed by macromolecules in gels or the solutions of polymers. The requirement to develop ecologically clean productions will give an opportunity of wide application of nanoreactors in chemistry and metallurgy. In this case, for the obtaining of metallic nanoparticles and nanowires in carbon shells, it is expedient to evaluate the possibilities of reduction-oxidation couple reactions with the participation of metal ions and organic compounds. The results obtained are experimentally confirmed[1-4].

10.2 EXPERIMENTAL AND DISCUSSION

For the synthesis of nanoparticles and nanowires from the mixture of metal salts and polyvinyl alcohol (PVA), the aqueous solutions of salts were mixed in a certain ratio with the aqueous solution of PVA. The average molar ratio of PVA in the mixture was five. The experiments were carried out on the glass substrates; after the obtained mixtures had been dried, they formed colored transparent films. On some samples, the films were broken due to the large surface tension. The films were heated at 250°C until their color, composition and morphology changed. To control the process, a complex of methods was used, that is, visual spectrophotometry, optical microscopy, X-ray photoelectron spectroscopy, and atomic power microscopy.

When PVA was added to the powders of metal chlorides, the color of the mixture changed: the mixture of copper chloride became yellow-green, the cobalt chloride mixture—blue, and nickel chloride—pale-green. Observing the color changes we can conclude that when PVA interacts with metal chlorides the complex compounds are formed.

Among the aforesaid metals iron is the most active. Brown-red inclusions on the photograph prove the formation of complex iron compounds. In addition, in all the photographs depicting the mixtures containing metal chlorides, we can observe a net of weaves, which are most likely the reflections of nanostructures. The application of optical microscopy and spectrophotometry allows characterizing kinetic parameters of coordination reactions, and also the peculiarities of corresponding reactions because of the possible soiling of raw materials. This is very important for the creation of control conditions during this stage of technological process.

The control of coordination reaction parameters and soiling level gives the possibility evaluate the quality of nanoproduct obtained. As it is shown at other stages of technological process, the quality of nanoproducts and their future application depend upon the characteristics of initial raw materials and intermediate product which is formed at the first stage.

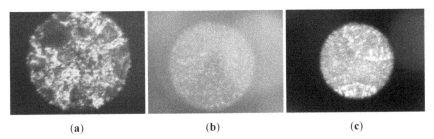

(a)   (b)   (c)

FIGURE 10.1   Photographs of the samples containing PVA and copper chloride (**a**), cobalt chloride (**b**), and nickel chloride (**c**).

In order to compare these structures, the morphology of the films changing over a certain range of temperatures was investigated with the help of atomic power microscopy (Figure 10.1).

When the nanoproduct pictures obtained by atomic power microscopy and optical microscopy are compared with TEM micrograph of the nanoproduct treated thermally and with aqueous solution for the matrix removal, some compliance between them can be noticed. The nanoproduct represents interweaving tubules containing Cu(I) and Cu(II). In Figure 10.2, there are also optical effects indicating light polarization at light transmission through the films owing to the defects appearing during the formation of complex compounds at the initial stage of the process.

Due to the fact that metal ions are active, in the polymer medium they immediately appear in the environment of the PVA molecules and form bonds with hydroxyl groups of this polymer. Polyvinyl alcohol replicates the structure of the particle that it surrounds; however, due to the tendency of the molecules of the metal salts or other metal compounds to combine, PVA seems to envelope the powder particles, and therefore the forms of the obtained nanostructures can be different. The optical microscopy method allows determining the structure of nanostructures at the early stage.

When the samples are heated, metal-containing nanotubes form as a result of dehydration. These processes are thoroughly described [5–6]. Dehydration leads to the darkening of the film. After the samples have been heated, the remaining net of weaves can be seen on the photograph, that is, the structure morphology has remained. To some extent, this fact indicates that the initially formed structure of matrices is inherited. The methods of optical spectroscopy and X-ray photoelectron spectroscopy allow determining the interaction energy of chemical particles in nanoreactors with active centers of nanoreactor walls, which stimulate reduction-oxidation processes.

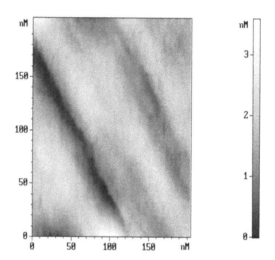

FIGURE 10.2   Micrographs of the surface geometry of PVA film with nickel chloride.

After the heating following the definite program (100-200-300-400°C) the dark brittle products are formed, which are washed out with hot water, dried and milled into dark powders. Nanoproducts obtained are investigated by X-ray photoelectron spectroscopy, transmission and scanning electron microscopy and also electron microdiffraction. For the same samples the Raman spectra are investigated. The main nanoproducts are the tubules (diameter—20–30 nm), welded with each other (Figure 10.3). The output of this nanoproduct in the isolated masses after drying exceeds 90 percent. The yield of carbon-metal-containing nanotubules in account of carbon in PVA equals 85–90 percent.

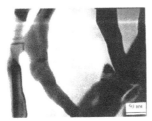

FIGURE 10.3 Micrographs of tubules.

Depending upon the nature of the metal salt and electrochemical potential of the metal, different metal reduction nanoproducts in the carbon shells differing in shape are formed. Based on this result we may speak about a new scientific branch—nanometallurgy.

Based on the results of theoretical and experimental investigations the following technological scheme (Figure 10.4) is proposed:

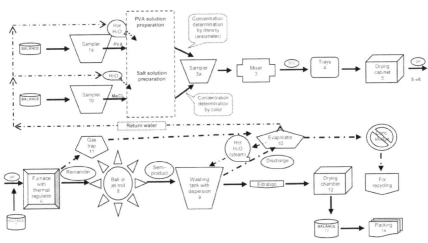

FIGURE 10.4 Technological scheme of nanoproduct production.

The technology proposed was realized with the production of nanoproduct the TEM microphotograph of which is presented below (Figure 10.5):

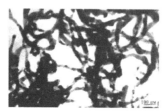

FIGURE 10.5   TEM image of nanoproduct obtained in the technology proposed.

Analogous nanoproduct (Ross 4 [6]) is obtained from polymeric materials by high-energy synthesis (Figure 10.6).

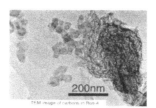

FIGURE 10.6   TEM image of nanoproduct "Ros-4" [6].
However this nanoproduct (Ros-4) does not contain metals and its price is $25/g.

10.3   CONCLUSION

The possible ways for obtaining metallic nanostructures in carbon shells have been determined. The investigation results allow speaking about the possibility of the isolation of metallic and metal-containing nanoparticles in the carbon shells differing in shape and structure. However, still there are problems related to the calculation and experiment because using the existing investigation methods it is difficult to unambiguously estimate the geometry and energy parameters of nanoreactors under the condition of "erosion" of their walls during the formation of metallic nanostructures in them. The technology of metallic nanoparticles and nanowires in carbon shells is proposed.

KEYWORDS

- **Nanoproduct**
- **Nanoreactors**
- **Nanowires**
- **Production technology**
- **Redox processes**

REFERENCES

1. Kodolov, V. I.; Kuznetsov, A. P.; and Nicolaeva, O. A.; et al. Surface and Interface Analysis, **2001**, *32,* 10–14.
2. Kodolov, V. I.; Shabanova, I. N.; and Makarova, L. G.; et al. *J. Struct. Chem.* **2001,** *42(2),* 260–264, (In Russian).
3. Kodolov, V. I.; Khokhriakov, N. V.; Nikolaeva, O. A.; and Volkov, V. L.; *Chem. Phys. Mesoscopy.* **2001,** *3(1),* 53–65, (In Russian).
4. Volkova, E. G.; Volkov, A. Yu.; and Murzakaev, A. M.; et al. *Phys. Metals Metallography.* **2003,** *95(4),* 342–345.
5. Didik, A. A.; Kodolov, V. I.; and Volkov, A. Yu.; et al. *Inorg. Mat.* **2003,** *39(6),* 693–697.
6. Information of Rosseter Holdings Ltd, in http://www.e-nanoscience.com/products.html; http://www.e-nanoscience.com/process.html; and http://www.e-nanoscience.com/prices.html

CHAPTER 11

REDOX PROCESSES IN POLYMERIC MATRIXES NANOREACTORS: A RESEARCH NOTE

V. I. KODOLOV[1,2], V. V. TRINEEVA[1,3], and YU. M. VASIL'CHENKO[1,2]

[1]Basic Research – High Educational Center of Chemical Physics and Mesoscopy, Udmurt Scientific Center, Ural Division, Russian Academy of Sciences, Russia

[2]M. T. Kalashnikov Izhevsk State Technical University, Russia

[3]Institute of Mechanics, Ural Division, Russian Academy of Sciences, Russia

CONTENTS

11.1 INTRODUCTION

Previously [1, 2] the models of nanoreactors in which carbon- or metal-containing nanostructures can be obtained were considered. The synthesis takes place due to the redox reaction between metal-containing substances and nanoreactor walls which are the macromolecules of polymer matrix with functional groups participating in the interaction with metal-containing compound or metal ion. If nanoreactors are nanopores or cavities in polymer gels which are formed in the process of solvent removal from gels and their transformation into xerogels or during the formation of crazes in the process of mechanical-chemical processing of polymers and inorganic phase in the presence of active medium, the process essence is as follows:

The nanoreactor is formed in the polymeric matrix which, by geometry and energy parameters, corresponds to the transition state of the reagents participating in the reaction. Then the nanoreactor is filled with reactive mass comprising reagents and solvent. The latter is removed and only the reagents oriented in a particular way stay in the nanoreactor and, if the sufficient energy impulse is available, for instance, energy isolated at the formation of coordination bonds between the fragments of reagents and functional groups in nanoreactor walls interact with the formation of necessary products. The initially appeared coordination bonds destruct together with the nanoproduct formation.

11.1.1. THEORETICAL IDEAS ABOUT THE INTERACTIONS IN NANOREACTORS OF POLYMERIC MATRIXES

The difference of potentials between the interacting particles and object walls stimulating these interactions is the driving force of self-organization processes (formation of nanoparticles with definite shapes). The potential jump at the boundary "nanoreactor wall—reacting particles" is defined by the wall surface charge and reacting layer size. If we consider the redox process as the main process preceding the nanostructure formation, the work for charge transport corresponds to the energy of nanoparticle formation process in the reacting layer. Then the equation of energy conservation for nanoreactor during the formation of nanoparticle mol will be as follows:

$$nF\Delta\phi = RT \ln\left(\frac{N_p}{N_r}\right) \tag{11.1}$$

where n—number reflecting the charge of chemical particles moving inside "the nanoreactor"; F—faraday number; $\Delta\varphi$—difference of potentials between the nanoreactor walls and flow of chemical particles; R—gas constant; T—process temperature; N_p—mol share of nanoparticles obtained; N_r—mol share of initial reagents from which the nanoparticles are obtained.

Using the above equation we can determine the values of equilibrium constants when reaching the certain output of nanoparticles, sizes of nanoparticles and shapes of nanoparticles formed with the appropriate equation modification. The internal cavity sizes or nanoreactor reaction zone and its geometry significantly influence the sizes and shapes of nanostructures.

The sequence of the processes is conditioned by the composition and parameters (energy and geometry) of nanoreactors. To accomplish such processes it is advisable to preliminarily select the polymeric matrix containing the nanoreactors in the form of nanopores or crazes as process appropriate. Such selection can be realized with the help of computer chemistry. Further the computational experiment is carried out with the reagents placed in the nanoreactor with the corresponding geometry and energy parameters. Examples of such computations were given in [3]. The experimental confirmation of polymer matrix and nanoreactor selection to obtain carbon- or metal-containing nanostructures was given in [4, 5]. Avrami equations are widely used for such processes which usually reflect the share of a new phase produced.

As follows from Avrami equation –

$$1 - \upsilon = \exp\left[-k\tau^n\right],$$

(11.2)

where υ—crystallinity degree, τ—duration, k—value corresponding to specific process rate, n— number of degrees of freedom changing from 1 to 6, the factor under the exponential is connected with the process rate with the duration (time) of the process. Under the conditions of the isothermal growth of the ordered system "embryo", it can be accepted that the nanoreactor activity will be proportional to the process rate in relation to the flowing process.

It was proposed to use the thermodynamics of small systems and Avrami equations to describe the formation processes of carbon nanostructures during recrystallization (graphitization) [6, 7]. These equations are successfully applied [8] to forecast permolecular structures and prognosticate the conditions on the level of parameters resulting in the obtaining of nanostructures of definite size and shape. The equation was also used to forecast the formation of fibers [9]. The application of Avrami equations in the processes of nanostructure formation: (a) embryo formation and crystal growth in polymers [8]—

$$(1 - \upsilon) = \exp\left[-k\tau^n\right],$$

(11.3)

where υ—crystallinity degree, τ—duration, k—value corresponding to the process specific rate, n—number of the degrees of freedom changing from 1 to 6; b) graphitization process with the formation of carbon nanostructures [6, 7] –

$$\upsilon = 1 - \exp\left[-B\tau^n\right],\qquad(11.4)$$

where υ—volume share that was changed, τ—duration, B—index connected with the process rate, n—value determining the process directedness; (c) process of fiber formation [9] –

$$\omega = 1 - \exp\left[-z\tau^n\right],\qquad(11.5)$$

where ω—share of the fiber formed, τ—process duration, z—statistic sum connected with the process rate constant, n—number of the degrees of freedom (for the fiber n equals 1).

Instead of k, B or z, based on the previous considerations, the activity of nanoreactors can be used (a).

At the same time, the metal orientation proceeds in interface regions and nanopores of polymeric phase which conditions further direction of the process to the formation of metal/carbon nanocomposite. In other words, the birth and growth of nanosize structures occur during the process in the same way as known from the macromolecule physics [6], in which Avrami equations are successfully used. The application of Avrami equations to the processes of nanostructure formation was previously discussed in the papers dedicated to the formation of ordered shapes of macromolecules [8], formation of carbon nanostructures by electric arc method [6], obtaining of fiber materials [9].

11.1.2 MODIFIED AVRAMI EQUATION FOR PROCESSES IN NANOREACTORS

Then the share of the product being formed (W) in nanoreactor will be expressed by the following equation –

$$W = 1 - \exp\left(-a\tau^n\right) = 1 - \exp\left[-\left(\frac{\varepsilon_S}{\varepsilon_V}\right)\tau^n\right] = 1 - \exp\left\{-\left[\left(\frac{\varepsilon_S^0 d}{\varepsilon_V^0}\right)\frac{S}{V}\right]\tau^n\right\}\qquad(11.6)$$

where a—nanoreactor activity, $a = \varepsilon_S/\varepsilon_V$; ε_S—surface energy reflecting the energy of interaction of reagents with nanoreactor walls, $\varepsilon_S = \varepsilon^0_S dS$; ε_V—nanoreactor volume energy, $\varepsilon_V = \varepsilon^0_V V$; $\varepsilon^0_S d$—multiplication of surface layer energy by its thickness, ε^0_V—energy of nanoreactor volume unit, S—surface of nanoreactor walls, V—nanoreactor volume.

When the metal ion moves inside the nanoreactor with redox interaction of ion (mol) with nanoreactor walls, the balance setting in the pair "metal containing—polymeric phase" can apparently be described with the following equation –

$$zF\Delta\phi = RT\ln K = RT\ln\left(\frac{N_p}{N_r}\right) = RT\ln(1-W), \qquad (11.7)$$

where z—number of electrons participating in the process; $\Delta\phi$—difference of potentials at the boundary "nanoreactor wall—reactive mixture"; F—Faraday number; R—universal gas constant; T—process temperature; K—process balance constant; N_p—number of moles of the product produced in nanoreactor; N_r—number of moles of reagents or atoms (ions) participating in the process which filled the nanoreactor; W—share of nanoproduct obtained in nanoreactor.

In turn, the share of the transformed components participating in phase interaction can be expressed with the equation which can be considered as a modified Avrami equation –

$$W = 1 - \exp\left[-\tau^n \exp\left(\frac{zF\Delta\phi}{RT}\right)\right], \qquad (8)$$

where τ—duration of the process in nanoreactor; n—number of degrees of freedom changing from 1 to 6. When "n" equals 1, one-dimensional nanostructures are obtained (linear nanostructures, nanofibers). If "n" equals 2 or changes from 1 to 2, flat nanostructures are formed (nanofilms, circles, petals, wide nanobands). If "n" changes from 2 to 3 and more, spatial nanostructures are formed as "n" also indicates the number of degrees of freedom. The selection of the corresponding equation recording form depends on the nanoreactor (nanostructure) shape and sizes and defines the nanostructure growth in the nanoreactor.

In case of nanostructure interaction with the medium molecules, the medium self-organization is effective with the proximity or concordance of the oscillations of separate fragments of nanostructures and chemical bonds of the medium molecules. This hypothesis of nanostructure influence on media self-organization was already discussed [10–13]. At the same time, it is possible to use quantum-chemical computational experiment with the known software products.

According to the paper [14] the process vibration nature is discovered. This fact corroborates by IR spectroscopic investigations of nanostructures aqueous soles. Self-organization processes in media and compositions can be compared with the processes of crystalline phase origin and growth. At the same time, the growth can be one-, two- and three-dimensional. For such processes Avrami equations are widely used which usually reflect the share of the new phase appearing. In this case, the degree of nanostructure influence on active media and compositions is defined by the number of nanostructures, their activity in this composition and interaction duration. The temperature growth during the formation of new-phases in self-organizing medium prevents the process development.

Different shapes of nanostructures appear during the formation of lamellar or linear substances. For instance, graphite, clay, mica, asbestos, many silicates have lamellar structure and, consequently, they contain nanostructures with the shape of rotation bodies. It can be explained by the facts that bands of lamellar structure are rolled into clews and spirals with further sewing between them and formation of spheres, cylinders, ellipsoids, cones, etc. Nanostructures formed can be classified based on complexity. The formation of film structures resembling petals that can be put together into segments and semispheres afterwards is possible for carbon-containing structures that are formed mainly from hexagons with a certain number of pentagons and insignificant number of heptagons. Under certain energy actions the distortion of coplanar (flat) aromatic rings is known, when π-electrons are shifted and charges are separated on the ring, the ring polarity goes up. For polymeric chains and bands formed from macromolecules the possibilities of formation of super molecular structures, corresponding to the possibility of taking a definite shape, can be predicted with the help of Avrami's equation [8] for one-, two-, and three-dimensional crystallization of macromolecules. Here, energy exchange during the formation of ordered nanostructures and without energy exchange with the surroundings, i.e. under athermal and thermal embryo formation, is considered.

As mentioned before, at two-dimensional growth circles that form petals and further more complex nanostructures resembling flowers can be formed [15]. When being rolled, the circles formed can produce distorted semispheres or can serve as a basis for the transformation into "a beady",

During the redox process connected with the coordination process, the character of chemical bonds changes. Therefore correlations of wave numbers of the changing chemical bonds can be applied as the characteristic of the nanostructure formation process in nanoreactor –

$$W = 1 - \exp\left[-\tau^n \left(\frac{v_s}{v_f} \right) \right], \tag{11.9}$$

where v_s corresponds to wave numbers of initial state of chemical bonds, and v_f—wave numbers of chemical bonds changing during the process.

The share of nanostructures (W) during the redox process can depend on the potential of nanoreactor walls interaction with the reagents, as well as the number of electrons participating in the process. At the same time, metal ions in the nanoreactor are reduced and its internal walls are partially oxidized (transformation of hydrocarbon fragments into carbon ones). Then the isochoric-isothermal potential (ΔF) is proportional to the product $zF\Delta\varphi$ and Avrami equation will look as follows –

$$W = 1 - k_1 \exp\left[-\tau^n \exp\left(\frac{zF\Delta\phi}{RT}\right)\right], \tag{11.10}$$

where k_1—proportionality coefficient taking into account the temperature factor; n—index of the process directedness to the formation of nanostructures with certain shapes; z—number of electrons participating in the process; $\Delta\varphi$—difference of potentials at the boundary "nanoreactor wall—reactive mixture"; F—faraday number; R—universal gas constant.

Then the equation for defining the share (W) of nanostructures formed can be written down as follows by analogy with the aforesaid equations:

$$1 - W = \exp\left[-\beta a \tau^n\right], \tag{11.11}$$

where β—coefficient taking into account the changes in the activity in the process of nanoproduct formation.

Substituting the value of a, we get the dependence of the share of nanostructures formed upon the ratio of the energy of nanoreactor internal surface to its volume energy –

$$W = 1 - \exp\left[-\beta\left(\frac{\varepsilon_S}{\varepsilon_V}\right)\tau^n\right] = 1 - \exp\left[-\beta\left(\frac{\varepsilon_S^0}{\varepsilon_V^0}\frac{S}{V}\right)\tau^n\right], \tag{11.12}$$

If the nanoreactor internal walls become the shells of nanostructures during the process, the nanostructures so obtained are the nanoreactor mirror reflections.

The formation processes of metal-containing nanostructures in carbon or carbon-polymeric shells in nanoreactors can be related to one type of reaction series using the terminology of the theory of linear dependencies of free energies (LFE) [16]. Then it is useful to introduce definite critical values for the volume, surface energy of nanoreactor internal walls, as well as the temperature critical value. When the ration $\lg k/k_c$ is proportional $-\Delta\Delta F/RT$, the ratio W/W_c can be transformed into the following expression:

$$\frac{W}{W_c} = b\exp\left\{-\left(\frac{k}{k_c}\right)\left(\frac{\tau}{\tau_c}\right)^n\right\} = b\exp\left\{-\left(\frac{\tau}{\tau_c}\right)^n \exp\left(-\frac{\Delta\Delta F}{RT}\right)\right\} =$$

$$b\exp\left\{\left(\frac{\tau}{\tau_c}\right)^n\left[\exp k_T \cdot k_{VS}\left(\frac{\varepsilon_V}{\varepsilon_{Vk}} - \frac{\varepsilon_S}{\varepsilon_{Sk}}\right)\frac{Q}{T}\right]\right\}$$

(11.13)

where values with index "c" are correspond to the critical (or standard) values, b—proportionality coefficient considering the temperature factor, k_{vs}—coefficient considering correlations $\varepsilon_v/\varepsilon_{Vk}$ and $\varepsilon_s/\varepsilon_{Sk}$, ε_v and ε_{Vk}—volume energies of nanoreactor and "equilibrium" nanoreactor calculated via the ratios of their volumes; ε_s and ε_{Sk}—surface energy and its equilibrium value, T and θ—temperature of the process and temperature of the equilibrium process; τ—time required to develop the process of nanostructure formation; n—index of the process directedness to the formation of nanostructures of definite shapes. The values of volume and surface energies are given after the transformation of ΔΔF in accordance with [17], in which the physical sense of Taft constants is substantiated using the indicated energies.

At the same time the share of nanostructures (W) during the redox process can depend upon the potential of nanoreactor wall interaction with reagents, as well as the number of electrons participating in the process. The metal ions in nanoreactor are reduced and its internal walls are partially oxidized (transformation of hydrocarbon fragments into carbon ones).

Then the Helmholtz thermodynamic potential (ΔF) is proportional to the product of zFΔφ and Avrami equations will be expressed by the following formulae in accordance with the above models, one of them can be as follows:

$$W = 1 - k\exp\left[-\tau^n \exp\left(\frac{zF\Delta\phi}{RT}\right)\right],$$

(11.14)

where k—proportionality coefficient considering the temperature factor, n—index of the process directedness to the formation of nanostructures of definite shapes; z—number of electrons participating in the process; Δφ—difference of potentials on the border "nanoreactor wall—reaction mixture"; F—Faraday number; R—universal gas constant. When n equals 1, one-dimensional nanostructures are obtained (linear nanosystems and narrow bands). If n equals 2 or changes from 1 to 2, narrow flat nanostructures are formed (nanofilms, circles, petals, broad nanobands). If n changes from 2 to 3 and over, spatial nanostructures are formed, since n also means the number of degrees of freedom. If in this equation we take k as 1 and consider the process in which copper is reduced with simultaneous formation of nanostructures of a definite shape, the share of such formations or transformation degree can be connected with the process duration.

EXAMPLES OF THE MODIFIED AVRAMI EQUATIONS APPLICATION

The experimental modeling of obtaining nanofilms after the alignment of copper compounds with polyvinyl alcohol at 200°C revealed that optimal duration when the share of nanofilms approaches 100% equals 2.5 hours. This corresponds to the calculated value based on the aforesaid Avrami equation. The calculations are made supposing the formation of copper nanocrystals on the nanofilms. It is pointed out that copper ions are predominantly reduced to metal. Therefore it was accepted for the calculations that n equals 2 (two-dimensional growth), potential of redox process during the ion reduction to metal ($\Delta\varphi$) equals 0.34 V, temperature (T) equals 473 K, Faraday number (F) corresponds to 26.81 (A×hour/mol), gas constant R equals 2.31 (W×hour/mol×degree). The analysis of the dimensionality shows the zero dimension of the ratio $\dfrac{zF\Delta\phi}{RT}$. The calculations are made when changing the process duration with a half-hour increment:

Duration, hours	0.5	1.0	1.5	2.0	2.5
Content of nanofilms, %	22.5	63.8	89.4	98.3	99.8

If nanofillms are scrolled together with copper nanowires, β is taken as equaled to 3, the temperature increases up to 400°C, the optimal time when the transformation degree reaches 99.97%, corresponds to the duration of 2 hours, thus also coinciding with the experiment. According to the calculation results if following the definite conditions of the system exposure, the duration of the exposure has the greatest influence on the value of nanostructure share. The selection of the corresponding equation form depends upon the shape and sizes of nanoreactor (nanostructure) and defines the nanostructure growth in nanoreactor or the influence distribution of the nanostructure on the structurally changing medium. With one-dimensional growth and when the activation zero is nearly zero, the equation for the specific rate of the influence distribution via the oscillations of one bond can be written down as follows:

$$W = 1 - \exp\left[-\beta v \tau^n\right],\tag{15}$$

where v—oscillation frequency of the bond through which the nanostructure influences upon the medium, β - coefficient considering the changes in the bond oscillation frequency in the process. In the case discussed the parameter βv can be represented as the ratio of frequencies of bond oscillations v_{is}/v_{fs}, that are changing during the process. At the same time v_{is} corresponds to the frequency of skeleton oscillations of C–C bond at 1,100 cm^{-1}, v_{fs}—symmetrical skeleton oscillations of C=C bond at 1,050 cm^{-1}. In this case the equation looks as follows:

$$W = 1 - \exp\left\{-\tau^n \cdot \frac{v_{is}}{v_{fs}}\right\}$$

(11.16)

For the example discussed the content of nanofilms in % will be changing together with the changes in the duration as follows:

Duration, hours	0.5	1.0	1.5	2.0	2.5
Content of nanofilms, %	23.0	64.9	90.5	98.5	99.9

By the analogy with the above calculations the parameters a in the equation (5) should be considered as a value that reflects the transition from the initial to final state of the system and represents the ratios of activities of system states. Under the aforesaid conditions the linear sizes of copper (from ion radius to atom radius) and carbon-carbon bond (from C–C to C=C) are changing during the process. Apparently the structure of copper ion and electron interacts with electrons of the corresponding bonds forming the layer with linear sizes $r_i + l_{C-C}$ in the initial condition and the layer with the size $r_a + l_{C=C}$ in the final condition. Then the equation for the content of nanofilms can be written down as follows:

$$W = 1 - \exp\left\{-\tau^n \cdot \frac{r_a + l_{C=C}}{r_i + l_{C-C}}\right\}$$

(11.17)

At the same time r_i for Cu^{2+} equals 0.082 nm, r_a for four-coordinated copper atom corresponds to 0.113 nm, bond energy C–C equals 0.154 nm, and C=C bond—0.142 nm. Representing the ratio of activities as the ratio of corresponding linear sizes and taking the value n as equaled to 2, at the same time changing τ in the same intervals as before, we get the following change in the transformation degree based on the process duration—

Duration, hours	0.5	1.0	1.5	2.0	2.5
Content of nanofilms, %	23.7	66.0	91.2	98.7	99.9

Thus with the help of Avrami equations or their modified analogs we can determine the optimal duration of the process to obtain the required result. It opens up the possibility of defining other parameters of the process and characteristics of nanostructures obtained (by shape and sizes).

The influence of nanostructures on the media and compositions can be assessed with the help of quantum-chemical experiment and Avrami equations.

Modified Avraami equations were tested to prognosticate the duration of the processes of obtaining metal/carbon nanofilms in the system "Cu—PVA" at 200°C [18]. The calculated time (2.5 hours) correspond to the experimental duration of obtaining carbon nanofilms on copper clusters.

The influence of nanostructures on the media and compositions was discussed based on quantum-chemical modeling [3]. After comparing the energies of interaction of fullerene derivatives with water clusters, it was found that the increase in the interactions in water medium under the nanostructure influence is achieved only with the participation of hydroxyfullerene in the interaction. The energy changes reflect the oscillatory process with periodic boosts and attenuations of interactions. The modeling results can identify that the transfer of nanostructure influence onto the molecules in water medium is possible with the proximity or concordance of oscillations of chemical bonds in nanostructure and medium. The process of nanostructure influence onto media has an oscillatory character and is connected with a definite orientation of particles in the medium in the same way as reagents orientate in nanoreactors of polymeric matrixes. To describe this process, it is advisable to introduce such critical parameters as critical content of nanoparticles, critical time and critical temperature [1]. The growth of the number of nanoparticles (n) usually leads to the increase in the number of interaction (N). Also such situation is possible when with the increase of n critical value, N value gets much greater than the number of active nanoparticles. If the temperature exceeds the critical value, this results in the distortion of self-organization processes in the composition being modified and decrease in nanostructure influence onto media.

KEYWORDS

- **Avrami equations**
- **Computational experiment.**
- **Metal/Carbon Nanocomposites**
- **Nanoreactor**
- **Polymeric matrixes**
- **Redox process**
- **Self-organization**

REFERENCES

1. Kodolov, V. I.; and Trineeva, V. V.; Perspectives of idea development about nanosystems self-organization in polymeric matrixes. In: A. M. Lipanov, Ed., The problems of nanochemistry for the creation of new materials, Institute for Engineering of Polymer Materials and Dyes, Torun, **2012**, 75–100.

2. Shabanova, I. N.; Kodolov, V. I.; Terebova, N. S.; Trineeva, V. V.; X Ray electron spectroscopy in investigation of Metal/Carbon nanosystems and nanostructured materials. Udmurt State University, Moscow-Izhevsk, **2012**.

3. Khokhriakov, N. V.; and Kodolov, V. I.; Quantum-chemical modeling of nanostructure formation," Nanotechnics, 2, 2005, pp. 108–112.

4. Kodolov, V. I.; Didik, A. A.; and Volkov, Yu. A.; Volkova, E. G.; Low-temperature synthesis of copper nanoparticles in carbon shells. HEIs' news. *Chemi. Chem. Eng.* **2004**, *47*(1), 27–30.

5. Lipanov, A. M.; Kodolov, V. I.; and Khokhriakov et al., N. V. Challenges in creating nanoreactors for the synthesis of metal nanoparticles in carbon shells. *Alternat. Energ. Ecol.* **2005**, *2*(22), 58–63.

6. Fedorov, V. B.; Khakimova, D. K.; Shipkov, N. N.; and Avdeenko, M. A.; To thermodynamics of carbon materials," Doklady AS USSR, **1974**, *219*, 3, 596–599.

7. Fedorov, V. B.; Khakimova, D. K.; and Shorshorov, M. H. et al.; "To kinetics of graphitation," Doklady AS USSR, Vol. 222. No 2, **1975**; pp. 399–402.

8. Vunderlikh, B.; "Physics of macromolecules, vol. 2," Mir, Moscow, **1979**.

9. Serkov, A. T.; Theory of chemical fiber formation. In: A.T. Serkov, Ed., Himiya, Moscow, **1975.**

10. Kodolov, V. I.; Khokhriakov, N. V.; and Kuznetsov, A. P.; et al, Perspectives of nanostructure and nanosystem application when creating composites with predicted behavior. In: A. A. Berlin, Ed., Space challenges in XXI century, vol. 3, Novel materials and technologies for space rockets and space development, Torus press, Moscow, **2007**; pp. 201–205.

11. Krutikov, V. A.; Didik, A. A.; and Yakovlev, G. I.; et al, Composite material with nanoreinforcement. *Alter. Energ. Ecol.* *4*(24), **2005**, pp. 36–41.

12. Kodolov, V. I.; Khokhriakov, N. V.; and Kuznetsov, A. P.; To the issue of the mechanism of nanostructure influence on structurally changing media during the formation of "intellectual" composites. *Nanotechnics.* **2006**, 3(7), 27–35.

13. Khokhriakov, N. V.; and Kodolov, V. I.; Influence of hydroxyfullerene on the structure of water. *Int. J. Quantum Chem.* **2011**, *111*(11), 2620–2624.

14. "Synthesis, functional properties and applications," Proceedings of Conf. NATO-ASI, July-August, 2002.

15. Palm, V. A.; Basics of quantitative theory of organic reactions. Himiya, Leningrad, **1967**.

16. Kodolov, V. I.; On modeling possibility in organic chemistry. *Org. React,* 2(4) **1965**, 11–18.

17. Kodolov, V. I.; Khokhriakov, N. V.; Trineeva, V. V.; and Blagodatskikh, I. I. Activity of nanostructures and its display in nanoreactors of polymeric matrixes and in active media. *Chem. Phys. & Mesoscopy.* **2008**, *10*(4), 448–460.

18. Kodolov, V. I.; Commentary to the paper (2008) by V. I. Kodolov et al., *Chem. Phys. Mesoscopy.* **2009**, *11*(1) 134–136.

CHAPTER 12

CONDITIONS FOR CARBON BLACK ACCUMULATION AT THE INTERFACE IN HETEROGENEOUS BINARY POLYMER BLENDS

A. E. ZAIKIN, V. V. MOLOKIN, and S. A. BOGDANOVA

Department of Plastics Technology, Kazan State Technological University, Kazan, Russia

CONTENTS

12.1   INTRODUCTION

The phenomenon of carbon black (CB) particles gathering at the interface in binary heterogeneous polymer blends is not only of fundamental interest, but has a practical aspect as a method for improving the electric conductivity of composite polymer materials [1–12] and the mechanical properties of polymer blends with a low interfacial adhesion [13–15]. The works examining the causes of this phenomenon are few in number. Most of authors only note that some part of a filler tends to accumulate at the interface. Contradictory assumptions of the conditions necessary for such localization are made.

Some workers [1–2] argue that a necessary condition for such localization is the essential difference between the energies of interactions of polymer components with a surface of powder particles. The others [5–8], in contrast, suppose that this phenomenon takes place only when the polymeric components of a blend have low and approximately equal energies of the interaction with the filler surface. Some authors [3] hold that a filler is driven out to the interface due to the crystallization of polymer components of a mixture.

Sumita et al. [9] reason, that the causes of filler localizing at the interface reduce to the well-known thermodynamic conditions for solid particles to reside at the interface between two nonmiscible liquids.

12.2   EXPERIMENTAL

The polymers used are characterized in Table 12.1.

High-pressure polyethylenes (PE) with different values of the melt-flow index (MFI) also were used. The viscosity values of PE are given in Figure 12.4. Poly(methyl methacrylate) (PMMA) of various viscosity was prepared by fractionating the initial polymer by molecular weight. The fractionation was carried out by stepwise polymer sedimentation from a chloroform solution using hexane. The viscosity values of the obtained PMMA samples are given in Figure 12.4.

Carbon black under study (no. 254) had specific surface area 250 m^2/g, average size and density of primary aggregates 28 nm and 1.8 g/cm^3, correspondingly, and specific volume electric resistance 1.5×10^{-3} Ohm×cm (with a density of 0.5 g/cm^3).

Low-molecular weight liquids were purified and distilled. The values of heat of carbon black wetting by liquids {ΔH} were measured by means of a DAK1-1A calorimeter at 298 K.

Polymers were mixed with carbon black in melt at 433 ± 5 K (for PP at 453 K) using laboratory rolls. The mixing was carried out in two stages. At the first stage the whole portion of CB was mixed with one of polymers, and then the second polymer was added. The mixing time at each stage was 5 min.

TABLE 12.1   The characteristic of used polymers

Polymer	Density (g/cm³)	Viscosity Pa·s (413 K, 15 s⁻¹)	Molecular Weight × 10⁻³	Comments
High Pressure Polyethylene (PE)	0.922	2,900	37 (η)	MFI = 2.1 g/10 min (at 463 K; 2.16 kg)
Polypropylene (PP)	0.91	2,500 (at 463 K)	-	MFI = 3.2 /10 min (at 503 K; 2.16 kg)
Polyurethane (PU)*	1.21	3,500	-	OH: NCO = 1.01
Polyisobutylene (PIB)	0.91	11,300	118 (η)	
cis-1,4-Polybutadiene (PBD)	0.92	-	104 (η)	Mooney—48 (373 K)
Polystyrene (PS)	1.05	9,800 (433 K)	190 (w)	-
Polymethylmethacrylate (PMMA)	1.19	11,200 (at 433 K)	150 (η)	-
Copolymer of Ethylene and Vinylacetate (EVA)	0.950	980	15.5 (w)	Containing 28.9 wt.% Vinyl-acetate
Copolymer of Butadiene and Acrylonitrile (BNR)	0.986	Moony—54 (at 373 K)	220 (w)	Containing 40 wt.% Acrylonitrile
Polyvinylacetate (PVA)	1.19	-	140 (η)	-
Polydimethylsiloxane (PDMS)	0.98	-	560 (η)	-
Polychloroprene (PCP)	1.22	-	170 (η)	Mooney—62 (373 K)

MFI **is** the melt-flow index, η is the viscosity-average molecular mass, w is the weight-average molecular mass.
*PU was prepared on the basis of polyoxytetraethylene glycol, 4,4—diphenylmethane diisozyanat and 1,4—buthane diol

Polymer samples for measurements of the specific volume electric resistance (ρ) were prepared by molding using a hydraulic press at 443 ± 3 K (for PP at 463 K) under a pressure of 30 MPa for 300 ± 5 sec. If ρ was less than 1×10^6 Ohm×cm, it was measured potentiometrically at 293 K (ISO 1853-75) using stripes 100 mm long, 10 mm wide and 1.2 mm thick. The potential difference was recorded for a 20 mm site so that contact resistance was eliminated. The results scattering between parallel experiments was ± 18 percent. For the electric resistance exceeding 1×10^6 Ohm×cm ρ was measured according to the ISO 2878-78 procedure on plates 1.2 mm thick with an area of 16 cm². In this case stainless steel electrodes were pressed over the whole surface to both sides of a plate. The measurements were made using an E6-13 teraohmmeter with a 10 V potential difference between the electrodes. The scattering in ρ values between parallel experiments was ±31 percent.

The carbon black distribution in polymer blends was examined by means of optical microscopy in transmitted light on thin (1–5 µm thick) sample slices according to the procedure described in [11].

The bond strength between a filler and polymer was assessed by the exfoliation force (F) of the polymer under study from filler particles fixed in a matrix of another polymer, polypropylene [14]. In this order carbon black was stirred into molten PP, and a plate was molded from this composition. The surface of the plate was treated with an abrasive to remove a polymer layer and expose the carbon black surface. Then the plate was coated from solution by a layer of the polymer under study. After removing a solvent the force of the coating polymer exfoliation from the plate surface was measured.

12.3   RESULTS AND DISCUSSION

It is known that carbon-black filled polymers conduct electric current only with the concentration of carbon black exceeding the threshold of percolation (φ_p). In heterogeneous polymer blends carbon black is distributed nonuniformly between polymer phases. If the concentrations of carbon black in both phases of a blend are lower than φ_p, the blend can conduct electric current only subject to the condition that the part of carbon black is localized at the interface and its concentration here reaches the percolation threshold [12]. So if the concentration of carbon black in both phases of a blend is only slightly lower than φ_p, even a minor accumulation of carbon black at the interface confers conductivity on the polymer blend. This enables the extent of carbon black aggregation at the interface to be judged by the conductivity value.

A variety of blends (Table 12.2) was analyzed under the condition that the carbon black is localized in a single polymer phase and its concentration in this phase is slightly below φ_p. To confine carbon black within one of the phases of a blend, the sequence of carbon black mixing with polymers was altered. Carbon black was first introduced into one of polymer components of a blend, and only then another component was added. A filler is known to retain almost entirely in that phase of a heterogeneous blend where it was introduced initially [1, 2, 10, 11, 12].

The localization of carbon black at the interface takes place not in all of the blends (Table 12.2). For most of blends the presence or absence of this phenomenon depends on the sequence of components mixing.

All the blends in Table 12.2 may be divided into three groups based on the presence of the superadditive electrical conductivity and the effect of blending sequence on electric conductivity.

TABLE 12.2 Electrical conductivity of the (P1 + CB) + P2 systems

Phase P1 (50 vol. %)	Phase P2 (50 vol. %)	φ_p CB for a Phase (P1, vol. %)	The Contents CB in a Phase (P1, vol. %)	ρ of a Phase (P1, Ohm·cm)	ρ of a Phase (P2, Ohm·cm)	ρ of a Blend (Ohm·cm)
Blends in which CB localization at the interface takes place only from one of two phases.						
PE + CB	PU	6.8	5.3	$>1 \cdot 10^{12}$	$3 \cdot 10^{10}$	$1.3 \cdot 10^{4}$
PU + CB	PE	10	8.5	$7 \cdot 10^{7}$	$>1 \cdot 10^{12}$	$9 \cdot 10^{8}$
PE + CB	BNR	6.8	5.3	$>1 \cdot 10^{12}$	$3 \cdot 10^{9}$	$5.9 \cdot 10^{4}$
BNR + CB	PE	16.5	15.5	$4 \cdot 10^{7}$	$>1 \cdot 10^{12}$	$2 \cdot 10^{9}$
PE + CB	PCP	6.8	5.3	$>1 \cdot 10^{12}$	$3 \cdot 10^{10}$	$2 \cdot 10^{4}$
PCP + CB	PE	18	16	$3 \cdot 10^{8}$	$>1 \cdot 10^{12}$	$1 \cdot 10^{11}$
PE + CB	PMMA	6.8	5.3	$>1 \cdot 10^{12}$	$>1 \cdot 10^{12}$	$2 \cdot 10^{4}$
PMMA + CB	PE	15.5	14	$>7 \cdot 10^{11}$	$>1 \cdot 10^{12}$	$>1 \cdot 10^{12}$
PS + CB	PMMA	11.5	9.3	$>1 \cdot 10^{12}$	$>1 \cdot 10^{12}$	$3.2 \cdot 10^{3}$
PMMA + CB	PS	15.5	14	$>1 \cdot 10^{12}$	$>1 \cdot 10^{12}$	$>1 \cdot 10^{12}$
EVA + CB	PS	5	2.7	$1.5 \cdot 10^{11}$	$>1 \cdot 10^{12}$	$>1 \cdot 10^{12}$
PS + CB	EVA	11.5	9.3	$>1 \cdot 10^{12}$	$7 \cdot 10^{11}$	$7.2 \cdot 10^{2}$
PE + CB	PDMS	6.8	5.3	$>1 \cdot 10^{12}$	$>1 \cdot 10^{12}$	$3 \cdot 10^{4}$
PDMS + CB	PE	4	3	$>1 \cdot 10^{12}$	$>1 \cdot 10^{12}$	$>1 \cdot 10^{12}$
PS + CB	PDMS	11.5	9.3	$>1 \cdot 10^{12}$	$>1 \cdot 10^{12}$	$2.2 \cdot 10^{3}$
PDMS + CB	PS	4	3	$>1 \cdot 10^{12}$	$>1 \cdot 10^{12}$	$>1 \cdot 10^{12}$
PIB + CB	PDMS	8.6	6.5	$>1 \cdot 10^{12}$	$>1 \cdot 10^{12}$	$2 \cdot 10^{7}$
PDMS + CB	PIB	4	3	$>1 \cdot 10^{12}$	$>1 \cdot 10^{12}$	$>1 \cdot 10^{12}$
Blends in which CB localization at the interface takes place from both phases.						
PE + CB	PS	6.8	5.3	$>1 \cdot 10^{12}$	$>1 \cdot 10^{12}$	$6 \cdot 10^{9}$
PS + CB	PE	11.5	9.3	$>1 \cdot 10^{12}$	$>1 \cdot 10^{12}$	$4 \cdot 10^{4}$
PP + CB	PS	5.3	4.2	$>1 \cdot 10^{12}$	$>1 \cdot 10^{12}$	$6.3 \cdot 10^{8}$

TABLE 12.2 *(Continued)*

Phase P1 (50 vol. %)	Phase P2 (50 vol. %)	φ_p CB for a Phase (P1, vol. %)	The Contents CB in a Phase (P1, vol. %)	ρ of a Phase (P1, Ohm·cm)	ρ of a Phase (P2, Ohm·cm)	ρ of a Blend (Ohm·cm)
PS + CB	PP	4.2	3.6	$>1 \cdot 10^{12}$	$>1 \cdot 10^{12}$	$6.9 \cdot 10^3$
PE + CB	EVA	6.8	5.3	$>1 \cdot 10^{12}$	$8 \cdot 10^{11}$	$1 \cdot 10^7$
EVA + CB	PE	5	2.7	$1.5 \cdot 10^{11}$	$>1 \cdot 10^{12}$	$5 \cdot 10^8$
Blends in which CB localization at the interface is very small.						
PE + CB	PIB	6.8	5.3	$>1 \cdot 10^{12}$	$>1 \cdot 10^{12}$	$>1 \cdot 10^{12}$
PIB + CB	PE	8.6	6.5	$>1 \cdot 10^{12}$	$>1 \cdot 10^{12}$	$>1 \cdot 10^{12}$
PE + CB	PP	5.9	4.8	$>1 \cdot 10^{12}$	$>1 \cdot 10^{12}$	$>1 \cdot 10^{12}$
PP + CB	PE	5.3	4.2	$>1 \cdot 10^{12}$	$>1 \cdot 10^{12}$	$>1 \cdot 10^{12}$
PE + CB	PBD	6.8	5.3	$>1 \cdot 10^{12}$	$>1 \cdot 10^{12}$	$>1 \cdot 10^{12}$
PBD + CB	PE	9	8	$>1 \cdot 10^{12}$	$>1 \cdot 10^{12}$	$>1 \cdot 10^{12}$

The first group consists of blends where the effect of superadditive electric conductivity is observed only with a certain sequence of components blending.

The second group involves blends where a lowered ρ value is observed only with preliminary carbon black introduction into either of two polymer components, but the sequence of components mixing strongly affects the degree of ρ lowering.

The blends which do not conduct electric current with any sequence of components blending constitute the third group.

The study of the compositions by optical microscopy shows that increased electric conductivity takes place only for those blends and mixing procedures for which the carbon black localization at the interface is observed (Figure 12.1).

It is reasonable to propose that the process of filler localization at the interface between polymeric phases as well as between low-molecular weight phases is fully controlled by the thermodynamics of the competitive wetting of a solid particle by these phases, which is consistent with Sumita [9].

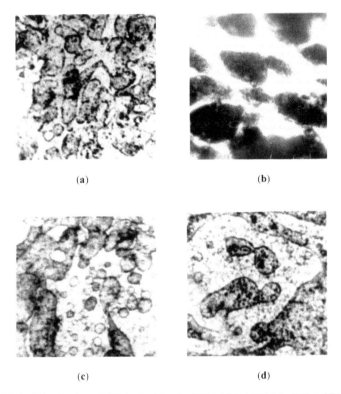

(a)   (b)

(c)   (d)

FIGURE 12.1   Distribution of CB in the blends: (PE + CB) + PU (**a**); (PU + CB) + PE (**b**), (PS + CB) + PE (**c**), (PE + CB) + PDMS (**d**) filled with 2.5 vol. % CB (Optical micrograph). The ratio of polymers is 1:1. Taking the components in brackets means their premixing. Magnification: 1,100.

The phenomenon of solid particles aggregation at the interface between low-molecular weight liquid phases has been much studied, for example, for emulsion stabilized by high-dispersity powders or for powder flotation [16]. The thermodynamic condition of localization of solid high-dispersity particles between low-molecular weight liquid phases stems from the Young law [16]:

$$-1 < (\sigma_{13} - \sigma_{23}) / \sigma_{12} < 1 \qquad (12.1)$$

Where σ_{13} is the interfacial tension between the first liquid and the particle surface, σ_{23} is the interfacial tension between the second liquid and the particle surface, and σ_{12} is the interfacial tension between the liquids.

Such a localization is thermodynamically efficient and will take place with any angle of solid surface wetting by liquid phases except for the angle equal to zero. With the unsatisfied condition (12.1), the particles of a filler would be fully wetted by one of the liquid phases (the phenomenon of spreading) and could not gather at the interface.

The validity of the last proposal is supported by the fact that carbon black localization at the interface is most pronounced for polymer blends with a high surface tension between polymers, that is, in those polymer pairs where thermodynamic gain of particle transfer to the interface is essential [17].

The applicability of the condition (12.1) for polymer blends is difficult to test because of lacking data on interfacial tension between a filler and polymer.

Nevertheless, there is an experimental observation contradictory to the condition above. This is dependence of carbon black localization at the interface on the sequence of components mixing. Indeed, the satisfied condition (12.1) inevitably results in the satisfied thermodynamic conditions of the transfer of particles to the interface both from the first phase ($\sigma_{13} > \sigma_{23} - \sigma_{12}$), and from the second phase ($\sigma_{23} > \sigma_{13} - \sigma_{12}$) [16].

Therefore, from the thermodynamic viewpoint, with the unsatisfied condition (12.1) the aggregation of solid particles at the interface must be observed for any sequence of components mixing. So, for example, the localization of carbon black at the interface of two low-molecular weight liquids does not depend on the sequence of components mixing. Altering the sequence of components mixing can affect only the thermodynamic efficiency of process of filler redistribution from the bulk to the interface, but not the parameters in Eq. (12.1). It should be noted that in most of publications cited above carbon black was introduced into the blend of polymers, and the problem of the influence of components blending sequence on the localization of carbon black at the interface was not considered [2, 4, 5–10].

To reveal the causes and conditions necessary for carbon black to localize at the interface there was a need in evaluating the energy of adsorption interaction of polymers with the filler surface.

However, now there is no simple and reliable method to assess the efficiency of interaction of polymers with a surface of high-dispersity powders.

At the same time it is well understood that the energy of adsorption interaction of low-molecular weight analogs of polymers with a solid surface which depends on the surface energy of a solid body and the chemical nature of an adsorbent is comparable with such interaction for polymers [18, 19]. This makes it possible to judge qualitatively the relative energy efficiency of adsorption interaction of corresponding polymers with the carbon black surface by the interaction energies of low-molecular weight analogs. The efficiency of the interaction of liquid low-molecular weight analogs of polymers with the carbon black surface was estimated by the value of wetting heat (ΔH) (Figure 12.2).

The studies show (Figure 12.2) that ΔH of CB is in a certain manner dependent on the polarity ξ. The polarity of polymers and liquids was calculated as follows:

$$\xi = (\delta_p^2 + \delta_h^2)/\delta^2 \qquad (12.2)$$

Where, δ_p and δ_h are polar and hydrogenous components of the solubility parameter of a liquid, δ is the solubility parameter of a liquid [20, 21].

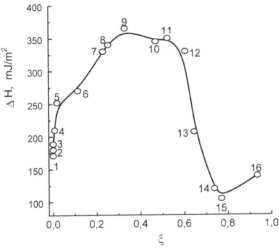

FIGURE 12.2 The heat of CB No. 254 wetting by liquids versus the polarity of a liquid. Carbon tetrachloride (1), n-hexane (2), 2,2,4—trimethylpentane (3), ethylbenzene (4), toluene (5), 1—chlorbutane (6), MMA (7), tetrahydrofuran (8), ethylacetate (9), hexanol—1 (10), dimethylformamide (11), acetonitrile (12), ethanol (13), ethylene glycol (14), glycerin (15), water (16).

With increasing ξ the wetting heat of CB No. 254 passed a maximum at ξ = 0.33. This value is consistent with the magnitudes of surface polarity for two other types of CB calculated on the basis of Hansen's three-dimensional solubility parameter. So, for two different types of CB ξ was equal to 0.43 and 0.38 [20, 21]. The ΔH increase up to a certain ξ value is due to the formation of not only dispersion bonds but hydrogen and polar bonds with oxygen- containing groups on the CB surface as well. The further fall in ΔH is obviously connected with the decreasing energy of dispersion interaction between liquids. The fact that the dependence of interaction of organic liquids with the surface of some metal oxides on the polarity of a liquid passes an extremum was noticed by the researchers [22] (the interaction was assessed by the impregnation of powders by liquids).

The data obtained indicate that the increase in the polarity of polymers up to ξ = 0.33 – 0.35 must result in rising energy of their adsorption interaction with CB.

Besides, the relative efficiency of the interaction between polymer and a filler was judged indirectly by the specific force of exfoliation (F) of polymer from a filler [14]. The permissibility and reliability of such estimation for elastic polymers is evidenced by the fact that F of these polymers rises linearly with the increasing CB volume fraction in PP [14].

The F values show that with increasing polarity of polymer (Table 12.3) its adhesion to a filler rises. This correlates well on a qualitative level with the estimates of their interaction with CB from the wetting heat data (Table 12.3).

According to the results of these measurements, PE, PIB, PBD, PP, PS, and PDMS have the lowest and closest magnitudes of the energy of adsorption interaction with the CB and aerosol surface. The rest of polymers, judging by wetting heat and exfoliation force (Table 12.3), arrange in the following series according to the increasing efficiency of their interaction with CB and aerosol: EVA < BNR < PCP < PU < PVA.

TABLE 12.3 The characteristics of the interaction of polymers and their low-molecular weight analogs with CB No. 254

Polymer			Low-Molecular Weight Analog		
Notation	x	DF,* kN/m	Name	x	DH, mJ/ m^2
	[17, 20, 21, 23]			[20, 21, 23]	
PE	0	0.025	n—hexane	0	170
PIB	0	0.02	2,2,4-trimethyl-pentane	0	175
PS	0.012	0.015	Ethylbenzene	0.0076	205
PBD	0.003	0.15	Hexene-1	-	180
PDMS	0.04**	0.03	Cyclic tetramer of dimethylsiloxane	0.04	200
EVA	-	0.3	-	-	-
BNR	-	0.46	-	-	-
PCP	0.11**	0.65	1-chlorobutane	0.11	270
PU	-	1.17	-	-	-
PVA	0.33	2	Ethylacetate	0.33	365

*DF is F gain for the CB content increase in the substratum (PP) from 0 to 42 vol. %
**calculated from the ratio of polar and dispersion components of surface tension [17].

The analysis shows that when an essential difference in adhesion of polymer components of a blend to CB takes place (PE + PU, PE + BNR, PE + PMMA, PE + PCP, PS + PMMA), CB moves to the interface only from the phase characterized by lower adhesion to CB.

However, this rule is invalid for explaining the effect of the sequence of components blending on CB localization at the interface in the blends of polymers having close energies of adhesion to a filler, such as PE + PDMS, PS + PDMS, PIB + PDMS, PE + PS, PS + PP.

The results obtained show that in these blends the removal of a filler from the bulk to the interface occurs from the phase of polymer characterized by higher value of cohesion energy (Table 12.4) (from PS to the interface with PO and PDMS, from PO to the interface with PDMS). As for the phase with low cohesion energy, such redistribution here either is absent (from PDMS to the interface with PE, PS, PIB) or is insignificant if the cohesion energy difference for polymer components is small (from PE or PP to the interface with PS).

TABLE 12.4   The characteristics of cohesion energy of polymers at 433 K [17, 23]

Polymer	s (mN/m)	Polymer	s (mN/m)
PDMS	13.0	EVA	26.2
PP	20*	PVA	26.3
PU	23 **	PS	29.9
PIB	24.8	PMMA	30.5
PE	25.1	PCP	31.5
PE	24.0*	-	-

*at 463 K
**given for the low-energy block of copolymer.

It is known [16] that the wetting force of liquids increases with increase in their adhesion to the surface as well as with decrease in their cohesion energy. Taking into account Dupre's equation, inequality Eq. (12.1) may be written in the following form:

$$-1 < \left(\sigma_1 - \sigma_2 + w_{a2} - w_{a1}\right)/\sigma_{12} < 1 \qquad (12.3)$$

Where, σ_1 and σ_2 are the surface tension values of the first and the second liquids at the interface with air, correspondingly; w_{a1} и w_{a2} are the adhesion energies of the first and the second liquids to the surface of a particle.

The liquid with greater adhesion to a surface and lower surface tension has a preference in wetting this surface. When the energies of polymer-surface interaction are close, the polymer with lower surface tension (smaller cohesion) exhibits better wetting force [16].

This suggests that the migration of carbon black toward the interface in most of blends proceeds effectively from the polymer phase with lower capability of wetting the carbon black surface. The redistribution of carbon black from the phase of polymer with higher wetting force to the interface takes place only for a small difference in wetting force of polymer components and is less pronounced.

This conclusion determines the conditions necessary for a filler to localize at the interface, but cannot explain the causes of such strong influence of the sequence of components mixing on a process, and hence, does not reveal the essence of a phenomenon.

It is impossible to understand the effect of blending sequence on filler localization at the interface from the single viewpoint of thermodynamics of particle wetting by two liquids. It is due to the fact that in polymer mixtures, for well-known reasons, the most energetically efficient distribution of a filler described by Eq. (12.1) is not achieved.

The study of a filler redistribution toward the interface shows that this process for all blends is completed after 2–3 min of mixing and does not depend on the sequence of components adding (Figure 12.3). The increasing time of mixing to 30 min does not lead to the filler localization at the interface in those cases, where it has not taken place within 3 min after the start of mixing. To the contrary, where the mixing was prolonged to 10 min, ρ for many blends slightly rises.

The essential difference in electrical conductivity values for blends of identical formula but different sequence of preparation, and the stability of conductivity during the mixing process suggest that the concentration of a filler at the interface is governed by the equilibrium between the number of particles arriving at the interface and those removed back to the phase. Indeed, taking into account that, although the redistribution of a filler from a phase to the interface gives no energy gain and is associated with overcoming a high activation barrier of macromolecule desorption, it does occur. So, according to the statistics, the concentration of filler particles at the interface must rise in time, but it remains constant (Figure 12.3).

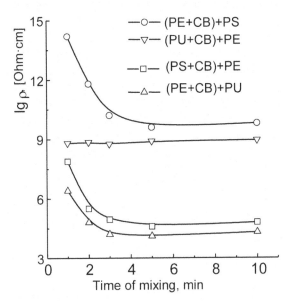

FIGURE 12.3 ρ of polymer blends versus time of mixing. The CB content: 5 vol. % in the PE phase, 8.5 vol. % in the PU phase, 9.3 vol. % in the PS phase. The ratio of polymers is 1:1.

A great difference in electrical conductivity of blends prepared by two different mixing procedures indicates that in a blend with low conductivity there exists a potential for greater number of filler particles to be localized at the interface than it is observed. However, this possibility is not realized. This can be explained only by the equilibrium existing between the number of particles arriving at the interface and those leaving the interface for a phase.

Let us analyze this equilibrium in more detail.

Firstly specify that the localization of a dispersed particle at the interface means that the macromolecules of both polymers are adsorbed on this particle. The velocity of particle transfer from a phase to the interface is determined by the number of their successful collisions dependent on the filler concentration in a phase. The collision is said to be successful when a part of macromolecules of the preliminarily filled polymer previously adsorbed on a filler is substituted by the macromolecules of another polymer component. The replacement of a very small number of macromolecules may be treated as a transfer of a filler particle to the interface.

The replacement is associated with overcoming an energy barrier of the desorption of macromolecules from a solid surface, which can be rather high for some macromolecules [24]. However, at the surface of a filler there always exists some part of macromolecules having a relatively low adsorption energy (small number of contacts with a surface). These macromolecules can be with relative ease substituted after a collision of particles for macromolecules of another polymer. High shear stresses acting during mixing promote a macromolecule to overcome the activation barrier of desorption.

Obviously, the average activation energy of desorption is higher for macromolecules of the polymer having the higher wetting force. Consequently, the rate of filler particles migration to the interface must be lower from a phase with higher wetting force than from a phase with lower wetting force.

However, with a satisfied condition (12.1), after a long period of mixing the interface would be filled with particles without regard to the phase which they left. Nevertheless, the experimental data are contradictory to this concept (Figure 12.3). Therefore, the essential role in establishing the equilibrium concentration at the interface is played by the rate of filler particles leaving the interface and arriving at a phase.

Consider the factors controlling the rate of particles leaving the interface for a phase.

The removal of a particle from the interface is a result of the shear stress exerted on the particle by the polymer environment and is described by the Stokes law [25]. The shear stress affecting a particle is in direct proportion to its size, the viscosity of a polymer medium, and the shear rate. In certain situations this shear can abstract a particle from the interface and return it to one of polymer phases. The adhesion of a particle (W_a) to the unfilled polymer phase counteracts its removal from the interface. Only those particles stay at the interface for which the force of binding to the opposite phase exceeds the force of their separation from this phase. The ratio of these forces will determine the equilibrium concentration of a filler at the interface.

Let us estimate these forces.

In both phases of a heterogenous polymer blend the shear stress is the same, so the same is the force of particle removal from the interface. Hence, the local equilibrium concentration of particles at the interface will be determined by their adhesion to a phase. In this case it is important to specify to what of two polymer phases the adhesion is considered. It is known [1] that the transfer of filler particles from one phase of polymer blend to another is observed very rarely. Therefore, in most cases the particles of a filler after abstraction from the interface return back in that phase from which they came to the interface. Otherwise a fast transfer of a filler would be observed from a phase of lower wetting force to opposite polymer phase. So the force holding particles at the interface is a result of their adhesion to the unfilled polymer phase. Besides, this suggests that dispersed particles at the interface occupy nonequilibrium positions and are confined predominantly in the preliminarily filled phase.

The energy of particle adhesion to a phase is determined by the following expressions [16]:

$$Wa31 = \sigma_{23} - \sigma_{13} + \sigma_{12} \qquad (12.4)$$

$$Wa32 = \sigma_{13} - \sigma_{23} + \sigma_{12} \qquad (12.5)$$

Where, $Wa31$ is the work of particle adhesion to the phase 1; $Wa32$ is the work of particle adhesion to the phase 2.

The adhesion of filler particles to the phase of polymer with the greater wetting force is higher than to the phase with smaller wetting force.

Analyzing the ratio between the rates of particles arriving at and leaving the interface, it may be concluded that the smaller is the wetting force of a polymer phase where particles are located, the greater is their concentration at the interface. This is consistent with experimental observations (Table 12.2, Figure 12.3).

However, the proposed mechanism of establishing the equilibrium concentration of carbon black at the interface between polymers does not explain why carbon black does not migrate to the interface from one of two phases whereas such transfer occurs from another phase. Indeed, if the condition (12.1) is satisfied, the transfer of carbon black from either of two phases to the interface is thermodynamically efficient. Even with a low adhesion of a filler to the interface the local concentration of a filler at the interface would slightly exceed its concentration in phase. However, such an excess is not observed for a number of polymer pairs attributed to the first group according to the data in Table 12.2.

A broad spectrum of the energies of adsorption of macromolecules on a solid surface suggests that under certain conditions the localization of a filler at the interface is feasible even if the condition (12.1) is not fulfilled. This may take place if the location of the whole of a filler within a phase of one of polymers and not at the interface is thermodynamically efficient ($\sigma_{13} > \sigma_{12} + \sigma_{23}$ or $\sigma_{23} > \sigma_{12} + \sigma_{13}$) and a filler was introduced initially into a phase of another polymer. In this case the transfer of CB from one phase to another is thermodynamically favorable. It is necessary for a filler particle to transfer from one phase to another that the macromolecules of polymer previously adsorbed on its surface should be fully replaced by the macromolecules of the second polymer. However, such a replacement is unlikely because of a very high adsorption energy of some part of macromolecules. The experimental data confirm that such replacement is very rare to occur [1, 10, 11]. The partial replacement of macromolecules is more probable since there is a portion of macromolecules with low adsorption energy [24]. Such partial replacement just implies the localization of a filler at the interface.

Since the interfacial tension between polymers is minor [17] and the adhesion of polymers to a filler differs essentially (Table 12.3), it may be supposed that the filler localizing between polymer phases by the latter mechanism is most probable. The filler localization at the interface in all blends assigned to the first group according to Table 12.2 is likely to follow the last mentioned scheme. In those blends the local concentration of a filler at the interface is also governed by the equilibrium between the number of particles arriving at the interface and leaving it for a phase.

From the above discussion it follows that with increasing difference in wetting forces of polymers the equilibrium concentration of CB at the interface must rise when a filler transfers to the interface from the phase of a lesser wetting force and must fall when it comes here from the phase of higher wetting force. Besides, the rise of interfacial tension between polymers must promote the increase in local concentration of CB at the interface when it is redistributed from any phase.

Taken together the experimental observations (Table 12.2) confirm the validity of the latter conclusion.

Thus, dependence of the interfacial filler concentration on the sequence of components mixing is due to the peculiarities of macromolecule adsorption on a solid surface. Because of a high activation energy of desorption from a solid surface for a major part of macromolecules [24], the redistribution of dispersed particles from one polymer phase to another practically does not occur and those cannot occupy equilibrium position at the interface. Under these circumstances when a difference in wetting forces of phases takes place, the sequence of blending has a determining effect on the possibility and extent of the localization of dispersed particles at the interface. Let us consider the blends assigned to the third group according to Table 12.2. In the PE + PIB, PE + PP, PE + PBD blends the polymer components have close values of cohesion energy and similar interaction intensity with the carbon black surface and hence, close wetting force. Besides, these pairs exhibit very low interfacial energy [17]. Consequently, even though the condition (12.1) is fulfilled, the equilibrium CB concentration at the interface in these polymer blends will only slightly exceed the concentration in phase. The experimental data are fully consistent with this assumption. So, if CB is confined only in a single polymer component and its concentration here is 1.5 vol. % below φ_p, the concentration of CB at the interface does not reach the percolation threshold (Table 12.2). However, when the concentration of CB in phase is 0.3 percent below φ_p, the PE + PP and PE + PIB blends conduct electric current and have ρ equal to $8 \cdot 10^3$ and $2 \cdot 10^4$ Ohm·cm correspondingly. Because of close wetting forces of the polymers in these blends, the migration of a filler to the interface is possible from both phases. Owing to this fact, when CB is preliminarily introduced into both polymer components and CB migrates to the interface from both phases, the fall in electric resistance of these blends below the "additive" values is observed even if the CB concentration in both phases is 1.5 percent below the percolation threshold.

The essential role played by shear stress in the process of filler localizing at the interface suggests that the process is strongly affected by the viscosity of polymer components. Particular attention was given to the systems (PE + CB) + PMMA and (PS + CB) + PE, a case of the most interest since a filler is here redistributed from a phase of polymer with lesser wetting force. In these systems the viscosity of PMMA and of PE correspondingly was varied (Figure 12.4). The variation of viscosity within 300 to 1,500–1,700 Pa·s range for PMMA and from 250 up to 1,500 Pa·s for PE only moderately increases ρ of the (PS + CB) + PE blend (Figure 12.4, the viscosity is given for the conditions close to those of mixing: shear rate is 100 s^{-1}). When PMMA and PE viscosity exceeds the above limits, ρ of blends rises essentially.

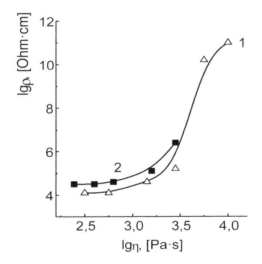

FIGURE 12.4 ρ of the (PE + CB) + PMMA (1) and (PS + CB) + PE (2) systems versus the effective viscosity of PMMA and PE correspondingly. Concentration of CB: 5.4 vol. % in the PE phase (1), 9.3 vol. % in the PS phase (2).

Such viscosity effect can be understood from the viewpoint of kinetics. The wetting of a filler by high-viscosity polymers occurs predominantly under the action of external forces straining polymer and indenting the filler particles into polymer bulk. When the polymers differ greatly in viscosity, the rate of strain will be far less for a high-viscosity polymer phase than for that of low-viscosity. So the high-viscosity component has no time to wet the filler particles surrounded by the low viscosity, easy to deform polymer. As a result, the CB localization at the interface is not observed. Besides, the rise of the viscosity of either of two polymer components increases viscosity of the system as a whole, and in the blending conditions this causes the shear stress and the force of particle break-off from the interface to rise. As a result, the equilibrium concentration of particles at the interface decreases.

The ratio of polymer viscosity values, under which a substantial drop in local interfacial CB concentration is observed, differs for various polymer pairs. The greater is the difference in wetting forces of polymers, the more times the viscosity of the second (having higher wetting force) polymer, can exceed the viscosity of the preliminarily filled polymer. So, this ratio is about four for the PE + PMMA blend, and about two for the PS+PE blend. The best conditions for particles to localize at the interface are provided when viscosity of the second polymer is slightly lower than that of a preliminarily filled polymer component.

12.4 CONCLUSIONS

(i) The possibility of dispersed particles to be redistributed from a polymer phase to the interface between two polymers and the extent to which this event proceeds is substantially determined by the relative capacity of polymer components to wet the surface of filler particles and by the sequence of blending the components. When a difference in the wetting force of polymers increases, the concentration of dispersed particles at the interface rises in the event that they have preliminarily been introduced into the phase of polymer with a lesser wetting force, and reduces where they have preliminarily been introduced into the phase of polymer with a higher wetting force.

(ii) The localization of a filler at the interface in polymer blends is also possible in the case when the residence of particles at the interface is thermodynamically inefficient, but when their transfer from a filled polymer phase to unfilled one gives energy gain.

(iii) The local concentration of a filler at the interface is determined by the equilibrium established in the process of blending between the number of filler particles arriving to the interface from a polymer phase and those moving in the opposite direction under the action of external mechanical forces. This concentration rises with increasing energy of particle adhesion to the phase of unfilled polymer and with decreasing shear stress experienced by a particle.

KEYWORDS

- **Carbon black (CB)**
- **Heterogeneous polymer blends**
- **Thermodynamics**

REFERENCES

1. Marsh, P. A.; Voet, A.; Price, L. D.; and Mullens, T. J.; *Rubber Chem. Technol*. **1968**, *41(2)*, 344–355.
2. Sircar, A. K.; *Rubber Chem. Technol*. **1981**, *54(4)*, 820–834.
3. Lipatov, Yu. S.; Mamunya, E. P.; Gladyreva, I. A.; and Lebedev, E. V.; *Vysokomolekulyarnye Soedineniya*. **1983**, *A25(7)*, 1483–1489.
4. Lipatov, Yu. S.; Mamunya, E. P.; Lebedev, E. V.; and Gladyreva, I. A.; *Kompozitsionnye Polimernye Materiaiy*. **1983**, *17*, 9–14.
5. Gubbels, F.; et al. *Macromolecules*. **1994**, *27(7)*, 1972–1974.
6. Gubbels, F.; et al. *Macromolecules*. **1995**, *28(4)*, 1559–1566.
7. Soares, B. G.; Gubbels, F.; Jerome, R.; Teyssie, Ph.; Vanlathen, E.; and Deltour, R.; *Polym. Bull*. **1995**, *35(1–2)*, 223–228.
8. Soares, B. G.; Gubbels, F.; Jerome, R.; Vanlathen, E.; and Deltour, R.; *Rubber Chem. Technol*. **1997**, *70(1)*, 60–70.
9. Sumita, M.; Sakata, K.; Asai, S.; Miyasaka, K.; and Nakagawa, H.; *Polym. Bull*. **1991**, *25(2)*, 265–271.
10. Sumita, M.; Sakata, K.; Hayakawa, Y.; Asai, S.; Miyasaka, K.; and Tanemura, M.; *Colloid Polym. Sci*. **1992**, *270(2)*, 134–139.
11. Pavliy, V. G.; Zaikin, A. E.; Kuznetsov, E. V.; and Michailova, L. N.; *Izvestiya Vysshikh Uchebnykh Zavedeniy. Khimiya i khim. Telhnologiya*. **1986**, *29*, vip. 5, 84.
12. Zaikin, A. E.; Mindubaev, P. Yu.; and Arkhireev, V. P.; *Polym. Sci*. **1999**, *B41(1–2)*, 15–19.
13. Pavliy, V. G.; Zaikin, A. E.; and Kuznetsov, E. V.; *Vysokomolekulyarnye Soedineniya*. **1987**, *A29(3)*, 447–450.
14. Zaikin, A. E.; Galikhanov, M. F.; and Arkhireev, V. P.; *Mekhanika Kompozitsionnykh Materialov i Konstruktsiy*. **1998**, *4(3)*, 55–61.
15. Zaikin, A. E.; Galikhanov, M. F.; Zverev, A. V.; and Arkhireev, V. P.; *Polym. Sci*. **1998**, *A40(5)*, 847.
16. Adamson, A.; The Physical Chemistry of Surface. New York: Wiley; **1982**, 4th ed.
17. Wu, S.; In: "Polymer Blends." eds. Paul, D. R.; Newmen, S.; Chapter 6, New York: Academic Press; **1978**, *1*.
18. Kraus, G.; In: "Reinforcement of Elastomers." eds. Kraus, G.; New York: Wiley Interscience; **1965**.
19. Dannenberg, E. M.; *Rubber Chem. Technol*. **1975**, *48(2)*, 410.
20. Hansen, C. M.; *J. Paint. Technol*. **1967**, *39(505)*, 104–113.
21. Hansen, C. M.; *J. Paint. Technol*. **1967**, *39(511)*, 505–510.

22. Stepin, S. N.; Bogachev, F. V.; and Svetlacov, N. V.; *Zhurnal Prikladnoy Khimii*. **1991**, *10*, 2107–2110.
23. Kinlock, E.; Adhesion and adhesives. Science and Technology. New York L.: University Press; 441.
24. Flir, G.; and Liklema, J.; Adsorption from Solution at the Solid/Liquid. eds. Parfitt, G. D.; and Rochester, C. H.; London: Academic; **1983**.
25. McKelvey, J. M.; Polymer Processing, Chapter 12, New York: John Wiley; **1963**.

CHAPTER 13

PERFORMANCE ANALYSIS OF MULTILAYER INSULATIONS INCRYOGENIC APPLICATIONS

ARASHESMAILI* and MEHDI MAEREFAT

Tarbiat Modares University, Faculty of Technical and Engineering, Mechanical Engineering Section, Energy Conversion Group; *E-mail: arash.esmaili@gmail.com

CONTENTS

13.1 INTRODUCTION

Multilayer insulations (MLI) consisted of some layers including heat shields and spacers. Heat transfer in such insulations is governed by conduction and radiation mechanisms, simultaneously.

Multilayer insulations have been used in cryogenics first. These insulations have been embedded in cryogen storage tanks. For better understanding, we should know major heat transfer mechanisms in such insulations. Schematic of main heat transfer mechanism in multilayer insulations is shown in Figure 13.1.

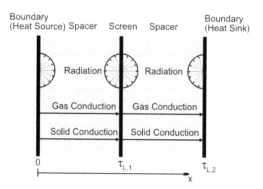

FIGURE 13.1 Schematic of main heat transfer mechanisms in multilayer insulations [1]

As shown in Figure 13.1, the main heat transfer mechanisms are radiation, gas conduction and solid conduction. In most of cases, we neglect convection. Thermal protection systems (TPS) in cryogen storage tanks are consisted of a two-layered wall with vacuumed inside. This will decreases gas conduction and convection significantly. For reduction of radiation, low emissivity heat shields are used. Polymeric or Ceramic-based spacers are inserted between the shields [1].

13.2 PREVIOUS RESEARCHES

Some experimental and theoretical studies have been done for modeling the heat transfer in multilayer insulations. Estimation of heat transfer for low temperature insulations in aerospace application has been done [2]. Effective thermal conductivity of seven configuration of multilayer insulation in temperature of 427°C and pressure of 5×10^{-5} mm of Hg has been measured. Radiation modeling in that study has been done using optically thin approximation and theoretical results has been compared to experimental data [3]. In 1968, latter study has been continued with temperatures up to 1,000°C and new mathematical formulation has been developed. In that formulation gas conduction has been neglected and radiation modeling has been done using optically thick approximation [4].

Applicable study for high temperature multilayer insulation has been continued by German researchers in early 1990s. They survived the feasibility of using of such insulations in space transportation vehicles. They neglect solid conduction and used optically thick approximation for radiation modeling [5]. Also a research has been done about application of MLI in fuel cell. The researchers used optically thick approximation and compared the results with experimental data [1]. A numerical solution presented for heat transfer in MLI. In preceding study optically thick approximation has been used and effective thermal conductivity for fibrous[1] insulations has been obtained [6].

In present study two-flux approximation is used for radiation modeling. This approximation unlike the others is not limited to specific ranges of optical thickness. This feature will become more important when we know optical thickness of each layer of multilayer insulation may vary from optical thickness of the other layers. So if we use optically thick or thin approximations, optical thickness of two different layers may fall into both thick and thin ranges.

Another benefit of using of two-flux approximation is that the radiativeproperties of the boundaries are spotted in formulation. The other approximations neglect the radiative properties of the boundaries in their formulation.

13.3   THEORETICAL ANALYSIS

Heat transfer mechanisms in MLI are solid conduction, gas conduction and radiation. Radiation itself may be in forms of absorption, scattering and emission.

One dimensional conservation of energy in an insulation leads to following PDE [7];

$$\rho c \frac{\partial T}{\partial t} = \frac{\partial q''}{\partial x} \qquad (13.1)$$

where ρ is density, c is specific heat, T is temperature, t is time and q'' is total heat flux that is summation of radiative and conductive heat fluxes;

$$q'' = k \frac{\partial T}{\partial x} + q_r'' \qquad (13.2)$$

Fibrous insulation is kind of MLI with no heat shield.

where k is thermal conductivity affected by both solid conduction and gas conduction. Merging Eqs. 13.1 and 13.2 we have;

$$\rho c \frac{\partial T}{\partial t} = \frac{\partial}{\partial x}(k\frac{\partial T}{\partial x}) - \frac{\partial q_r{}''}{\partial x}$$

(13.3)

Thermal conductivity and specific heat coefficient both are functions of temperature.

13.3.1 APPROXIMATE SOLUTIONS FOR RADIATION MODELING

Since governing equations of MLI have no exact solution, numerical solution along approximate solution should be employed. The most common approximate solutions are;

13.3.1.1 OPTICALLY THICK APPROXIMATION

This approximation is valid for the cases that optical thickness is very greater than unity. For example when we have a high density and/or high thickness fibrous insulation we may use this approximation. In this case characteristic length is so greater than mean free path of photons and every element of medium is only affected by neighboring elements. So the radiation problem leads to the diffusion one and radiative heat flux will be obtained from Eq. 13.4 [7];

$$q_r{}'' = -\frac{16\sigma T^3}{3\beta}\frac{\partial T}{\partial x}$$

(13.4)

and Eq. 13.3 become;

$$\rho c \frac{\partial T}{\partial t} = \frac{\partial}{\partial x}\{(k + \frac{16\sigma}{3\beta}T^3)\frac{\partial T}{\partial x}\}$$

(13.5)

Since this formulation has no radiative properties from bounding walls, then it is not valid in boundaries and it is valid only inside the participating medium. Local effective thermal conductivity will be defined in the form of Eq. 13.6;

Optically thick approximation is also known as diffusion approximation [8].

13.3.1.2 OPTICALLYTHINAPPROXIMATION

This approximation is valid for the cases that optical thickness is very smaller than unity. Characteristic length is so smaller than mean free path of photons and every emitted photon from medium elements will go through medium without any restriction caused by absorption or scattering mechanisms. Gradient of radiative flux in this case will lead to zero;

(13.7)

and radiative flux will become;

(13.8)

Therefore, the radiation problem reduces to a simple conduction problem with radiative boundary conditions. This approximation is valid for multilayer insulations with thin spacers [9].

13.3.1.3   TWO-FLUX APPROXIMATION

This approximation is valid for all ranges of optical thickness and radiative flux will be obtained from set of ODEs [10];

$$\frac{\partial q_r''}{\partial \tau} = (1 - \omega)(4\pi I_b - G) \tag{13.9}$$

$$\frac{\partial G}{\partial \tau} = -3q_r'' \tag{13.10}$$

where G is incident radiation, I_b is radiation intensity of black body and q_r'' is radiative flux. In a medium with index of refraction of unity we have:

$$I_b = \frac{e_b}{\pi} = \frac{\sigma T^4}{\pi} \tag{13.11}$$

The Eqs. 13.8 and 13.9 are in optical coordinates, where $\tau = \beta x$, so by turning them back into Cartesian coordinates and using Eq. 13.10, we lead to distribution of incident radiation in the medium;

$$-\frac{1}{3\beta^2(1 - \omega)}\frac{\partial^2 G}{\partial x^2} + G = 4\sigma T^4 \tag{13.12}$$

$$-\frac{2}{3\beta(\frac{\varepsilon_1}{2 - \varepsilon_1})}\frac{\partial G}{\partial x} + G = 4\sigma T_1^4 (@ x = 0) \tag{13.13}$$

Equation 13.11 should be solved with equations of conservation of energy, simultaneously [10]. Boundary conditions for Eq. 13.11 are:

$$-\frac{2}{3\beta(\frac{\varepsilon_1}{2-\varepsilon_1})}\frac{\partial G}{\partial x}+G=4\sigma T_1^4\ (@\ x=0) \tag{13.13}$$

$$-\frac{2}{3\beta(\frac{\varepsilon_2}{2-\varepsilon_2})}\frac{\partial G}{\partial x}+G=4\sigma T_2^4\ (@\ x=L) \tag{13.14}$$

Having incident radiation distribution, radiative flux and its gradient could be calculated using Eqs. 13.14 and 13.15;

$$q_r^{''}=-\frac{1}{3\beta}\frac{\partial G}{\partial x} \tag{13.15}$$

$$\frac{\partial q_r^{''}}{\partial x}=\beta(1-\omega)(4\sigma T^4-G) \tag{13.16}$$

As mentioned before in present study two-flux approximation is used for radiation modeling, because first, it is not limited to any ranges of optical thickness, and second, it has radiation properties of solid walls in its formulation.

13.4   NUMERICAL SOLUTION

Two-flux equations has been discretized using finite difference method, but discretization of energy equation is done by finite volume method, because first this method has intrinsic conservation [11], and second in this casefinite difference do not lead to good results. The discretized form of energy conservation is shown in Eq. 16;

$$(\rho c_{j-1}\frac{\Delta x_{j-1}}{2}+\rho c_j\frac{\Delta x_j}{2})\frac{T_j^{n+1}-T_j^n}{\Delta t}=$$
$$\frac{k_{j-1}}{\Delta x_{j-1}}(T_{j-1}^n-T_j^n)+\frac{k_j}{\Delta x_j}(T_{j-1}^n-T_j^n)+\frac{1}{3\beta_{j-1}\Delta x_{j-1}}(G_{j-1}^n-G_j^n)-\frac{1}{3\beta_j\Delta x_j}(G_j^n-G_{j+1}^n) \tag{13.17}$$

Subscripts indicate spatial index and superscripts indicate temporal index. Temperature-dependent properties are obtained by averaging between neighboring nodes;

$$k_{j-1}=k(\frac{T_{j-1}+T_j}{2}) \tag{13.18}$$

$$k_j = k(\frac{T_j + T_{j+1}}{2})$$ (13.19)

These formulations also can be applied on specific heat and attenuation coefficients. The discretized form of two-flux equations is shown in equations 19-21:

$$[2 + 3\beta_1\Delta x\{\beta_1\Delta x(1-\omega_1) + \frac{\varepsilon_1}{2-\varepsilon_1}\}]G_1 - 2G_2 = 12\beta_1\Delta x\sigma T_1^4\{\beta_1\Delta x(1-\omega_1) + \frac{\varepsilon_1}{2-\varepsilon_1}\}$$ (13.20)

$$-g_{j-1} + \left[2 + 3\beta_j^2\Delta x^2\left(1-\omega_j\right)\right]G_j - G_{j+1} = 12\beta_j^2\Delta x^2\left(1-\omega_j\right)\sigma T_j^4$$ (13.21)

$$-2G_{n-1} + [2 + 3\beta_n\Delta x\{\beta_n\Delta x(1-\omega_n) + \frac{\varepsilon_n}{2-\varepsilon_n}\}]G_n - 2G_2 = 12\beta_n\Delta x\sigma T_n^4\{\beta_n\Delta x(1-\omega_n) + \frac{\varepsilon_n}{2-\varepsilon_n}\}$$ (13.22)

Equations 13.19 and 13.21 are for boundaries and Eq. 13.20 is for intermediate nodes. Whole domain is divided to n-1 finite volumes. Set of equations above, establishes tridiagonal algebra. The value for incident radiation is calculated in every step and inserted into energy equation.

13.5 RESULTS AND DISCUSSIONS

In current study a formulation presented for combined conduction-radiation modeling. This formulation has been validated with results of Ozisik solution. He presented a numerical solution for one-dimensional combined conduction-radiation in a slab [10]. Ozisik used normal-expansion mode for radiation modeling that is less cost-effective and hard to be applied on multilayer geometries, but it is more accurate. For comparison we the parametersshould be nondimensionalizedas shown in Table 13.1.

TABLE 13.1 Dimensionless parameters in Ozisik solution

$\theta = \dfrac{T}{T_1}$	$t^* = \dfrac{k}{\rho c}\beta^2 t$	$\tau = \beta x$	$N_r = \dfrac{k\beta}{4\sigma T_1^3}$	$G^* = \dfrac{G}{4\sigma T_1^3}$	$q_r^* = \dfrac{q_r''}{4\sigma T_1^3}$

N_r is ratio of conductive flux to radiative flux. Dimensionless energy equation is shown in Eq. (13.22)

$$\frac{\partial\theta}{\partial t^*} = \frac{\partial}{\partial\tau}(\frac{\partial\theta}{\partial\tau} - \frac{1}{N_r}q_r^*)$$ (13.23)

And total heat flux is:

$$q^* = \left(-\frac{\partial \theta}{\partial \tau} + \frac{1}{N_r} q_r^* \right) \tag{13.24}$$

A comparison between results of two formulations is shown in Figure 13.2:

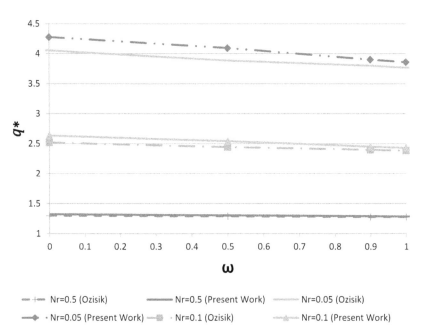

FIGURE 13.2   Comparison of dimensionless heat fluxes obtained from numerical solution and Ozisik solution

It is obvious from Figure 13.2 that with increase of N_r the total heat fluxes decreases and the results of present work are in agreement with Ozisik results.

Variation of dimensionless temperature respect to dimensionless lengths is shown in Figure 13.3. In this case temperature difference between hot boundary and cold boundary is 100 K and sample thickness is 26.6 cm.

$$\Phi = \frac{T(x) - T(L)}{T(0) - T(L)} \tag{13.25}$$

$$x^* = \frac{x}{L} \tag{13.26}$$

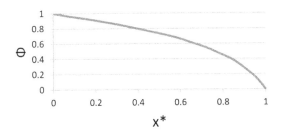

FIGURE 13.3 Variation of dimensionless temperature respect to dimensionless length

As shown in Figure 13.3, the variation of temperature is nonlinear and near cold boundary the gradient is greater.

$$q_s'' = \frac{q_s''}{q_T''} \tag{13.27}$$

$$q_r'' = \frac{q_r''}{q_T''} \tag{13.28}$$

In Figure 13.4 the contribution of radiative and conductive heat fluxes is shown. By marching from hot boundary ($X^*=0$) to cold boundary ($X^* = 1$) the contribution of radiative flux (Eq. 13.27) decreases and the contribution of conductive flux (Eq. 13.26)increases. In $X^* = 0.9$ radiative flux equals conductive flux.

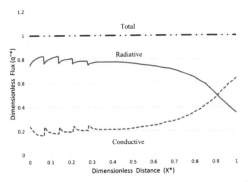

FIGURE 13.4 Contribution of radiative and conductive heat fluxes in dimensionless length of insulation

13.5 CONCLUSION

Thermal performance analysis of a multilayer insulation has been done. From temperature distribution along insulation thickness, it can be concluded that thermal gradients near cold boundary are greater and also in this region conductive heat flux is dominated, so in case of optimal design the spacers in vicinity of cold boundary should be made of materials with lower thermal conductivity. On the other side radiative heat flux is dominated near hot boundary and layer density should be greater and/or heat shields should be coated with low emissive materials such as gold.

KEYWORDS

- **Thermal Properties**
- **Combined Conduction-Radiation**
- **Cryogenics**
- **Cryogen storage tanks**
- **Multilayer insulations**

REFERENCES

1. Spinnler, Markus.; Winter, Edgar R. F.; Viskanta, Raymond.; Studies on high temperature multilayer thermal insulation. *Int. J. Heat. Mass. Transfer.* **2004**, *47*, 1305–1312.
2. Doenecke, J.; "Survey and Evaluation of Multilayer Insulation Heat Transfer Measurements,"SAE Technical Paper 932117, 23rd International Conference on Environmental Systems, Colorado Springs, Colorado, July; **1993.**
3. Cunnington, G. R.; Ziennan, C. A.; Funai, A. I.; and Lindahn, A.; Performance of Multilayer Insulation Systems for Temperatures to 700K, NASA CR-907, October; **1967.**
4. DeWitt, W. D.; Gibbon, R. L.; and Reid, R. L.; Multi-foil Type Thermal Insulation, Proceedings of Intersociety Energy Conversion Engineering Conference (ECEC), pp. 263–271; **1968.**
5. Gallert, H.; and Keller, K.; Metallic thermal protection concept for hypersonic vehicles. *J. Aircraf.* **1991**, *28*(6), 410–416.
6. Zhao Shu-yuan, Temperature and pressure dependent effective thermal conductivity of fibrous insulation. *Int. J. Therm. Sci.* **2009.**
7. Sparrow, E. M.; and Cess, R. D.; Radiation heat transfer, Augmented Edition, McGraw- Hill: **Washington. D.C.; 1978.**
8. Pawel, R. E.; McElroy, D. L.; Weaver, F. J.; and Graves, R. S.; High temperature thermal conductivity of a fibrous Alumina Ceramic. Ed. David Yarbrough, Thermal Conductivity, 19, Plenum Press; **1985** pp. 301–313.
9. Stauffer, T.; Jog, M.; and Ayyaswamy, P.; The effective thermal conductivity of multi foil insulation as a function of temperature and pressure. AIAA Paper 92-2939, AIAA 27th Thermophysics Conference, Nashville, Tennessee; July **1992.**

10. Ozisik, M. N.; Radiative Transfer and Interactions with Conduction and Convection. John Wiley & Sons, Inc.; **1973**.
11. Anderson, D. A.; Tannehill, J. C.; and Pletcher, R. H.; Computational fluid mechanics and heat transfer. Hemisphere Publishing Corporation; **1984**.
12. Siegel, R.; Transient Thermal Analysis for Heating a Translucent Wall with Opaque Radiation Barriers. *J. Thermophys. Heat. Transf.* **1999**, *13*, 277–284.

CHAPTER 14

THE EFFECT PARTICLE SIZE OF MICROHETEROGENOUS CATALYST TICL$_4$–AL(ISO-C$_4$H$_9$)$_3$ ON THE BASIC PATTERNS OF ISOPRENE POLYMERIZATION

ELENA M. ZAKHAROVA[1], VADIM Z. MINGALEEV[1], and VADIM P. ZAKHAROV[2]

[1]Institute of Organic Chemistry, Ufa Scientific Center of Russian Academy of Sciences, pr. Oktyabrya 71, Ufa, Bashkortostan, 450054, Russia

[2]Bashkir State University, Zaki Validi str. 32, Ufa, 450076 Bashkortostan, Russia, E-mail: zaharovvp@mail.ru

CONTENTS

14.1 INTRODUCTION

The formation of highly stereoregular polymers under the action of microhetero-geneous Ziegler–Natta catalysts is accompanied by broadening of the polymer MWD [1, 2]. This phenomenon is related to the kinetic heterogeneity of active sites (AC) [1, 3, 4]. The possible existence of several kinetically nonequivalent AC of polymerization correlates with the nonuniform particle size distribution of a catalyst [4]. At present time much attention is given to study the influence of microheterogeneous catalysts particle size on the properties of polymers [5, 6]. However, almost no detailed study of the effect particle size catalysts on their kinetic heterogeneity.

The microheterogeneous catalytic system based on $TiCl_4$ and $Al(iso-C_4H_9)_3$ that is widely used for the production of the cis-1,4-isoprene. Our study has shown [7] that the targeted change of the solid phase particle size during the use of a tubular turbulent reactor at the stage of catalyst exposure for many hours is an effective method for controlling the polymerization process and some poly-mer characteristics of isoprene. We suppose that the key factor is the interrela-tion between the reactivity of isoprene polymerization site and the size of cata-lyst particles on which they localize.

The aim of this study was to investigate the interrelation between the particle size of a titanium catalyst and its kinetic heterogeneity in the polymerization of isoprene.

14.2 EXPERIMENTAL

Titanium catalytic systems (Table 14.1) were prepared through two methods. **Method 1.** At 0 or –10°C in a sealed reactor 30–50 mL in volume with a calcu-lated content of toluene, calculated amounts of $TiCl_4$ and $Al(iso-C_4H_9)_3$ toluene solutions (cooled to the same temperature) were mixed. The molar ratio of the components of the catalyst corresponded to its maximum activity in isoprene polymerization. The resulting catalyst was kept at a given temperature (Table 14.1) for 30 min under constant stirring.

TABLE 14.1   Titanium catalytic systems and their fractions used for isoprene polymerization

Catalyst	Labels	Molar ratio of catalyst components			T, °C	Method	Range of particle diameters in fractions of titanium catalysts, µm		
		Al/Ti	DPO/Ti	PP/Al			Fraction I	Fraction II	Fraction III
$TiCl_4$–Al(i-$C_4H_9)_3$	C-1	1	-	-	0	1	0.7–4.5	0.15–0.65	0.03–0.12
						2	–	0.20–0.7	0.03–0.18

TiCl$_4$–Al(i-C$_4$H$_9$)$_3$ –DPO	C-2	1	0.15	-	0	1	0.7–4.5	0.15–0.65	0.03–0.12
						2	–	0.15–0.68	0.03–0.12
TiCl$_4$–Al(i-C$_4$H$_9$)$_3$ –DPO–PP	C-3	1	0.15	0.15	0	1	–	0.12–0.85	0.03–0.10
						2	–	0.15–0.80	0.03–0.12
TiCl$_4$–Al(i-C$_4$H$_9$)$_3$ –DPO–PP	C-4	1	0.15	0.15	-10	1	–	0.12–0.45	0.03–0.10
						2	–	0.12–0.18	0.04–0.11
Averaged ranges, µm							0.7–4.5	0.15–0.69	0.03–0.14

Note: DPO – diphenyloxide, PP – piperylene

Method 2. After preparation and exposure of titanium catalysts via method 1, the system was subjected to a hydrodynamic action via single circulation with solvent through a six-section tubular turbulent unit of the diffuser-confuser design [8] for 2–3 s.

The catalyst was fractionated through sedimentation in a gravitational field. For this purpose calculated volumes of catalysts prepared through methods 1 and 2 were placed into a sealed cylindrical vessel filled with toluene. In the course of sedimentation, the samples were taken from the suspension column at different heights, a procedure that allowed the separation of fractions varying in particle size.

The titanium concentrations in the catalyst fractions were determined via FEK colorimeter with a blue light filter in a cell with a 50 mm thick absorbing layer. A K$_2$TiF$_6$ solution containing 1×10^{-4} g Ti/mL was used as a standard. The catalyst particle size distribution was measured via the method of laser diffraction on a Shimadzu Sald-7101 instrument.

Before polymerization, isoprene was distilled under a flow of argon in the presence of Al(iso-C$_4$H$_9$)$_3$ and then distilled over a TiCl$_4$–Al(iso-C$_4$H$_9$)$_3$ catalytic system, which provided a monomer conversion of 5–7%. The polymerization on fractions of the titanium catalyst was conducted in toluene at 25°C under constant stirring. In this case, the calculated amounts of solvent, monomer, and catalyst were successively placed into a sealed ampoule 10–12 mL in volume. The monomer and catalyst concentrations were 1.5 and 5×10^{-3} mol/L, respectively. The polymerization was terminated via the addition of methanol with 1% ionol and 1% HCl to the reaction mixture. The polymer was repeatedly washed with pure methanol and dried to a constant weight. The yield was estimated gravimetrically.

The MWD of polyisoprene was analyzed via GPC on a Waters GPC-2000 chromatograph equipped with three columns filled with a Waters microgel (a pore size of 103–106 A) at 80°C with toluene as an eluent. The columns were preliminarily calibrated relative to Waters PS standards with a narrow MWD ($M_w/M_n = 1.01$). The analyses were conducted on a chromatograph, which allows calculations with allowance for chromatogram blurring. Hence, the need for additional correction of chromatograms was eliminated.

The microstructure of polyisoprene was determined via high-resolution 1H NMR spectroscopy on a Bruker AM-300 spectrometer (300 MHz).

The MWD of cis-1,4-polyisoprene obtained under the aforementioned experimental conditions, $q_w(M)$, were considered through the equation

$$q_w(M) = \int_0^\infty \Psi(\beta) M\beta^2 \exp(-M\beta) d\beta$$

where β is the probability of chain termination and $\psi(\beta)$ is the distribution of active site over kinetic heterogeneity, M is current molecular weight.

As was shown previously [9] Eq. (14.1) is reduced to the Fredholm integral equation of the first kind, which yields function $\psi(\beta)$ after solution via the Tikhonov regularization method. This inverse problem was solved on the basis of an algorithm from [9]. As a result, the function of the distribution over kinetic heterogeneity in $\psi(\ln\beta)–\ln M$ coordinates with each maximum related to the functioning of AC of one type was obtained.

14.3 RESULTS

After mixing of the components of the titanium catalyst, depending on its formation conditions, particles 4.5 μm to 30 nm in diameter, which are separated into three arbitrary fractions, are formed (Table 14.1).

During the formation of catalyst C-1 via method 1, the fraction composed of relatively coarse particles, fraction I, constitutes up to 85 percent (Figure 14.1). In method 2, the hydrodynamic action on the titanium catalyst formed under similar conditions results in an increase in the content of fraction II. Analogous trends are typical of catalyst C-2. The catalyst modification with piperylene additives, catalyst C-3, is accompanied by the disappearance of fraction I and an increase in the content of fraction II (Figure 14.1), as was found during the hydrodynamic action on C-1. The hydrodynamic action on a two-component catalyst is equivalent to the addition of piperylene to the catalytic system. The preparation of catalytic complex C-3 via method 2 results in narrowing of the particle size distribution of fraction II owing to disintegration of particles 0.50–0.85 μm in diameter (Figure 14.1). The reduction of the catalyst exposure temperature to -10°C (catalyst C-4) is accompanied by further disintegration of fraction II (Figure 14.1). In this case the content of particles 0.19–0.50 μm in diameter decreases to 22 percent with predominance of particles 0.15–0.18 μm in diameter. The formation of C-4 via method 2 results in additional dispersion of particles of fraction II, with the content of particles 0.15–0.18 μm in diameter attaining 95 percent.

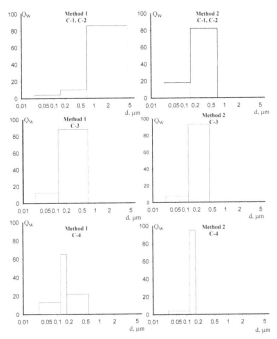

FIGURE 14.1 Fractional compositions of titanium catalyst C-1–C-2 (Table 14.1)

The content of the finest catalyst particles in the range 0.03–0.14 μm, fraction III, attains 5–12 percent and is practically independent of the catalyst formation conditions. The most considerable changes are shown by particles 0.18–4.50 μm in diameter. Particles of fraction I are easily dispersed as a result of the hydrodynamic action in turbulent flows and the addition of catalytic amounts of piperylene, and their diameter becomes equal to that of particles from fraction II. The decrease in the catalyst exposition temperature from 0 to −10°C with subsequent hydrodynamic action leads to a more significant reduction of particle size and the formation of a narrow fraction.

Isolated catalyst fractions differing in particle size were used for isoprene polymerization. The cis-1,4-polymer was obtained for all fractions, regardless of their formation conditions. The contents of cis-1,4 and 3,4 units were 96–97 and 3–4 percent, respectively. Coarse particles (fraction I) are most active in isoprene polymerization (method 1) on different fractions of C-1 (Figure 14.2).

As the particle size of C-1 decreases, its activity drops significantly. The catalyst modification with diphenyloxide (C-2) has practically no effect on the fractional composition, but the activities of different catalyst fractions change. The most marked increase in activity was observed for fraction I. Catalyst C-3 prepared via method 1 comprises two fractions, with fraction II having the maximum activity. The decrease of the catalyst exposition temperature to −10°C (C-4) results in further increase in the rate of isoprene polymerization on particles of fraction II.

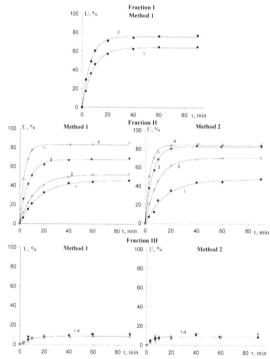

FIGURE 14.2 *cis*-1,4-Polyisoprene yields U vs. polymerization times τ in the presence of fractions particles of titanium catalysts (*1*) C-1, (*2*) C-2, (*3*) C-3, and (*4*) C-4.

The hydrodynamic action increases the content of fraction II in C-1, but its activity in isoprene polymerization does not increase (Figure 14.2). For C-2 the analogous change in the fractional composition is accompanied by an increased activity of fraction II (method 2). The addition of piperylene to C-3 results in a stronger effect on the activity of fraction II under the hydrodynamic action. The change of the hydrodynamic regime in the reaction zone does not affect the activity of fine particles of catalyst fraction III. Isoprene polymerization in the presence of fraction III always has a low rate and a cis-1,4-polyisoprene yield not exceeding 7–12 percent.

During polymerization with C-1 composed of fraction I (method 1), the weight-average molecular mass of polyisoprene increases with the process time (Table 14.2). Polyisoprene prepared with catalyst fraction II has a lower molecular mass. A more considerable decrease in the weight-average molecular mass is observed during isoprene polymerization on the finest particles of fraction III, with M_w being independent of the polymerization time. The width of the MWD for polyisoprene obtained on fraction I increases with polymerization time from 2.5 to 5.1. The polymer prepared in the presence of catalyst fraction III shows a narrow molecularmass distribution ($M_w/M_n \sim 3.5$). The addition of DPO is accompanied by an increase in the weight-average molecular mass of polyisoprene obtained in the presence of fraction I. Under a hydrodynamic action (method 2) on catalysts C-1 and C-2, polyisoprene with an increased weight-average molecular mass is formed. During the addition of piperylene to the catalyst (C-3), M_w of polyisoprene increases to values characteristic of C-2 formed via method 2. The formation of the catalytic system $TiCl_4–Al(iso-C_4H_9)_3–DPO–$piperylene via method 1 at $-10°C$ results in substantial increases in the activity of the catalyst and the reactivity of the active centers of fraction II. The polymerization of isoprene on catalyst fraction III, regardless of the conditions of catalytic system formation (electrondonor additives, exposure temperature, and hydrodynamic actions), yields a low molecular-mass polymer with a polydispersity of 3.0–3.5.

TABLE 14.2 Molecular-mass characteristics of *cis*-1,4-polyisoprene. 1 and 2 are the methods of catalyst preparation

C	τ min	Fraction I				Fraction II				Fraction III			
		$M_w \times 10^{-4}$		M_w/M_n		$M_w \times 10^{-4}$		M_w/M_n		$M_w \times 10^{-4}$		M_w/M_n	
		1	2	1	2	1	2	1	2	1	2	1	2
C-1	3	17.1		2.6		43.2	43.6	4.6	4.8	25.4	21.4	3.7	3.8
	20	56.3		4.2		46.5	57.5	4.0	5.9	23.7	22.6	3.4	3.7
	40	64.9		4.6		47.4	62.7	4.1	5.8	27.6	28.9	3.6	3.6
	90	72.2		5.1		54.3	61.4	4.3	6.2	26.8	22.4	3.7	3.6
C-2	3	24.3		3.3		39.5	54.8	4.3	4.9	21.1	22.3	3.8	3.7
	20	69.7		3.8		41.2	67.5	3.9	4.6	24.6	24.6	3.7	3.4
	40	78.4		3.7		42.6	72.8	4.4	4.3	26.3	22.4	3.3	3.5
	90	84.6		4.0		58.0	73.2	4.3	4.2	24.7	24.3	3.5	3.3

	3	66.2	60.5	3.9	3.8	22.4	20.0	3.8	3.6
C-3	20	75.4	71.1	3.7	3.5	28.1	21.3	3.7	3.7
	40	79.6	78.8	4.1	4.0	21.5	23.6	3.8	3.8
	90	81.6	79.6	3.9	3.8	23.2	26.8	3.6	3.3
	3	81.2	72.7	3.6	3.7	25.4	20.3	3.8	3.6
C-4	20	62.7	51.3	3.3	3.2	24.8	21.6	3.7	3.7
	40	57.6	44.3	3.1	3.2	23.2	26.8	3.6	3.6
	90	61.4	48.2	3.2	3.4	19.6	21.5	3.1	3.3

The solution of the inverse problem of the formation of the MWD in the case of cis-1,4-polyisoprene made it possible to obtain curves of the active site distribution over kinetic heterogeneity (Figures 14.3, 14.4. 14.5 and 14.6). As a result of averaging of the positions of all maxima three types of polymerization site that produce isoprene macromolecules with different molecular masses were found:

type A (lnM =10.7), type B (lnM = 11.6), and type C (lnM = 13.4).

The polymerization of isoprene in the presence of fractions I and II of C-1 (method 1) occurs on site of types B and C (Figure 14.3). The polymerization in the presence of fraction III proceeds on active site of type A only. The single site and low activity character of the catalyst composed of particles of fraction III is typical of all the studied catalysts and all the methods of their preparation. Thus, these curves are not shown in subsequent figures. Catalyst C-2 prepared via method 1 likewise features the presence of type B and type C site in isoprene polymerization on fractions I and II (Figure 14.4). The hydrodynamic action (method 2) on C-1 accompanied by dispersion of particles of fraction I does not change the types of site of polymerization on particles of fraction II (Figure 14.5). A similar trend is observed during the same action on a DPO-containing titanium catalyst in turbulent flows (Figure 14.5).

In the presence of piperylene, the main distinction of function $\psi(\ln\beta)$ relative to the distributions considered above is a significantly decreased area of the peak due to the active site of type B (Figure 14.6). The decrease of the catalyst exposition temperature to −10°C (C-4, method 1) allows the complete "elimination" of active site of type B (Figure 14.6). With allowance for the low content of fraction III, it may be concluded that, under these conditions, a single site catalyst is formed. In the case of the hydrodynamic action on C-4, the particles of fraction II contain active centers of type C with some shift of unimodal curve $\psi(\ln\beta)$ to smaller molecular masses (Figure 14.6).

14.3 DISCUSSION

The particles of the titanium catalyst 0.03–0.14 μm in diameter, regardless of the conditions of catalytic system preparation, feature low activity in polyisoprene synthesis, and the resulting polymer has a low molecular mass and a narrow MWD. The molecular-mass characteristics of polyisoprene and the activity of the catalyst comprising particles 0.15–4.50 μm in diameter depend to a great extent on its formation conditions.

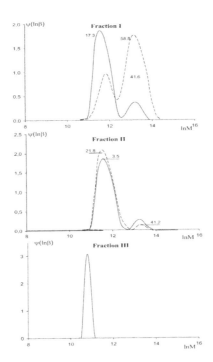

FIGURE 14.3 Active site distributions over kinetic heterogeneity during isoprene polymerization on fractions c-1. Method 1. Here and in FIGURES 4–5 numbers next to the curves are conversions (%).

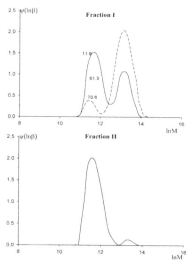

FIGURE 14.4 Active site distributions over kinetic heterogeneity during isoprene polymerization on fractions C-2. Method 1.

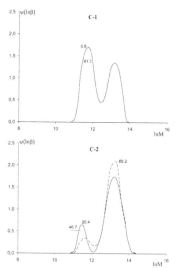

FIGURE 14.5 Active site distributions over kinetic heterogeneity during isoprene polymerization on particles of fraction II of C-1 and C-2. Method 2.

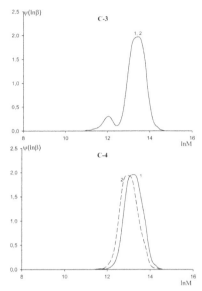

FIGURE 14.6 Active site distributions over kinetic heterogeneity during isoprene polymerization on particles of fraction II of C-3 and C-4. Methods 1—1, method 2—2.

As was shown previously [10, 11] the region of coherent scattering for particles based on $TiCl_3$ spans 0.003–0.1 µm, a range that corresponds to the linear size of the minimum crystallites. Coarser catalyst particles are aggregates of these minimum crystallites. This circumstance makes it possible to suggest that the fraction of catalyst particles 0.03–0.14 µm in diameter that was isolated in this study is a mixture of primary crystallites of β-$TiCl_3$ that cannot be separated via sedimentation. The fractions of catalyst particles with larger diameters are formed by stable aggregates of 2–1100 primary crystallites. There is sense in the suggestion that the elementary crystallites are combined into larger structures via additional Al–Cl bonds between titanium atoms on the surface of a minimum of two elementary crystallites, i.e. $(Ti)_1$–Cl–Al–Cl–$(Ti)_2$. Similar structures can be formed with the participation of AlR_2Cl and $AlRCl_2$, which are present in the liquid phase of the catalyst. Trialkylaluminum AlR_3 is incapable of this type of bonding. Thus, the structure of the most alkylated Ti atom (in the limit, a monometallic center of polymerization), which has the minimum reactivity, should be assigned to the active centers localized on particles 0.03–0.14 µm in diameter [9]. On particles 0.15–4.50 µm in diameter in clusters of primary crystallites, highactivity bimetallic centers with the minimum number of Ti–C bonds at a Ti atom are present. Thus, the experimental results obtained in this study show that the nature of the polymerization center resulting from successive parallel reactions between the pristine components of the catalytic system determines the size of the titanium catalyst particles and, consequently, their activity in isoprene polymerization.

14.4 CONCLUSION

We first examined the isoprene polymerization on the fractions of the titanium catalyst particles which were isolated by sedimentation of the total mixture. The results obtained allow to consider large particle as clusters which are composed of smaller particles. In the formation of these clusters is modified ligands available titanium atoms. In the process of polymerization or catalyst preparation the most severe effects are large particles (clusters). This result in the developing process later on substantially smaller as compared to initial size particles. These particles are fragments of clusters which are located over the active centers of polymerization. Note that the stereospecificity is not dependent on the size of the catalyst particles.

Hypothesis about clusters agrees well with the main conclusions of this paper:

I. Isolated the fraction of particles of titanium catalyst $TiCl_4-Al(iso-C_4H_9)_3$: I—0.7–4.5 µm, II—0.15–0.68 µm, III – 0.03–0.13 µm. With decreasing particle size decreases the rate of polymerization the molecular weight and width of the molecular weight distribution. Hydrodynamic impact leads to fragmentation of large particles of diameter greater than 0.5 µm.

II. Isoprene polymerization under action of titanium catalyst is occurs on three types active sites: type A – lnM = 10.7; type B – lnM = 11.6; type C—lnM = 13.4. Fractions I and II particles contain the active site of type B and C. The fraction III titanium catalyst is represented by only one type of active sites producing low molecular weight polymer (lnM = 10.7).

III. The use of hydrodynamic action turbulent flow, doping DPO and piperylene, lowering temperature of preparation of the catalyst allows to form single site catalyst with high reactivity type C (lnM = 13.4), which are located on the particles of a diameter of 0.15–0.18 µm.

ACKNOWLEDGMENTS

This study was financially supported by the Council of the President of the Russian Federation for Young Scientists and Leading Scientific Schools Supporting Grants (project no. MD-4973.2014.8)

KEYWORDS

- Active sites
- Isoprene polymerization
- Particles size effect.
- Single site catalysts
- Ziegler-Natta catalyst

REFERENCES

1. Kissin, Yu. V.; *J. Cataly.* **2012**, *292*, 188–200.
2. Hlatky, G. G.; *Chem. Rev.* **2000**, *100*, 1347–1376.
3. Kamrul Hasan, A. T. M.; Fang, Y.; Liu, B.; and Terano, M.; *Polymer.* **2010,** *51*, 3627–3635.
4. Schmeal, W. R.; and Street, J. R.; *J. Polym. Sci. Polym. Phys. Edition* **1972**, *10*, 2173–2183.
5. Ruff, M.; and Paulik, C.; *Macromol. React. Eng.* **2013**, *7*, 71–83.
6. Taniike, T.; Thang, V. Q.; Binh, N. T.; Hiraoka, Y.; Uozumi, T.; Terano, M.; *Macromol. Chem. Phys.* **2011**, *212*, 723–729.
7. Morozov, Yu. V.; Nasyrov, I. Sh.; Zakharov, V. P.; Mingaleev, V. Z.; and Monakov, Yu. B.; *Russ. J. Appl. Chem.* **2011**, *84*, 1434–1437.
8. Zakharov, V. P.; Berlin, A. A.; Monakov Yu. B.; and Deberdeev, R. Ya.; Physicochemical Fundamentals of Rapid Liquid Phase Processes, Moscow: Nauka, **2008**, 348 p.
9. Monakov, Y. B.; Sigaeva N. N.; and Urazbaev, V. N.; Active Sites of Polymerization. Multiplicity: Stereospecific and Kinetic Heterogeneity. Leiden: Brill Academic, **2005**; 397 p.
10. Grechanovskii, V. A.; Andrianov, L. G.; Agibalova, L. V.; Estrin, A. S.; and Poddubnyi, I. Ya.; *Vysokomol. Soedin., Ser. A.* **1980**, *22*, 2112–2120.
11. Guidetti, G.; Zannetti., R.; Ajò, D.; Marigo, A.; Vidali, M.; *Eur. Polym. J.* **1980**, *16*, 1007–1015.

CHAPTER 15

INTERNAL STRUCTURE AND THE EQUILIBRIUM CONFIGURATION (SHAPE) OF SEPARATE NONINTERACTING NANOPARTICLES BY THE MOLECULAR MECHANICS AND DYNAMICS METHODS

A. V. VAKHRUSHEV and A. M. LIPANOV

Institute of Mechanics, Ural Branch of the Russian Academy of Sciences, T. Baramsinoy 34, Izhevsk, Russia E-mail: postmaster@ntm.udm.ru

CONTENTS

15.1   INTRODUCTION

The properties of a nanocomposite are determined by the structure and properties of the nanoelements, which form it. One of the main tasks in making nanocomposites is building the dependence of the structure and shape of the nanoelements forming the basis of the composite on their sizes. This is because with an increase or a decrease in the specific size of nanoelements (nanofibers, nanotubes, nanoparticles, etc.), their physical-mechanical properties such as coefficient of elasticity, strength, deformation parameter, and so on, are varying over one order [1–5].

The calculations and experiments show that this is primarily due to a significant rearrangement (which is not necessarily monotonous) of the atomic structure and the shape of the nanoelement. The experimental investigation of the above parameters of the nanoelements is technically complicated and laborious because of their small sizes. In addition, the experimental results are often inconsistent. In particular, some authors have pointed to an increase in the distance between the atoms adjacent to the surface in contrast to the atoms inside the nanoelement, while others observe a decrease in the aforementioned distance [6].

Thus, further detailed systematic investigations of the problem with the use of theoretical methods, that is, mathematical modeling are required.

The atomic structure and the shape of nanoelements depend both on their sizes and on the methods of obtaining which can be divided into two main groups:

(i) Obtaining nanoelements in the atomic coalescence process by "assembling" the atoms and by stopping the process when the nanoparticles grow to a desired size (the so-called "bottom-up" processes). The process of the particle growth is stopped by the change of physical or chemical conditions of the particle formation, by cutting off supplies of the substances that are necessary to form particles, or because of the limitations of the space where nanoelements form.

(ii) Obtaining nanoelements by breaking or destruction of more massive (coarse) formations to the fragments of the desired size (the so-called "up down" processes).

In fact, there are many publications describing the modeling of the "bottom-up" processes [7–8], while the "up down" processes have been studied very little. Therefore, the objective of this work is the investigation of the regularities of the changes in the structure and shape of nanoparticles formed in the destruction ("up down") processes depending on the nanoparticle sizes, and building up theoretical dependences describing the above parameters of nanoparticles.

When the characteristics of powder nanocomposites are calculated it is also very important to take into account the interaction of the nanoelements since the changes in their original shapes and sizes in the interaction process and during the formation (or usage) of the nanocomposite can lead to a significant change in its properties and a cardinal structural rearrangement. In addition, the experimental investigations show the appearance of the processes of ordering and self-assembling leading to a more organized form of a nanosystem [9–15]. In general, three main processes can be distinguished: the first process is due to the regular structure formation at the interaction of the nanostructural elements with the surface where they are situated; the second one arises from the interaction of the nanostructural elements with one another; the third process takes place because of the influence of the ambient medium surrounding the nanostructural elements. The ambient medium influence can have "isotropic distribution" in the space or it can be presented by the action of separate active molecules connecting nanoelements to one another in a certain order. The external action significantly changes the original shape of the structures formed by the nanoelements. For example, the application of the external tensile stress leads to the "stretch" of the nanoelement system in the direction of the maximal tensile stress action; the rise in temperature, vice versa, promotes a decrease in the spatial anisotropy of the nanostructures [10]. Note that in the self-organizing process, parallel with the linear moving, the nanoelements are in rotary movement. The latter can be explained by the action of moment of forces caused by the asymmetry of the interaction force fields of the nanoelements, by the presence of the "attraction" and "repulsion" local regions on the nanoelement surface, and by the "non-isotropic" action of the ambient as well.

The above phenomena play an important role in nanotechnological processes. They allow developing nanotechnologies for the formation of nanostructures by the self-assembling method (which is based on self-organizing processes) and building up complex spatial nanostructures consisting of different nanoelements (nanoparticles, nanotubes, fullerenes, supermolecules, etc.) [15]. However, in a number of cases, the tendency toward self-organization interferes with the formation of a desired nanostructure. Thus, the nanostructure arising from the self-organizing process is, as a rule, "rigid" and stable against external actions. For example, the "adhesion" of nanoparticles interferes with the use of separate nanoparticles in various nanotechnological processes, the uniform mixing of the nanoparticles from different materials and the formation of nanocomposite with desired properties. In connection with this, it is important to model the processes of static and dynamic interaction of the nanostructure elements. In this case, it is essential to take into consideration the interaction force moments of the nanostructure elements, which causes the mutual rotation of the nanoelements.

The investigation of the above dependences based on the mathematical modeling methods requires the solution of the aforementioned problem on the atomic level. This requires large computational aids and computational time, which makes the development of economical calculation methods urgent. The objective of this work was the development of such a technique.

This chapter gives results of the studies of problems of numeric modeling within the framework of molecular mechanics and dynamics for investigating the regularities of the amorphous phase formation and the nucleation and spread of the crystalline or hypocrystalline phases over the entire nanoparticle volume depending on the process parameters, nanoparticles sizes and thermodynamic conditions of the ambient. Also the method for calculating the interactions of nanostructural elements is offered, which is based on the potential built up with the help of the approximation of the numerical calculation results using the method of molecular dynamics of the pairwise static interaction of nanoparticles. Based on the potential of the pairwise interaction of the nanostructure elements, which takes into account forces and moments of forces, the method for calculating the ordering and self-organizing processes has been developed. The investigation results on the self-organization of the system consisting of two or more particles are presented and the analysis of the equilibrium stability of various types of nanostructures has been carried out. These results are a generalization of the authors' research in [16–24]. A more detailed description of the problem you can obtain in these works.

15.2   PROBLEM STATEMENT AND MODELING TECHNIQUE

The problem on calculating the internal structure and the equilibrium configuration (shape) of separate noninteracting nanoparticles by the molecular mechanics and dynamics methods has two main stages:

(i) The "initiation" of the task, that is, the determination of the conditions under which the process of the nanoparticle shape and structure formation begins.

(ii) The process of the nanoparticle formation.

Note that the original coordinates and initial velocities of the nanoparticle atoms should be determined from the calculation of the macroscopic parameters of the destructive processes at static and dynamic loadings taking place both on the nanoscale and on the macroscale. Therefore, in the general case, the coordinates and velocities are the result of solving the problem of modeling physical-mechanical destruction processes at different structural levels. This problem due to its enormity and complexity is not considered in this paper. The detailed description of its statement and the numerical results of its solution are given in the works of the authors [16–19].

The problem of calculating the interaction of ordering and self-organization of the nanostructure elements includes three main stages: the first stage is building the internal structure and the equilibrium configuration (shape) of each separate noninteracting nanostructure element; the second stage is calculating the pairwise interaction of two nanostructure elements; and the third stage is establishing the regularities of the spatial structure and evolution with time of the nanostructure as a whole.

Let us consider the above problems in sequence.

15.2.1 THE CALCULATION OF THE INTERNAL STRUCTURE AND THE SHAPE OF THE NONINTERACTING NANOELEMENT

The initialization of the problem is in giving the initial coordinates and velocities of the nanoparticle atoms

$$\vec{x}_i = \vec{x}_{i0}, \vec{V}_i = \vec{V}_{i0}, t = 0 \ \vec{x}_i \subset \Omega_k, \tag{15.1}$$

where $\vec{x}_{i0}, \vec{x}_i$ are original and current coordinates of the i-th atom; $\vec{V}_{i0}, \vec{V}_i$ are initial and current velocities of the i-th atom, respectively; Ω_k is an area occupied by the nanoelement.

The problem of calculating the structure and the equilibrium configuration of the nanoelement will be carried out with the use of the molecular dynamics method taking into consideration the interaction of all the atoms forming the nanoelement. Since, at the first stage of the solution, the nanoelement is not exposed to the action of external forces, it is taking the equilibrium configuration with time, which is further used for the next stage of calculations.

At the first stage, the movement of the atoms forming the nanoparticle is determined by the set of Langevin differential equations at the boundary conditions (15.1) [25]

$$m_i \cdot \frac{d\vec{V}_i}{dt} = \sum_{j=1}^{N_k} \vec{F}_{ij} + \vec{F}_i(t) - \alpha_i m_i \vec{V}_i, \qquad i = 1, 2, ..., N_k,$$

$$\frac{d\vec{x}_i}{dt} = \vec{V}_i, \tag{15.2}$$

where N_k is the number of atoms forming each nanoparticle; m_i is the mass of the i-th atom; α_i is the "friction" coefficient in the atomic structure; $\vec{F}_i(t)$ is a random set of forces at a given temperature which is given by Gaussian distribution.

The interatomic interaction forces usually are potential and determined by the relation

$$\vec{F}_{ij} = -\sum_{1}^{n} \frac{\partial \Phi(\vec{\rho}_{ij})}{\partial \vec{\rho}_{ij}} \;, i = 1, 2, ..., N_k, \;\; j = 1, 2, ..., N_k, \tag{15.3}$$

where $\vec{\rho}_{ij}$ is a radius vector determining the position of the i-th atom relative to the j-th atom; $\Phi(\vec{\rho}_{ij})$ is a potential depending on the mutual positions of all the atoms; n is the number of interatomic interaction types.

In the general case, the potential $\Phi(\vec{\rho}_{ij})$ is given in the form of the sum of several components corresponding to different interaction types:

$$\Phi(\vec{\rho}_{ij}) = \Phi_{cb} + \Phi_{va} + \Phi_{ta} + \Phi_{pg} + \Phi_{vv} + \Phi_{es} + \Phi_{hb}. \tag{15.4}$$

Here the following potentials are implied: Φ_{cv}—of chemical bonds; Φ_{va}—of valence angles; Φ_{ta}—of torsion angles; Φ_{pg}—of flat groups; Φ_{vv}—of van der Waals contacts; Φ_{es}—of electrostatics; Φ_{hb}—of hydrogen bonds.

The above addends have different functional forms. The parameter values for the interaction potentials are determined based on the experiments (crystallography, spectral, calorimetric, etc.) and quantum calculations [25].

Giving original coordinates (and forces of atomic interactions) and velocities of all the atoms of each nanoparticle in accordance with Eq. (15.2), at the start time, we find the change of the coordinates and the velocities of each nanoparticle atoms with time from the equation of motion Eq. (15.1). Since the nanoparticles are not exposed to the action of external forces, they take some atomic equilibrium configuration with time that we will use for the next calculation stage.

15.2.2 THE CALCULATION OF THE PAIRWISE INTERACTION OF THE TWO NANOSTRUCTURE ELEMENTS

At this stage of solving the problem, we consider two interacting nanoelements. First, let us consider the problem statement for symmetric nanoelements, and then for arbitrary shaped nanoelements.

First of all, let us consider two symmetric nanoelements situated at the distance S from one another (Figure 15.1) at the initial conditions

$$\vec{x}_i = \vec{x}_{i0}, \vec{V}_i = 0, \; t = 0, \; \vec{x}_i \subset \Omega_1 \bigcup \Omega_2, \tag{15.5}$$

where Ω_1, Ω_2 are the areas occupied by the first and the second nanoparticle, respectively.

We obtain the coordinates $\vec{x}_{i0}$ from Eq. (15.2) solution at initial conditions (15.1). It allows calculating the combined interaction forces of the nanoelements

$$\vec{F}_{b1} = -\vec{F}_{b2} = \sum_{i=1}^{N_1} \sum_{j=1}^{N_2} \vec{F}_{ij} , \qquad (15.6)$$

where i, j are the atoms and N_1, N_2 are the numbers of atoms in the first and in the second nanoparticle, respectively.

Forces $\vec{F}_{ij}$ are defined from Eq. (15.3).

In the general case, the force magnitude of the nanoparticle interaction $\left|\vec{F}_{bi}\right|$ can be written as product of functions depending on the sizes of the nanoelements and the distance between them:

$$\left|\vec{F}_{bi}\right| = \Phi_{11}(S_c) \cdot \Phi_{12}(D) \qquad (15.7)$$

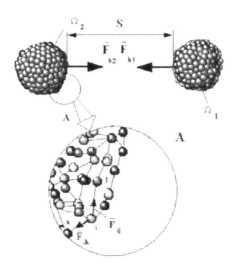

FIGURE 15.1   The scheme of the nanoparticle interaction; A—an enlarged view of the nanoparticle fragment.

The $\vec{F}_{bi}$ vector direction is determined by the direction cosines of a vector connecting the centers of the nanoelements.

Now, let us consider two interacting asymmetric nanoelements situated at the distance S_c between their centers of mass (Figure 15.2) and oriented at certain specified angles relative to each other.

In contrast to the previous problem, the interatomic interaction of the nano-elements leads not only to the relative displacement of the nanoelements but to their rotation as well. Consequently, in the general case, the sum of all the forces of the interatomic interactions of the nanoelements is brought to the principal vector of forces $\vec{F}_c$ and the principal moment $\vec{M}_c$

$$\vec{F}_c = \vec{F}_{b1} = -\vec{F}_{b2} = \sum_{i=1}^{N_1} \sum_{j=1}^{N_2} \vec{F}_{ij} ,$$ (15.8)

$$\vec{M}_c = \vec{M}_{c1} = -\vec{M}_{c2} = \sum_{i=1}^{N_1} \sum_{j=1}^{N_2} \vec{\rho}_{cj} \times \vec{F}_{ij} ,$$ (15.9)

where $\vec{\rho}_{cj}$ is a vector connecting points c and j.

The main objective of this calculation stage is building the dependences of the forces and moments of the nanostructure nanoelement interactions on the distance S_c between the centers of mass of the nanostructure nanoelements, on the angles of mutual orientation of the nanoelements $\Theta_1, \Theta_2, \Theta_3$ (shapes of the nanoelements) and on the characteristic size D of the nanoelement. In the general case, these dependences can be given in the form

$$\vec{F}_{bi} = \vec{\Phi}_F (S_c, \Theta_1, \Theta_2, \Theta_3, D),$$ (15.10)

$$\vec{M}_{bi} = \vec{\Phi}_M (S_c, \Theta_1, \Theta_2, \Theta_3, D),$$ (15.11)

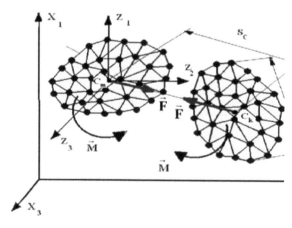

FIGURE 15.2 Two interacting nanoelements; $\vec{M}$, $\vec{F}$ are the principal moment and the principal vector of the forces, respectively.

For spherical nanoelements, the angles of the mutual orientation do not influence the force of their interaction; therefore, in Eq. (15.12), the moment is zero.

In the general case, functions (15.11) and (15.12) can be approximated by analogy with Eq. (15.8), as the product of functions $S_0, \Theta_1, \Theta_2, \Theta_3, D$, respectively. For the further numerical solution of the problem of the self-organization of nanoelements, it is sufficient to give the above functions in their tabular form and to use the linear (or nonlinear) interpolation of them in space.

15.3 PROBLEM FORMULATION FOR INTERACTION OF SEVERAL NANOELEMENTS

When the evolution of the nanosystem as whole (including the processes of ordering and self-organization of the nanostructure nanoelements) is investigated, the movement of each system nanoelement is considered as the movement of a single whole. In this case, the translational motion of the center of mass of each nanoelement is given in the coordinate system X_1, X_2 X_3, and the nanoelement rotation is described in the coordinate system Z_1, Z_2 Z_3, which is related to the center of mass of the nanoelement (Figure 15.2). The system of equations describing the above processes has the form

$$
\begin{cases}
M_k \dfrac{d^2 X_1^k}{dt^2} = \displaystyle\sum_{j=1}^{N_e} F_{X_1}^{kj} + F_{X_1}^{ke}, \\[4mm]
M_k \dfrac{d^2 X_2^k}{dt^2} = \displaystyle\sum_{j=1}^{N_e} F_{X_2}^{kj} + F_{X_2}^{ke}, \\[4mm]
M_k \dfrac{d^2 X_3^k}{dt^2} = \displaystyle\sum_{j=1}^{N_e} F_{X_3}^{kj} + F_{X_3}^{ke}, \\[4mm]
J_{Z_1}^k \dfrac{d^2 \Theta_1^k}{dt^2} + \dfrac{d\Theta_2^k}{dt} \cdot \dfrac{d\Theta_3^k}{dt}(J_{Z_3}^k - J_{Z_2}^k) = \displaystyle\sum_{j=1}^{N_e} M_{Z_1}^{kj} + M_{Z_1}^{ke}, \\[4mm]
J_{Z_2}^k \dfrac{d^2 \Theta_2^k}{dt^2} + \dfrac{d\Theta_1^k}{dt} \cdot \dfrac{d\Theta_3^k}{dt}(J_{Z_1}^k - J_{Z_3}^k) = \displaystyle\sum_{j=1}^{N_e} M_{Z_2}^{kj} + M_{Z_2}^{ke}, \\[4mm]
J_{Z_3}^k \dfrac{d^2 \Theta_3^k}{dt^2} + \dfrac{d\Theta_2^k}{dt} \cdot \dfrac{d\Theta_1^k}{dt}(J_{Z_2}^k - J_{Z_1}^k) = \displaystyle\sum_{j=1}^{N_e} M_{Z_3}^{kj} + M_{Z_3}^{ke},
\end{cases}
\qquad (15.12)
$$

where x_i^k, Θ_i^k are coordinates of the centers of mass and angles of the spatial orientation of the principal axes Z_1, Z_2, Z_3 of nanoelements; $F_{X_1}^{kj}, F_{X_2}^{kj}, F_{X_3}^{kj}$ are the interaction forces of nanoelements; $F_{X_1}^{ke}, F_{X_2}^{ke}, F_{X_3}^{ke}$ are external forces acting on nanoelements; N_e is the number of nanoelements; M_k is a mass of a nanoelement; $M_{Z_1}^{kj}, M_{Z_2}^{kj}, M_{Z_3}^{kj}$ is the moment of forces of the nanoelement interaction; $M_{Z_1}^{ke}, M_{Z_2}^{ke}, M_{Z_3}^{ke}$ are external moments acting on nanoelements; $J_{Z_1}^k, J_{Z_2}^k, J_{Z_3}^k$ are moments of inertia of a nanoelement.

The initial conditions for the system of Eqs. (15.13) and (15.14) have the form

$$\vec{X}^k = \vec{X}_0^k;\ \Theta^k = \Theta_0^k;\ \vec{V}^k = \vec{V}_0^k;\ \frac{d\Theta^k}{dt} = \frac{d\Theta_0^k}{dt}; t = 0, \tag{15.13}$$

15.4 NUMERICAL PROCEDURES AND SIMULATION TECHNIQUES

In the general case, the problem formulated in the previous sections has no analytical solution at each stage; therefore, numerical methods for solving are used, as a rule. In this work, for the first stages, the numerical integration of the equation of motion of the nanoparticle atoms in the relaxation process are used in accordance with Verlet scheme [26]:

$$\vec{x}_i^{n+1} = \vec{x}_i^n + \Delta t\ \vec{V}_i^n + \left((\Delta t)^2 / 2m_i\right)\left(\sum_{j=1}^{N_k} \vec{F}_{ij} + \vec{F}_i - \alpha_i m_i \vec{V}_i^n\right)^n, \tag{15.14}$$

$$\vec{V}_i^{n+1} = (1 - \Delta t\alpha_i)\vec{V}_i^n + (\Delta t / 2m_i)((\sum_{j=1}^{N_k} \vec{F}_{ij} + \vec{F}_i)^n + (\sum_{j=1}^{N_k} \vec{F}_{ij} + \vec{F}_i)^{n+1}), \tag{15.15}$$

where $\vec{x}_i^n$, $\vec{V}_i^n$ are a coordinate and a velocity of the i-th atom at the n-th step with respect to the time; Δt is a step with respect to the time.

The solution of the Eq. (15.13) also requires the application of numerical methods of integration. In the present work, Runge–Kutta method [27] is used for solving Eq. (15.13).

$$(X_i^k)_{n+1} = (X_i^k)_n + (V_i^k)_n \Delta t + \frac{1}{6}(\mu_{1i}^k + \mu_{2i}^k + \mu_{3i}^k)\Delta t, \tag{15.16}$$

$$(V_i^k)_{n+1} = (V_i^k)_n + \frac{1}{6}(\mu_{1i}^k + 2\mu_{2i}^k + 2\mu_{3i}^k + \mu_{4i}^k) \tag{15.17}$$

$$\mu_{1i}^k = \Phi_i^k(t_n;(X_i^k)_n,...;(V_i^k)_n...)\Delta t \ ,$$

$$\mu_{2i}^k = \Phi_i^k(t_n + \frac{\Delta t}{2};(X_i^k + V_i^k\frac{\Delta t}{2})_n,...;(V_i^k)_n + \frac{\mu_{1i}^k}{2},...)\Delta t \ ,$$

$$\mu_{3i}^k = \Phi_i^k(t_n + \frac{\Delta t}{2};(X_i^k + V_i^k\frac{\Delta t}{2} + \mu_{1i}^k\frac{\Delta t}{4})_n,...;(V_i^k)_n + \frac{\mu_{2i}^k}{2},...)\Delta t \ , \quad (15.18)$$

$$\mu_{4i}^k = \Phi_i^k(t_n + \Delta t;(X_i^k + V_i^k\Delta t + \mu_{2i}^k\frac{\Delta t}{2})_n,...;(V_i^k)_n + \mu_{2i}^k,...)\Delta t \ .$$

$$\Phi_i^k = \frac{1}{M_k}(\sum_{j=1}^{N_e} F_{X_3}^{kj} + F_{X_3}^{ke}) \quad (15.19)$$

$$(\Theta_i^k)_{n+1} = (\Theta_i^k)_n + (\frac{d\Theta_i^k}{dt})_n\Delta t + \frac{1}{6}(\lambda_{1i}^k + \lambda_{2i}^k + \lambda_{3i}^k)\Delta t \quad (15.20)$$

$$(\frac{d\Theta_i^k}{dt})_{n+1} = (\frac{d\Theta_i^k}{dt})_n + \frac{1}{6}(\lambda_{1i}^k + 2\lambda_{2i}^k + 2\lambda_{3i}^k + \lambda_{4i}^k) \quad (15.21)$$

$$\lambda_{1i}^k = \Psi_i^k(t_n;(\Theta_i^k)_n,...;(\frac{d\Theta_i^k}{dt})_n...)\Delta t \ ,$$

$$\lambda_{2i}^k = \Psi_i^k(t_n + \frac{\Delta t}{2};(\Theta_i^k + \frac{d\Theta_i^k}{dt}\frac{\Delta t}{2})_n,...;(\frac{d\Theta_i^k}{dt})_n + \frac{\lambda_{1i}^k}{2},...)\Delta t \ ,$$

$$\lambda_{3i}^k = \Psi_i^k(t_n + \frac{\Delta t}{2};(\Theta_i^k + \frac{d\Theta_i^k}{dt}\frac{\Delta t}{2} + \lambda_{1i}^k\frac{\Delta t}{4})_n,...;(\frac{d\Theta_i^k}{dt})_n + \frac{\lambda_{2i}^k}{2},...)\Delta t \ , \quad (15.22)$$

$$\lambda_{4i}^k = \Psi_i^k(t_n + \Delta t;(\Theta_i^k + \frac{d\Theta_i^k}{dt}\Delta t + \lambda_{2i}^k\frac{\Delta t}{2})_n,...;(\frac{d\Theta_i^k}{dt})_n + \lambda_{2i}^k,...)\Delta t$$

$$\Psi_1^k = \frac{1}{J_{Z_1}^k}(-\frac{d\Theta_2^k}{dt}\cdot\frac{d\Theta_3^k}{dt}(J_{Z_3}^k - J_{Z_2}^k) + \sum_{j=1}^{N_e} M_{Z_1}^{kj} + M_{Z_1}^{ke}) \ ,$$

$$\Psi_2^k = \frac{1}{J_{Z_2}^k}(-\frac{d\Theta_1^k}{dt}\cdot\frac{d\Theta_3^k}{dt}(J_{Z_1}^k - J_{Z_3}^k) + \sum_{j=1}^{N_e} M_{Z_2}^{kj} + M_{Z_2}^{ke}) \ , \quad (15.23a)$$

$$\Psi_3^k = \frac{1}{J_{Z_3}^k}(-\frac{d\Theta_1^k}{dt}\cdot\frac{d\Theta_2^k}{dt}(J_{Z_1}^k - J_{Z_2}^k) + \sum_{j=1}^{N_e} M_{Z_3}^{kj} + M_{Z_3}^{ke})$$

,

were $i=1,2,3;\quad k=1,2,\ldots N_e$

15.5   RESULTS AND DISCUSSIONS

Let us consider the realization of the above procedure taking as an example the calculation of the metal nanoparticle.

The potentials of the atomic interaction of Morse Eq. (15.23b) and Lennard-Johns Eq. (15.24) were used in the following calculations:

$$\Phi(\vec{\rho}_{ij})_m = D_m\ (\exp(-2\lambda_m(|\vec{\rho}_{ij}| - \rho_0)) - 2\exp(-\lambda_m(|\vec{\rho}_{ij}| - \rho_0))),\qquad (15.23b)$$

$$\Phi(\vec{\rho}_{ij})_{LD} = 4\varepsilon\left[\left(\frac{\sigma}{|\vec{\rho}_{ij}|}\right)^{12} - \left(\frac{\sigma}{|\vec{\rho}_{ij}|}\right)^{6}\right],\qquad (15.24)$$

where $D_m, \lambda_m, \rho_0, \varepsilon, \sigma$ are the constants of the materials studied.

For sequential and parallel solving the molecular dynamics equations, the program package developed at Applied Mechanics Institute, the Ural Branch of the Russian Academy of Sciences, and the advanced program package NAMD developed at the University of Illinois and Beckman Institute (USA) by the Theoretical Biophysics Group were used. The graphic imaging of the nanoparticle calculation results was carried out with the use of the program package VMD.

15.5.1   STRUCTURE AND FORMS OF NANOPARTICLES

At the first stage of the problem, the coordinates of the atoms positioned at the ordinary material lattice points (Figure 15.3, Eq. (15.1)) were taken as the original coordinates. During the relaxation process, the initial atomic system is rearranged into a new "equilibrium" configuration (Figure 15.3, Eq. (15.2)) in accordance with the calculations based on Eqs. (15.6), (15.7), (15.8), and (15.9), which satisfies the condition when the system potential energy is approaching the minimum (Figure 15.3, the plot).

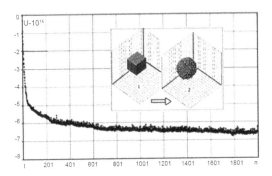

FIGURE 15.3 The initial crystalline (1) and cluster (2) structures of the nanoparticle consisting of 1,331 atoms after relaxation; the plot of the potential energy U [J] variations for this atomic system in the relaxation process (n—number of iterations with respect to the time).

After the relaxation, the nanoparticles can have quite diverse shapes: globe-like, spherical centered, spherical eccentric, spherical icosahedral nanoparticles, and asymmetric nanoparticles (Figure 15.4).

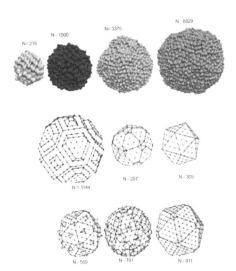

FIGURE 15.4 Nanoparticles of diverse shapes, depending on the number of atoms they consist of.

In this case, the number of atoms N significantly determines the shape of a nanoparticle. Note, that symmetric nanoparticles are formed only at a certain number of atoms. As a rule, in the general case, the nanoparticle deviates from the symmetric shape in the form of irregular raised portions on the surface. Besides, there are several different equilibrium shapes for the same number of atoms. The plot of the nanoparticle potential energy change in the relaxation process (Figure 15.5) illustrates it.

As it follows from this figure, the curve has two areas: the area of the decrease of the potential energy and the area of its stabilization promoting the formation of the first nanoparticle equilibrium shape (1). Then, a repeated decrease in the nanoparticle potential energy and the stabilization area corresponding to the formation of the second nanoparticle equilibrium shape are observed (2). Between them, there is a region of the transition from the first shape to the second one (P). The second equilibrium shape is more stable due to the lesser nanoparticle potential energy. However, the first equilibrium shape also "exists" rather long in the calculation process. The change of the equilibrium shapes is especially characteristic of the nanoparticles with an "irregular" shape. The internal structure of the nanoparticles is of importance since their atomic structure significantly differs from the crystalline structure of the bulk materials: the distance between the atoms and the angles change, and the surface formations of different types appear. In Figure 15.6, the change of the structure of a two-dimensional nanoparticle in the relaxation process is shown.

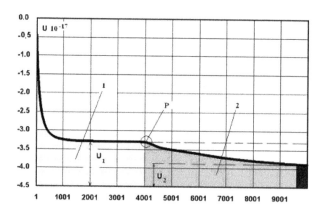

FIGURE 15.5   The plot of the potential energy change of the nanoparticle in the relaxation process.

(1)—a region of the stabilization of the first nanoparticle equilibrium shape; (2)—a region of the stabilization of the second nanoparticle equilibrium shape; (P)—a region of the transition of the first nanoparticle equilibrium shape into the second one.

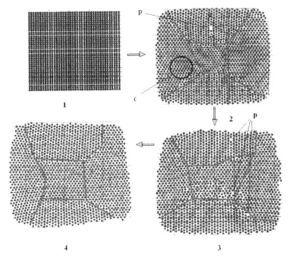

FIGURE 15.6 The change of the structure of a two-dimensional nanoparticle in the relaxation process: 1—the initial crystalline structure; 2, 3, 4—the nanoparticles structures which change in the relaxation process; p—pores; c—the region of compression.

The figure shows how the initial nanoparticle crystalline structure (1) is successively rearranging with time in the relaxation process (positions 2, 3, 4). Note that the resultant shape of the nanoparticle is not round, that is, it has "remembered" the initial atomic structure. It is also of interest that in the relaxation process, in the nanoparticle, the defects in the form of pores (designation "p" in the figure) and the density fluctuation regions (designation "c" in the figure) have been formed, which are absent in the final structure.

15.5.2 NANOPARTICLES INTERACTION

Let us consider some examples of nanoparticles interaction. Figure 15.7 shows the calculation results demonstrating the influence of the sizes of the nanoparticles on their interaction force. One can see from the plot that the larger nanoparticles are attracted stronger, that is, the maximal interaction force increases with the size growth of the particle. Let us divide the interaction force of the nanoparticles by its maximal value for each nanoparticle size, respectively. The obtained plot of the "relative" (dimensionless) force (Figure 15.8) shows that the value does not practically depend on the nanoparticle size since all the curves come close and can be approximated to one line.

Figure 15.9 displays the dependence of the maximal attraction force between the nanoparticles on their diameter that is characterized by nonlinearity and a general tendency toward the growth of the maximal force with the nanoparticle size growth.

The total force of the interaction between the nanoparticles is determined by multiplying of the two plots (Figures 15.8 and 15.9).

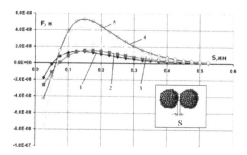

FIGURE 15.7   The dependence of the interaction force F [N] of the nanoparticles on the distance S [nm] between them and on the nanoparticle size: 1—d = 2.04; 2—d = 2.40; 3—d = 3.05; 4—d = 3.69; 5—d = 4.09 [nm].

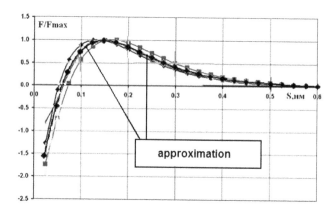

FIGURE 15.8   The dependence of the "relative" force F of the interaction of the nanoparticles on the distance S [nm] between them.

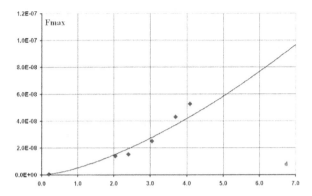

FIGURE 15.9 The dependence of the maximal attraction force F_{max} [N] the nanoparticles on the nanoparticle diameter d [nm].

Using the polynomial approximation of the curve in Figure 15.5 and the power mode approximation of the curve in Figure 15.6, we obtain

$$\overline{F} = (-1.13S^6 + 3.08S^5 - 3.41S^4 - 0.58S^3 + 0.82S - 0.00335)10^3, \qquad (15.25)$$

$$F_{max}\cdot = 0.5 \cdot 10^{-9} \cdot d^{1.499}, \qquad (15.26)$$

$$F = F_{max} \cdot \overline{F}, \qquad (15.27)$$

where d and S are the diameter of the nanoparticles and the distance between them [nm], respectively; F_{max} is the maximal force of the interaction of the nanoparticles [N].

Dependences Eqs. (15.25), (15.26), and (15.27) were used for the calculation of the nanocomposite ultimate strength for different patterns of nanoparticles' "packing" in the composite (Figure 15.10).

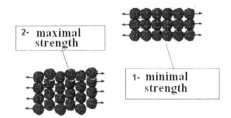

FIGURE 15.10 Different types of the nanoparticles' "packing" in the composite.

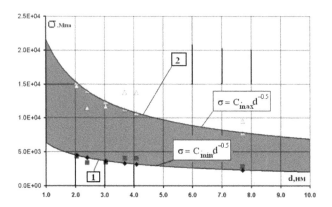

FIGURE 15.11   The dependence of the ultimate strength σ [MPa] of the nanocomposite formed by monodisperse nanoparticles on the nanoparticle sizes d [nm].

Figure 15.11 shows the dependence of the ultimate strength of the nanocomposite formed by monodisperse nanoparticles on the nanoparticle sizes. One can see that with the decrease of the nanoparticle sizes, the ultimate strength of the nanomaterial increases, and vice versa. The calculations have shown that the nanocomposite strength properties are significantly influenced by the nanoparticles' "packing" type in the material. The material strength grows when the packing density of nanoparticles increases. It should be specially noted that the material strength changes in inverse proportion to the nanoparticle diameter in the degree of 0.5, which agrees with the experimentally established law of strength change of nanomaterials (the law by Hall-Petch) [18]:

$$\sigma = C \cdot d^{-0.5}, \tag{15.28}$$

where $C = C_{max} = 2.17 \cdot 10^4$ is for the maximal packing density; $C = C_{min} = 6.4 \cdot 10^3$ is for the minimal packing density.

The electrostatic forces can strongly change force of interaction of nanoparticles. For example, numerical simulation of charged sodium (NaCl) nanoparticles system (Figure 15.12) has been carried out. Considered ensemble consists of eight separate nanoparticles. The nanoparticles interact due to van der Waals and electrostatic forces.

Results of particles center of masses motion are introduced at Figure 15.13 representing trajectories of all nanoparticles included into system. It shows the dependence of the modulus of displacement vector $|R|$ on time. One can see that nanoparticle moves intensively at first stage of calculation process. At the end of numerical calculation, all particles have got new stable locations, and the graphs of the radius vector $|R|$ become stationary. However, the nanoparticles continue to "vibrate" even at the final stage of numerical calculations. Nevertheless, despite of "vibration," the system of nanoparticles occupies steady position.

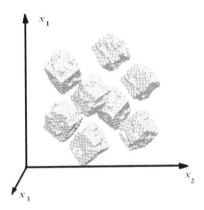

FIGURE 15.12   Nanoparticles system consists of eight nanoparticles NaCl.

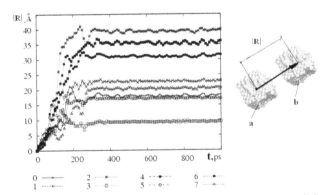

FIGURE 15.13   The dependence of nanoparticle centers of masses motion $|R|$ on the time t; a, b--the nanoparticle positions at time 0 and t, accordingly; 1-8—are the numbers of the nanoparticles.

However, one can observe a number of other situations. Let us consider, for example, the self-organization calculation for the system consisting of 125 cubic nanoparticles, the atomic interaction of which is determined by Morse potential (Figure 15.14).

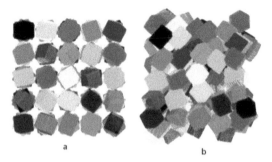

a

b

FIGURE 15.14   The positions of the 125 cubic nanoparticles: (a)—initial configuration and (b)—final configuration of nanoparticles.

As you see, the nanoparticles are moving and rotating in the self-organization process forming the structure with minimal potential energy. Let us consider, for example, the calculation of the self-organization of the system consisting of two cubic nanoparticles, the atomic interaction of which is determined by Morse potential [12]. Figure 15.15 displays possible mutual positions of these nanoparticles. The positions, where the principal moment of forces is zero, corresponds to pairs of the nanoparticles 2–3; 3–4; 2–5 (Figure 15.15) and defines the possible positions of their equilibrium.

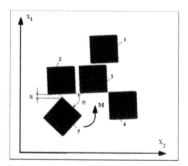

FIGURE 15.15   Characteristic positions of the cubic nanoparticles.

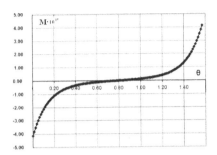

FIGURE 15.16 The dependence of the moment M [Nm] of the interaction force between cubic nanoparticles 1–3 (see in Figure 15.9) on the angle of their relative rotation θ [rad].

Figure 15.16 presents the dependence of the moment of the interaction force between the cubic nanoparticles 1–3 (Figure 15.15) on the angle of their relative rotation. From the plot follows that when the rotation angle of particle 1 relative to particle 3 is $\pi/4$, the force moment of their interaction is zero. At an increase or a decrease in the angle the force moment appears. In the range of $\pi/8 < \theta < 3\pi/4$ the moment is small. The force moment rapidly grows outside of this range. The distance S between the nanoparticles plays a significant role in establishing their equilibrium. If $S > S_0$ (where S_0 is the distance, where the interaction forces of the nanoparticles are zero), then the particles are attracted to one another. In this case, the sign of the moment corresponds to the sign of the angle θ deviation from $\pi/4$. At $S < S_0$ (the repulsion of the nanoparticles), the sign of the moment is opposite to the sign of the angle deviation. In other words, in the first case, the increase of the angle deviation causes the increase of the moment promoting the movement of the nanoelement in the given direction, and in the second case, the angle deviation causes the increase of the moment hindering the movement of the nanoelement in the given direction. Thus, the first case corresponds to the unstable equilibrium of nanoparticles, and the second case—to their stable equilibrium. The potential energy change plots for the system of the interaction of two cubic nanoparticles (Figure 15.17) illustrate the influence of the parameter S. Here, curve 1 corresponds to the condition $S < S_0$ and it has a well-expressed minimum in the $0.3 < \theta < 1.3$ region. At $\theta < 0.3$ and $\theta > 1.3$, the interaction potential energy sharply increases, which leads to the return of the system into the initial equilibrium position. At $S > S_0$ (curves 2–5), the potential energy plot has a maximum at the $\theta = 0$ point, which corresponds to the unstable position.

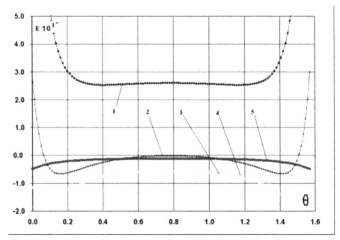

FIGURE 15.17   The plots of the change of the potential energy E [Nm] for the interaction of two cubic nanoparticles depending on the angle of their relative rotation θ [rad] and the distance between them (positions of the nanoparticles 1–3, Figure 15.9).

The carried out theoretical analysis is confirmed by the works of the scientists from New Jersey University and California University in Berkeley who experimentally found the self-organization of the cubic microparticles of plumbum zirconate-titanate (PZT) [28]: The ordered groups of cubic microcrystals from PZT obtained by hydrothermal synthesis formed a flat layer of particles on the air-water interface, where the particle occupied the more stable position corresponding to position 2–3 in Figure 15.15.

Thus, the analysis of the interaction of two cubic nanoparticles has shown that different variants of their final stationary state of equilibrium are possible, in which the principal vectors of forces and moments are zero. However, there are both stable and unstable stationary states of this system: nanoparticle positions 2–3 are stable, and positions 3–4 and 2–5 have limited stability or they are unstable depending on the distance between the nanoparticles.

Note that for the structures consisting of a large number of nanoparticles, there can be a quantity of stable stationary and unstable forms of equilibrium. Accordingly, the stable and unstable nanostructures of composite materials can appear. The search and analysis of the parameters determining the formation of stable nanosystems is an urgent task.

It is necessary to note, that the method offered has restrictions. This is explained by change of the nanoparticles form and accordingly variation of interaction pair potential during nanoparticles coming together at certain conditions.

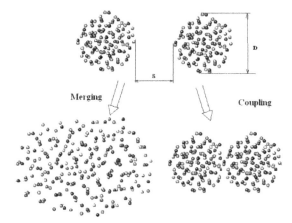

FIGURE 15.18   Different type of nanoparticles connection (merging and coupling).

The merge (accretion [4]) of two or several nanoparticles into a single whole is possible (Figure 15.18). Change of a kind of connection cooperating nanoparticles (merging or coupling in larger particles) depending on the its sizes, it is possible to explain on the basis of the analysis of the energy change graph of connection nanoparticles (Figure 15.19). From Figure 15.19 follows, that, though with the size increasing of a particle energy of nanoparticles connection E_{np} grows also, its size in comparison with superficial energy E_S of a particle sharply increases at reduction of the sizes nanoparticles. Hence, for finer particles energy of connection can appear sufficient for destruction of their configuration under action of a mutual attraction and merging in larger particle.

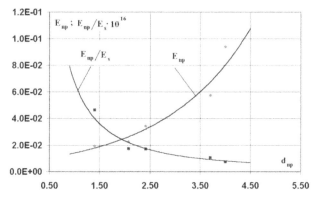

FIGURE 15.19   Change of energy of nanoparticles connection E_{np} [Nm] and E_{np} ration to superficial energy E_S depending on nanoparticles diameter d [nm]. Points designate the calculated values. Continuous lines are approximations.

Spatial distribution of particles influences on rate of the forces holding nano-structures, formed from several nanoparticles, also. On Figure 15.20 the chain nanoparticles, formed is resulted at coupling of three nanoparticles (from 512 atoms everyone), located in the initial moment on one line. Calculations have shown, that in this case nanoparticles form a stable chain. Thus, particles practically do not change the form and cooperate on small platforms.

In the same figure the result of connection of three nanoparticles, located in the initial moment on a circle and consisting of 256 atoms everyone is submitted. In this case particles incorporate among themselves "densely," contacting on a significant part of the external surface.

Distance between particles at which they are in balance it is much less for the particles collected in group ($L^0_{3np} < L^0_{2np}$) It confirms also the graph of forces from which it is visible, that the maximal force of an attraction between particles in this case (is designated by a continuous line) in some times more, than at an arrangement of particles in a chain (dashed line) $F_{3np} > F_{2np}$.

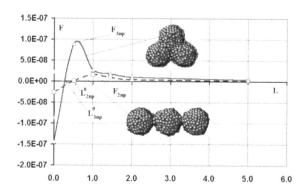

FIGURE 15.20   Change of force F [N] of 3 nanoparticles interaction, consisting of 512 atoms everyone, and connected among themselves on a line and on the beams missing under a corner of 120 degrees, accordingly, depending on distance between them L [nm].

Experimental investigation of the spatial structures formed by nanoparticles [4], confirm that nanoparticles gather to compact objects. Thus the internal nuclear structure of the connections area of nanoparticles considerably differs from structure of a free nanoparticle.

Nanoelements kind of interaction depends strongly on the temperature. In Figure 15.21 shows the picture of the interaction of nanoparticles at different temperatures (Figure 15.22) It is seen that with increasing temperature the interaction of changes in sequence: Coupling (1.2), merging (3.4). With further increase in temperature the nanoparticles dispersed.

FIGURE 15.21   Change of nanoparticles connection at increase in temperature.

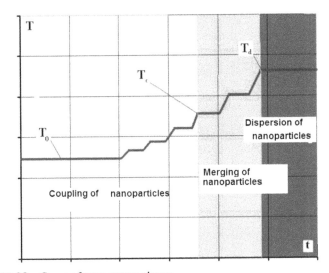

FIGURE 15.22   Curve of temperature change.

In the conclusion to the section we will consider problems of dynamics of nanoparticles. The analysis of interaction of nanoparticles among themselves also allows to draw a conclusion on an essential role in this process of energy of initial movement of particles. Various processes at interaction of the nanoparticles, moving with different speed, are observed: The processes of agglomerate formation, formation of larger particles at merge of the smaller size particles, absorption by large particles of the smaller ones, dispersion of particles on separate smaller ones or atoms.

For example, in Figure 15.23 the interactions of two particles are moving toward each other with different speed are shown. At small speed of moving is formed steady agglomerate (Figure 15.23).

In a Figure 15.23 (left) is submitted interaction of two particles moving toward each other with the large speed. It is visible, that steady formation in this case is not appearing and the particles collapse.

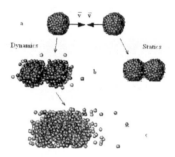

FIGURE 15.23 Pictures of dynamic interaction of two nanoparticles: (**a**) an initial configuration nanoparticles; (**b**) nanoparticles at dynamic interaction; and (**c**) the "cloud" of atoms formed because of dynamic destruction two nanoparticles.

Nature of interaction of nanoparticles, along with speed of their movement, essentially depends on a ratio of their sizes. In Figure 15.24 pictures of interaction of two nanoparticles of zinc of the different size are presented. As well as in the previous case of a nanoparticle move from initial situation (1) toward each other. At small initial speed, nanoparticles incorporate at contact and form a steady conglomerate (2).

FIGURE 15.24 Pictures of interaction of two nanoparticles of zinc: 1—initial configuration of nanoparticles; 2—connection of nanoparticles; 3, 4—absorption by a large nanoparticle of a particle of the smaller size; and 5—destruction of nanoparticles at blow.

At increase in speed of movement larger nanoparticle absorbs smaller, and the uniform nanoparticle (3, 4) is formed. At further increase in speed of movement of nanoparticles, owing to blow, the smaller particle intensively takes root in big and destroys its.

The given examples show that use of dynamic processes of pressing for formation of nanocomposites demands a right choice of a mode of the loading providing integrity of nanoparticles. At big energy of dynamic loading, instead of a nanocomposite with dispersion corresponding to the initial size of nanoparticles, the nanocomposite with much larger grain that will essentially change properties of a composite can be obtained.

CONCLUSIONS

In conclusion, the following basic regularities of the nanoparticle formation and self-organization should be noted.

(i) The existence of several types of the forms and structures of nanoparticles is possible depending on the thermodynamic conditions.

(ii) The absence of the crystal nucleus in small nanoparticles.

(iii) The formation of a single (ideal) crystal nucleus defects on the nucleus surface connected to the amorphous shell.

(iv) The formation of the polycrystal nucleus with defects distributed among the crystal grains with low atomic density and change interatomic distances. In addition, the grain boundaries are nonequilibrium and contain a great number of grain-boundary defects.

(v) When there is an increase in sizes, the structure of nanoparticles is changing from amorphous to roentgen-amorphous and then into the crystalline structure.

(vi) The formation of the defect structures of different types on the boundaries and the surface of a nanoparticle.

(vii) The nanoparticle transition from the globe-shaped to the crystal-like shape.

(viii) The formation of the "regular" and "irregular" shapes of nanoparticles depending on the number of atoms forming the nanoparticle and the relaxation conditions (the rate of cooling, first of all).

(ix) The structure of a nanoparticle is strained because of the different distances between the atoms inside the nanoparticle and in its surface layers.

(x) The systems of nanoparticles can to form stable and unstable nanostructures.

ACKNOWLEDGMENTS

This work was carried out with financial support from the Research Program of the Ural Branch of the Russian Academy of Sciences: the projects 12-P-12-2010, 12-C-1-1004, 12-T-1-1009 and was supported by the grants of Russian Foundation for Basic Research (RFFI) 04-01-96017-r2004ural_a; 05-08-50090-a; 07-01-96015-r_ural_a; 08-08-12082-ofi; 11-03-00571-a.

The author is grateful to his young colleagues doctor A. A. Vakhrushev and doctor A. Yu. Fedotov for active participation in the development of the software complex, calculations and analyzing of numerous calculation results.

The calculations were performed at the Joint Supercomputer Center of the Russian Academy of Sciences.

KEYWORDS

- **Dynamic interaction**
- **Equilibrium configuration**
- **Internal structure and shape**
- **Relaxation process**

REFERENCES

1. Qing-Qing, Ni, Yaqin, Fu, and Masaharu, Iwamoto, *"Evaluation of elastic modulus of nano particles in PMMA/silica nanocomposites."* J. Soc. Mater. Sci. Japan. **2004**, *53(9)*, 956–961.
2. Ruoff, R. S.; Nicola M. Pugno, *"Strength of nanostructures."* Mechanics of the 21st Century. Proceeding **of the 21-th International Congress of Theoretical and Applied Mechanics. Warsaw: Springer; 2004,** 303–311.
3. **Diao, J.; Gall, K.; and Dunn, M. L.;** *"Atomistic simulation of the structure and elastic properties of gold nanowires."* J. Mech. Phys. Solids. **2004**, *52(9)*, 1935–1962.
4. **Dingreville, R.; Qu, J.; and Cherkaoui, M.;** *"Surface free energy and its effect on the elastic behavior of nano-sized particles, wires and films."* J. Mech. Phys. Solids. **2004**, *53(8)*, 1827–1854.
5. **Duan, H. L.; Wang, J.; Huang, Z. P.; and Karihaloo, B. L.;** *"Size-dependent effective elastic constants of solids containing nano-inhomogeneities with interface stress."* J. Mech. Phys. Solids. **2005**, *53(7)*, 1574–1596.
6. Gusev, A. I.; and Rempel, A. A.; "Nanocrystalline Materials." Moscow: Physical Mathematical Literature; **2001**, (In Russian).
7. Hoare, M. R.; "Structure and dynamics of simple microclusters." *Ach. Chem. Phys.* **1987**, *40*, 49–135.
8. *Brooks, B. R.; Bruccoleri, R. E.; Olafson, B. D.; States, D. J.; Swaminathan, S.; and Karplus, M.; "CHARMM: A program for macromolecular energy minimization, and dynamics calculations." J. Comput. Chem.* **1983**, *4(2), 187–217.*

9. Friedlander, S. K.; *"Polymer-like behavior of inorganic nanoparticle chain aggregates."* *J. Nanopart. Res.* **1999**, *1*, 9–15.

10. Grzegorczyk, M.; Rybaczuk, M.; and Maruszewski, K.; *"Ballistic aggregation: An alternative approach to modeling of silica sol–gel structures."* *Chaos, Solitons Fractals.* **2004**, *19*, 1003–11.

11. Shevchenko, E. V.; Talapin, D. V. Kotov, N. A.; O'Brien, S.; and Murray, C. B.; *"Structural diversity in binary nanoparticle superlattices."* *Nature Lett.* **2006**, *439*, 55–59.

12. Kang, Z. C.; and Wang, Z. L.; *"On Accretion of nanosize carbon spheres."* *J. Phys. Chem.* **1996**, *100*, 5163–65.

13. Melikhov, I. V.; and Bozhevol'nov, V. E.; *"Variability and self-organization in nanosystems."* *J. Nanopart. Res.***2003**, *5*, 465–72.

14. Kim, D.; and Lu, W.; *"Self-organized nanostructures in multi-phase epilayers."* *Nanotechnology.* **2004**, *15*, 667–74.

15. Kurt E. Geckeler, *"Novel supermolecular nanomaterials: from design to reality."* Proceeding of the 12 Annual International Conference on Composites/Nano Engineering, Tenerife, Spain, August 1-6, CD Rom Edition, **2005**.

16. Vakhrouchev, A. V.; and Lipanov, A. M.; *"A numerical analysis of the rupture of powder materials under the power impact influence."* *Comp. Struct.* **1992**, *1/2(44)*, 481–86.

17. Vakhrouchev, A. V.; *"Modeling of static and dynamic processes of nanoparticles interaction."* CD-ROM Proceeding of the 21-th international congress of theoretical and applied mechanics, ID12054, Warsaw, Poland; **2004**.

18. Vakhrouchev, A. V.; *"Simulation of nanoparticles interaction."* Proceeding of the 12 Annual International Conference on Composites/Nano Engineering, Tenerife, Spain; August 1–6, CD Rom Edition, **2005**.

19. Vakhrouchev, A. V.; *"Simulation of nano-elements interactions and self-assembling."* *Model. Simulat. Mater. Sci. Eng.* **2006**, *14*, 975–991.

20. Vakhrouchev, A. V.; and Lipanov, A. M.; Numerical Analysis of the Atomic structure and Shape of Metal Nanoparticles. *Comput. Math. Math. Phys.* **2007**, *47(10)*, 1702–1711.

21. Vakhrouchev, A. V.; Modelling of the process of formation and use of powder nanocomposites. Composites with Micro and Nano-Structures. Computational Modeling and Experiments. Computational Methods in Applied Sciences Series. Barcelona, Spain: Springer Science; **2008**, *9*, 107–136.

22. Vakhrouchev, A. V.; Modeling of the nanosystems formation by the molecular dynamics, mesodynamics and continuum mechanics methods. *Multidis. Model. Mater. Struct.* **2009**, *5(2)*, 99–118.

23. Vakhrouchev, A. V.; Theoretical Bases of Nanotechnology Application to Thermal Engines and Equipment. Izhevsk: Institute of Applied Mechanics, Ural Branch of the Russian Academy of Sciences; **2008**, 212 p (In Russian).

24. Alikin, V. N.; Vakhrouchev, A. V.; Golubchikov, V. B.; Lipanov, A. M.; and Serebrennikov, S. Y.; Development and Investigation of the Aerosol Nanotechnology. Moscow: Mashinostroenie; **2010**, 196 p (In Russian).

25. Heerman, W. D.; *"Computer Simulation Methods in Theoretical Physics."* Berlin: Springer-Verlag; **1986.**

26. Verlet, L.; *"Computer 'experiments' on classical fluids I. Thermo dynamical properties of Lennard-Jones molecules,"* *Phys. Rev.* **1967**, *159*, 98–103.

27. Korn, G. A.; and Korn, M. T.; *"Mathematical Handbook."* New York: McGraw-Hill Book Company; **1968.**

28. Self-organizing of microparticles piezoelectric materials, News of chemistry, datenews.php. htm

CHAPTER 16

TRENDS IN MEMBRANE FILTRATION TECHNOLOGY: THEORY AND APPLICATION

SHIMA MAGHSOODLOU and AREZOO AFZALI

University of Guilan, Rasht, Iran

CONTENTS

16.1   MEMBRANES FILTRATION

Membrane filtration is a mechanical filtration technique which uses an absolute barrier to the passage of particulate material as any technology currently available in water treatment. The term "membrane" covers a wide range of processes, including those used for gas/gas, gas/liquid, liquid/liquid, gas/solid, and liquid/solid separations. Membrane production is a large-scale operation. There are two basic types of filters: depth filters and membrane filters.

Depth filters have a significant physical depth and the particles to be maintained are captured throughout the depth of the filter. Depth filters often have a flexuous three-dimensional structure, with multiple channels and heavy branching so that there is a large pathway through which the liquid must flow and by which the filter can retain particles. Depth filters have the advantages of low cost, high throughput, large particle retention capacity, and the ability to retain a variety of particle sizes. However, they can endure from entrainment of the filter medium, uncertainty regarding effective pore size, some ambiguity regarding the overall integrity of the filter, and the risk of particles being mobilized when the pressure differential across the filter is large.

The second type of filter is the membrane filter, in which depth is not considered momentous. The membrane filter uses a relatively thin material with a well-defined maximum pore size and the particle retaining effect takes place almost entirely at the surface. Membranes offer the advantage of having well-defined effective pore sizes, can be integrity tested more easily than depth filters, and can achieve more filtration of much smaller particles. They tend to be more expensive than depth filters and usually cannot achieve the throughput of a depth filter. Filtration technology has developed a well defined terminology that has been well addressed by commercial suppliers.

The term membrane has been defined in a number of ways. The most appealing definitions to us are the following:

"A selective separation barrier for one or several components in solution or suspension" [19]. "A thin layer of material that is capable of separating materials as a function of their physical and chemical properties when a driving force is applied across the membrane."

Membranes are important materials which form part of our daily lives. Their long history and use in biological systems has been extensively studied throughout the scientific field. Membranes have proven themselves as promising separation candidates due to advantages offered by their high stability, efficiency, low energy requirement and ease of operation. Membranes with good thermal and mechanical stability combined with good solvent resistance are important for industrial processes [1].

The concept of membrane processes is relatively simple but nevertheless often unknown. Membranes might be described as conventional filters but with much finer mesh or much smaller pores to enable the separation of tiny particles, even molecules. In general, one can divide membranes into two groups: porous and nonporous. The former group is similar to classical filtration with pressure as the driving force; the separation of a mixture is achieved by the rejection of at least one component by the membrane and passing of the other components through the membrane (see Figure 16.1). However, it is important to note that nonporous membranes do not operate on a size exclusion mechanism.

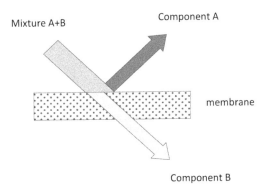

FIGURE 16.1 Basic principle of porous membrane processes.
Membrane separation processes can be used for a wide range of applications and can often offer significant advantages over conventional separation such as distillation and adsorption since the separation is based on a physical mechanism. Compared to conventional processes, therefore, no chemical, biological, or thermal change of the component is involved for most membrane processes. Hence membrane separation is particularly attractive to the processing of food, beverage, and bioproducts where the processed products can be sensitive to temperature (vs. distillation) and solvents (vs. extraction).

Synthetic membranes show a large variety in their structural forms. The material used in their production determines their function and their driving forces. Typically the driving force is pressure across the membrane barrier (see Table 16.1) [2–4]. Formation of a pressure gradient across the membrane allows separation in a bolter-like manner. Some other forms of separation that exist include charge effects and solution diffusion. In this separation, the smaller particles are allowed to pass through as permeates whereas the larger molecules (macromolecules) are retained. The retention or permeation of these species is ordained by the pore architecture as well as pore sizes of the membrane employed. Therefore based on the pore sizes, these pressure driven membranes can be divided into reverse osmosis (RO), nanofiltration (NF), ultrafiltration (UF), and microfiltration (MF), are already applied on an industrial scale to food and bioproduct processing [5–7].

TABLE 16.1   Driving forces and their membrane processes

Driving Force	Membrane Process
Pressure difference	Microfiltration, Ultrafiltration,
	Nanofiltration, Reverse osmosis
Chemical potential difference	Pervaporation, Pertraction, Dialysis, Gas separation, Vapour permeation, Liquid Membranes
Electrical potential difference	Electrodialysis, Membrane electrophoresis,
	Membrane electrolysis
Temperature difference	Membrane distillation

16.1.1   MICROFILTRATION MEMBRANES

Microfiltration membranes have the largest pore sizes and thus use less pressure. They involve removing chemical and biological species with diameters ranging between 100 to 10000 nm and components smaller than this, pass through as permeates. MF is primarily used to separate particles and bacteria from other smaller solutes [4].

16.1.2   ULTRAFILTRATION MEMBRANES

Ultrafiltration membranes operate within the parameters of the micro- and nano-filtration membranes. Therefore UF membranes have smaller pores as compared to MF membranes. They involve retaining macromolecules and colloids from solution which range between 2–100 nm and operating pressures between 1 and 10 bar. for example, large organic molecules and proteins. UF is used to separate colloids such as proteins from small molecules such as sugars and salts [4].

16.1.3   NANOFILTRATION MEMBRANES

Nanofiltration membranes are distinguished by their pore sizes of between 0.5–2 nm and operating pressures between 5 and 40 bar. They are mainly used for the removal of small organic molecules and di- and multivalent ions. Additionally, NF membranes have surface charges that make them suitable for retaining ionic pollutants from solution. NF is used to achieve separation between sugars, other organic molecules, and multivalent salts on the one hand from monovalent salts and water on the other. Nanofiltration, however, does not remove dissolved compounds [4].

16.1.4 REVERSE OSMOSIS MEMBRANES

Reverse osmosis membranes are dense semipermeable membranes mainly used for desalination of sea water [38]. Contrary to MF and UF membranes, RO membranes have no distinct pores. As a result, high pressures are applied to increase the permeability of the membranes [4]. The properties of the various types of membranes are summarized in Table 16.2.

TABLE 16.2 Summary of properties of pressure driven membranes [4]

	MF	UF	NF	RO
Permeability (L/h.m².bar)	1,000	10–1,000	1.5–30	0.05–1.5
Pressure (bar)	0.1–2	0.1–5	3–20	5–1,120
Pore Size (nm)	100–10,000	2–100	0.5–2	0.5
Separation Mechanism	sieving	sieving	Sieving, charge effects	Solution diffusion
Applications	Removal of bacteria	Removal of bacteria, fungi,virses	Removal of multivalen-tions	desalinatiob

The NF membrane is a type of pressure-driven membrane with properties in between RO and UF membranes. NF offers several advantages such as low operation pressure, high flux, high retention of multivalent anion salts and an organic molecular above 300, relatively low investment and low operation and maintenance costs. Because of these advantages, the applications of NF worldwide have increased [8]. In recent times, research in the application of nanofiltration techniques has been extended from separation of aqueous solutions to separation of organic solvents to homogeneous catalysis, separation of ionic liquids, food processing, and so on [9].

Figure 16.2 presents a classification on the applicability of different membrane separation processes based on particle or molecular sizes. RO process is often used for desalination and pure water production, but it is the UF and MF that are widely used in food and bioprocessing.

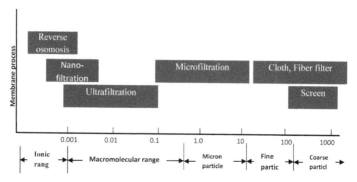

FIGURE 16.2   The applicability ranges of different separation processes based on sizes.

While MF membranes target on the microorganism removal, and hence are given the absolute rating, namely, the diameter of the largest pore on the membrane surface, UF/NF membranes are characterized by the nominal rating due to their early applications of purifying biological solutions. The nominal rating is defined as the molecular weight cutoff (MWCO) that is the smallest molecular weight of species, of which the membrane has more than 90 percent rejection (see later for definitions). The separation mechanism in MF/UF/NF is mainly the size exclusion, which is indicated in the nominal ratings of the membranes. The other separation mechanism includes the electrostatic interactions between solutes and membranes, which depends on the surface and physiochemical properties of solutes and membranes [5]. Also, The principal types of membrane are shown schematically in Figures 16.3 and 16.4 are described briefly below.

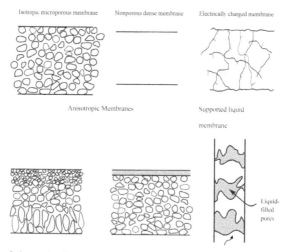

FIGURE 16.3   Schematic diagrams of the principal types of membranes.

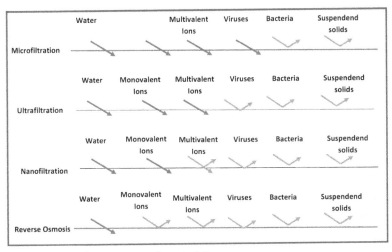

FIGURE 16.4   Membrane process characteristics.

16.2   THE RELATIONSHIP BETWEEN NANOTECHNOLOGY AND FILTRATION

Nowadays, nanomaterials have become the most interested topic of materials research and development due to their unique structural properties (unique chemical, biological, and physical properties as compared to larger particles of the same material) that cover their efficient uses in various fields, such as ion exchange and separation, catalysis, biomolecular isolation and purification as well as in chemical sensing [10]. However, the understanding of the potential risks (health and environmental effects) posed by nanomaterials hasn't increased as rapidly as research has regarding possible applications.

One of the ways to enhance their functional properties is to increase their specific surface area by the creation of a large number of nanostructured elements or by the synthesis of a highly porous material.

Classically, porous matter is seen as material containing three-dimensional voids, representing translational repetition, while no regularity is necessary for a material to be termed "porous." In general, the pores can be classified into two types: open pores which connect to the surface of the material, and closed pores which are isolated from the outside. If the material exhibits mainly open pores, which can be easily transpired, then one can consider its use in functional applications such as adsorption, catalysis and sensing. In turn, the closed pores can be used in sonic and thermal insulation, or lightweight structural applications. The use of porous materials offers also new opportunities in such areas as coverage chemistry, guest—host synthesis and molecular manipulations and reactions for manufacture of nanoparticles, nanowires and other quantum nano-structures. The International Union of Pure and Applied Chemistry (IUPAC) defines porosity scales as follows (Figure 16.5):

(i)   Microporous materials 0–2-nm pores
(ii) Mesoporous materials 2–50-nm pores
(iii) Macroporous materials >50-nm pores

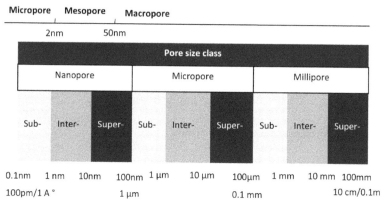

FIGURE 16.5 New pore size classification as compared with the current IUPAC nomenclature.

This definition, it should be noted, is somewhat in conflict with the definition of nanoscale objects, which typically have large relative porosities (>0.4), and pore diameters between 1 and 100 nm. In order to classify porous materials according to the size of their pores the sorption analysis is one of the tools often used. This tool is based on the fact that pores of different sizes lead to totally different characteristics in sorption isotherms. The correlation between the vapor pressure and the pore size can be written as the Kelvin equation:

$$r_p \left(\frac{p}{p_0} \right) = \frac{2 \gamma V_L}{RT \ln \left(\frac{p}{p_0} \right)} + t \left(\frac{p}{p_0} \right) \tag{16.1}$$

Therefore, the isotherms of microporous materials show a steep increase at very low pressures (relative pressures near zero) and reach a plateau quickly. Mesoporous materials are characterized by a so called capillary doping step and a hysteresis (a discrepancy between adsorption and desorption). Macroporous materials show a single or multiple adsorption steps near the pressure of the standard bulk condensed state (relative pressure approaches one) [10].

Nanoporous materials exuberate in nature, both in biological systems and in natural minerals. Some nanoporous materials have been used industrially for a long time. Recent progress in characterization and manipulation on the nanoscale has led to noticeable progression in understanding and making a variety of nanoporous materials: from the merely opportunistic to directed design. This is most strikingly the case in the creation of a wide variety of membranes where control over pore size is increasing dramatically, often to atomic levels of perfection, as is the ability to modify physical and chemical characteristics of the materials that make up the pores [11].

The available range of membrane materials includes polymeric, carbon, silica, zeolite and other ceramics, as well as composites. Each type of membrane can have a different porous structure, as illustrated in Figure 16.6. Membranes can be thought of as having a fixed (immovable) network of pores in which the molecule travels, with the exception of most polymeric membranes [12–13]. Polymeric membranes are composed of an amorphous mix of polymer chains whose interactions involve mostly Van der Waals forces. However, some polymers manifest a behavior that is consistent with the idea of existence of opened pores within their matrix. This is especially true for high free volume, high permeability polymers, as has been proved by computer modeling, low activation energy of diffusion, negative activation energy of permeation, solubility controlled permeation [14–15]. Although polymeric membranes have often been viewed as nonporous, in the modeling framework discussed here it is convenient to consider them nonetheless as porous. Glassy polymers have pores that can be considered as "frozen" over short times scales, while rubbery polymers have dynamic fluctuating pores (or more correctly free volume elements) that move, shrink, expand and disappear [16].

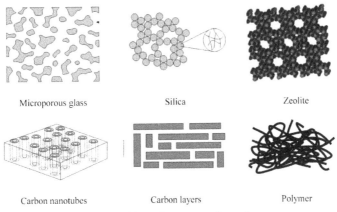

FIGURE 16.6 Porous structure within various types of membranes.

Three nanotechnologies that are often used in the filtering processes and show great potential for applications in remediation are:

(i) Nanofiltration (and its sibling technologies: reverse osmosis, ultrafiltration, and microfiltration), is a fully-developed, commercially-available membrane technology with a large number of vendors. Nanofiltration relies on the ability of membranes to discriminate between the physical size of particles or species in a mixture or solution and is primarily used for water pretreatment, treatment, and purification). There are almost 600 companies in worldwide which offering membrane systems.

(ii) Electrospinning is a process utilized by the nanofiltration process, in which fibers are stretched and elongated down to a diameter of about 10 nm. The modified nanofibers that are produced are particularly useful in the filtration process as an ultraconcentrated filter with a very large surface area. Studies have found that electrospun nanofibers can capture metallic ions and are continually effective through refiltration.

(iii) Surface modified membrane is a term used for membranes with altered makeup and configuration, though the basic properties of their underlying materials remain intact.

16.3   TYPES OF MEMBRANES

As it mentioned, membranes have achieved a momentous place in chemical technology and are used in a broad range of applications. The key property that is exploited is the ability of a membrane to control the permeation rate of a chemical species through the membrane. In essence, a membrane is nothing more than a discrete, thin interface that moderates the permeation of chemical species in contact with it. This interface may be molecularly homogeneous, that is completely uniform in composition and structure or it may be chemically or physically heterogeneous for example, containing holes or pores of finite dimensions or consisting of some form of layered structure. A normal filter meets this definition of a membrane, but, generally, the term filter is usually limited to structures that separate particulate suspensions larger than $1-10$ μm [17].

The preparation of synthetic membranes is however a more recent invention which has received a great audience due to its applications [18]. Membrane technology like most other methods has undergone a developmental stage, which has validated the technique as a cost-effective treatment option for water. The level of performance of the membrane technologies is still developing and it is stimulated by the use of additives to improve the mechanical and thermal properties, as well as the permeability, selectivity, rejection and fouling of the membranes [19]. Membranes can be fabricated to possess different morphologies. However, most membranes that have found practical use are mainly of asymmetric structure. Separation in membrane processes takes place as a result of differences in the transport rates of different species through the membrane structure, which is usually polymeric or ceramic [20].

The versatility of membrane filtration has allowed their use in many processes where their properties are suitable in the feed stream. Although membrane separation does not provide the ultimate solution to water treatment, it can be economically connected to conventional treatment technologies by modifying and improving certain properties [21].

The performance of any polymeric membrane in a given process is highly dependent on both the chemical structure of the matrix and the physical arrangement of the membrane [22]. Moreover, the structural impeccability of a membrane is very important since it determines its permeation and selectivity efficiency. As such, polymer membranes should be seen as much more than just sieving filters, but as intrinsic complex structures which can either be homogenous (isotropic) or heterogeneous (anisotropic), porous or dense, liquid or solid, organic or inorganic [22–23].

16.3.1   ISOTROPIC MEMBRANES

Isotropic membranes are typically homogeneous/uniform in composition and structure. They are divided into three subgroups, namely: microporous, dense and electrically charged membranes[20]. Isotropic microporous membranes have evenly distributed pores (Figure 16.7a) [27]. Their pore diameters range between 0.01–10 μm and operate by the sieving mechanism. The microporous membranes are mainly prepared by the phase inversion method albeit other methods can be used. Conversely, isotropic dense membranes do not have pores and as a result they tend to be thicker than the microporous membranes (Figure 16.7b). Solutes are carried through the membrane by diffusion under a pressure, concentration or electrical potential gradient. Electrically charged membranes can either be porous or nonporous. However in most cases they are finely microporous with pore walls containing charged ions (Figure 16.7c) [20, 28].

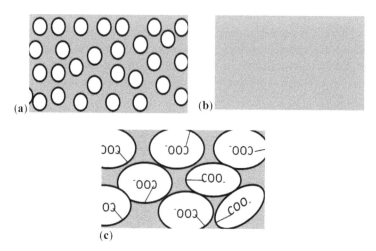

FIGURE 16.7 Schematic diagrams of isotropic membranes: (a) microporous; (b) dense; and (c) electrically charged membranes.

16.3.2 ANISOTROPIC MEMBRANES

Anisotropic membranes are often referred to as Loeb-Sourirajan, based on the scientists who first synthesized them [24–25]. They are the most widely used membranes in industries. The transport rate of a species through a membrane is inversely proportional to the membrane thickness. The membrane should be as thin as possible due to high transport rates are eligible in membrane separation processes for economic reasons. Contractual film fabrication technology limits manufacture of mechanically strong, defect-free films to thicknesses of about 20 μm. The development of novel membrane fabrication techniques to produce anisotropic membrane structures is one of the major breakthroughs of membrane technology. Anisotropic membranes consist of an extremely thin surface layer supported on a much thicker, porous substructure. The surface layer and its substructure may be formed in a single operation or separately [17]. They are represented by nonuniform structures which consist of a thin active skin layer and a highly porous support layer. The active layer enjoins the efficiency of the membrane, whereas the porous support layer influences the mechanical stability of the membrane. Anisotropic membranes can be classified into two groups, namely: (1) integrally skinned membranes where the active layer is formed from the same substance as the supporting layer and (2) composite membranes where the polymer of the active layer differs from that of the supporting sublayer [25] .In composite membranes, the layers are usually made from different polymers. The separation properties and permeation rates of the membrane are determined particularly by the surface layer and the substructure functions as a mechanical support. The advantages of the higher fluxes provided by anisotropic membranes are so great that almost all commercial processes use such membranes [17] (Figure 16.8).

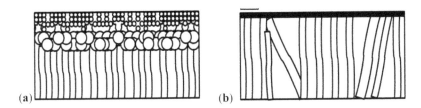

FIGURE 16.8   Schematic diagrams of anisotropic membranes: (a) Loeb-Sourirajan and (b) thin film composite membranes.

16.3.3   POROUS MEMBRANE

In Knudsen diffusion (Figure 16.9a), the pore size forces the penetrant molecules to collide more frequently with the pore wall than with other incisive species [26]. Except for some special applications as membrane reactors, Knudsen-selective membranes are not commercially attractive because of their low selectivity [27]. In surface diffusion mechanism (Figure 16.9b), the pervasive molecules adsorb on the surface of the pores so move from one site to another of lower concentration. Capillary condensation (Figure 16.9c) impresses the rate of diffusion across the membrane. It occurs when the pore size and the interactions of the penetrant with the pore walls induce penetrant condensation in the pore [28]. Molecular-sieve membranes in Figure 16.9(d) have gotten more attention because of their higher productivities and selectivity than solution-diffusion membranes. Molecular sieving membranes are means to polymeric membranes. They have ultra microporous (<7 Å) with sufficiently small pores to barricade some molecules, while allowing others to pass through. Although they have several advantages such as permeation performance, chemical and thermal stability, they are still difficult to process because of some properties like fragile. Also they are expensive to fabricate.

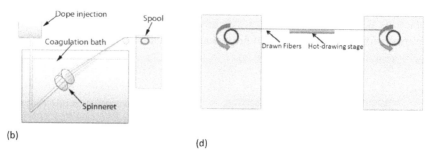

FIGURE 16.9   Schematic representation of membrane-based gas separations. (a) Knudsen-flow separation, (b) surface-diffusion, (c) capillary condensation, (d) molecular-sieving separation, and (e) solution-diffusion mechanism.

16.3.4 NONPOROUS (DENSE) MEMBRANE

Nonporous, dense membranes consist of a dense film through which permeants are transported by diffusion under the driving force of a pressure, concentration, or electrical potential gradient. The separation of various components of a mixture is related directly to their relative transport rate within the membrane, which is determined by their diffusivity and solubility in the membrane material. Thus, nonporous, dense membranes can separate permeants of similar size if the permeant concentrations in the membrane material differ substantially. Reverse osmosis membranes use dense membranes to perform the separation. Usually these membranes have an anisotropic structure to improve the flux [17].

The mechanism of separation by nonporous membranes is different from that by porous membranes. The transport through nonporous polymeric membranes is usually described by a solution—diffusion mechanism (Figure 16.9e). The most current commercial polymeric membranes operate according to the solution—diffusion mechanism. The solution—diffusion mechanism has three steps: (1) the absorption or adsorption at the upstream boundary, (2) activated diffusion through the membrane, and (3) desorption or evaporation on the other side. This solution—diffusion mechanism is driven by a difference in the thermodynamic activities existing at the upstream and downstream faces of the membrane as well as the intermolecular forces acting between the permeating molecules and those making up the membrane material.

The concentration gradient causes the diffusion in the direction of decreasing activity. Differences in the permeability in dense membranes are caused not only by diffusivity differences of the various species but also by differences in the physicochemical interactions of the species within the polymer. The solution—diffusion model assumes that the pressure within a membrane is uniform and that the chemical potential gradient across the membrane is expressed only as a concentration gradient. This mechanism controls permeation in polymeric membranes for separations.

16.4 CARBON NANOTUBES-POLYMER MEMBRANE

Iijima discovered carbon nanotubes (CNTs) in 1991 and it was really a revolution in nanoscience because of their distinguished properties. CNTs have the unique electrical properties and extremely high thermal conductivity [29–30] and high elastic modulus (>1 TPa), large elastic strain—up to 5 percent, and large breaking strain—up to 20 percent. Their excellent mechanical properties could lead to many applications [31]. For example, with their amazing strength and stiffness, plus the advantage of lightness, perspective future applications of CNTs are in aerospace engineering and virtual biodevices [32].

CNTs have been studied worldwide by scientists and engineers since their discovery, but a robust, theoretically precise and efficient prediction of the mechanical properties of CNTs has not yet been found. The problem is, when the size of an object is small to nanoscale, their many physical properties cannot be modeled and analyzed by using constitutive laws from traditional continuum theories, since the complex atomistic processes affect the results of their macroscopic behavior. Atomistic simulations can give more precise modeled results of the underlying physical properties. Due to atomistic simulations of a whole CNT are computationally infeasible at present, a new atomistic and continuum mixing modeling method is needed to solve the problem, which requires crossing the length and time scales. The research here is to develop a proper technique of spanning multiscales from atomic to macroscopic space, in which the constitutive laws are derived from empirical atomistic potentials which deal with individual interactions between single atoms at the microlevel, whereas Cosserat continuum theories are adopted for a shell model through the application of the Cauchy-Born rule to give the properties which represent the averaged behavior of large volumes of atoms at the macrolevel [33–34]. Since experiments of CNTs are relatively expensive at present, and often unexpected manual errors could be involved, it will be very helpful to have a mature theoretical method for the study of mechanical properties of CNTs. Thus, if this research is successful, it could also be a reference for the research of all sorts of research at the nanoscale, and the results can be of interest to aerospace, biomedical engineering [35].

Subsequent investigations have shown that CNTs integrate amazing rigid and tough properties, such as exceptionally high elastic properties, large elastic strain, and fracture strain sustaining capability, which seem inconsistent and impossible in the previous materials. CNTs are the strongest fibers known. The Young's Modulus of SWNT is around 1TPa, which is 5 times greater than steel (200 GPa) while the density is only $1.2 \sim 1.4$ g/cm^3. This means that materials made of nanotubes are lighter and more durable.

Beside their well-known extrahigh mechanical properties, single-walled carbon nanotubes (SWNTs) offer either metallic or semiconductor characteristics based on the chiral structure of fullerene. They possess superior thermal and electrical properties so SWNTs are regarded as the most promising reinforcement material for the next generation of high performance structural and multifunctional composites, and evoke great interest in polymer based composites research. The SWNTs/polymer composites are theoretically predicted to have both exceptional mechanical and functional properties, which carbon fibers cannot offer [36].

16.4.1 CARBON NANOTUBES

Nanotubular materials are important "building blocks" of nanotechnology, in particular, the synthesis and applications of CNTs [37–39]. One application area has been the use of carbon nanotubes for molecular separations, owing to some of their unique properties. One such important property, extremely fast mass transport of molecules within carbon nanotubes associated with their low friction inner nanotube surfaces, has been demonstrated via computational and experimental studies [40–41]. Furthermore, the behavior of adsorbate molecules in nanoconfinement is fundamentally different than in the bulk phase, which could lead to the design of new sorbents [42].

Finally, their one-dimensional geometry could allow for alignment in desirable orientations for given separation devices to optimize the mass transport. Despite possessing such attractive properties, several intrinsic limitations of carbon nanotubes inhibit their application in large scale separation processes: the high cost of CNT synthesis and membrane formation (by microfabrication processes), as well as their lack of surface functionality, which significantly limits their molecular selectivity [43]. Although outer-surface modification of carbon nanotubes has been developed for nearly two decades, interior modification via covalent chemistry is still challenging due to the low reactivity of the inner-surface. Specifically, forming covalent bonds at inner walls of carbon nanotubes requires a transformation from sp^2 to sp^3 hybridization. The formation of sp^3 carbon is energetically unfavorable for concave surfaces [44].

Membrane is a potentially effective way to apply nanotubular materials in industrial-scale molecular transport and separation processes. Polymeric membranes are already prominent for separations applications due to their low fabrication and operation costs. However, the main challenge for utilizing polymer membranes for future high-performance separations is to overcome the trad-eoff between permeability and selectivity. A combination of the potentially high throughput and selectivity of nanotube materials with the process ability and mechanical strength of polymers may allow for the fabrication of scalable, high-performance membranes [45–46].

16.4.2 *STRUCTURE OF CARBON NANOTUBES*

Two types of nanotubes exist in nature: multiwalled carbon nanotube)MWNTs(, which were discovered by Iijima in 1991 [39] and SWNTs, which were discov-ered by Bethune et al. in (1993) [47–48].

Single-wall nanotube has only one single layer with diameters in the range of 0.6–1 nm and densities of 1.33–1.40 g/cm^3[49] MWNTs are simply com-posed of concentric SWNTs with an inner diameter is from 1.5 to 15 nm and the outer diameter is from 2.5 to 30 nm [50]. SWNTs have better defined shapes of cylinder than MWNT, thus MWNTs have more possibilities of structure defects and their nanostructure is less stable. Their specific mechanical and electronic properties make them useful for future high strength/modulus materials and nanodevices. They exhibit low density, large elastic limit without breaking (of up to 20–30% strain before failure), exceptional elastic stiffness, greater than 1,000 GPa and their extreme strength which is more than twenty times higher than a high-strength steel alloy. Besides, they also posses superior thermal and elastic properties: thermal stability up to 2,800°C in vacuum and up to 750°C in air, thermal conductivity about twice as high as diamond, electric current carry-ing capacity 1,000 times higher than copper wire [51]. The properties of CNTs strongly depend on the size and the chirality and dramatically change when SWCNTs or MWCNTs are considered [52].

CNTs are formed from pure carbon bonds. Pure carbons only have two cova-lent bonds: sp^2 and sp^3. The former constitutes graphite and the latter constitutes diamond. The sp^2 hybridization, composed of one s orbital and two p orbitals, is a strong bond within a plane but weak between planes. When more bonds come together, they form six-fold structures, like honeycomb pattern, which is a plane structure, the same structure as graphite [53].

Graphite is stacked layer by layer so it is only stable for one single sheet. Wrapping these layers into cylinders and joining the edges, a tube of graphite is formed, called nanotube [54].

Atomic structure of nanotubes can be described in terms of tube chirality, or helicity, which is defined by the chiral vector, and the chiral angle, θ. Figure 16.10 shows visualized cutting a graphite sheet along the dotted lines and rolling the tube so that the tip of the chiral vector touches its tail. The chiral vector, often known as the roll-up vector, can be described by the following equation [55]:

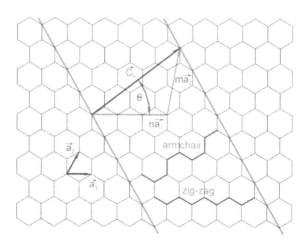

FIGURE 16.10 Schematic diagram showing how graphite sheet is "rolled" to form CNT.

$$C_h = na_1 + ma_2 \qquad (16.2)$$

As shown in Figure 16.10, the integers (n, m) are the number of steps along the carbon bonds of the hexagonal lattice. Chiral angle determines the amount of "twist" in the tube. Two limiting cases exist where the chiral angle is at 0° and 30°. These limiting cases are referred to as ziz-zag (0°) and armchair (30°), based on the geometry of the carbon bonds around the circumference of the nanotube. The difference in armchair and zig-zag nanotube structures is shown in Figure 16.11 In terms of the roll-up vector, the ziz-zag nanotube is (n, 0) and the armchair nanotube is (n, n). The roll-up vector of the nanotube also defines the nanotube diameter since the interatomic spacing of the carbon atoms is known [36].

Chiral vector C_h is a vector that maps an atom of one end of the tube to the other. C_h can be an integer multiple a_1 of a_2, which are two basis vectors of the graphite cell. Then we have $C_h = a_1 + a_2$, with integer n and m , and the constructed CNT is called a (n, m) CNT, as shown in Figure 16.12. It can be proved that for armchair CNTs n = m, and for zigzag CNTs m = 0. In Figure 16.12, the structure is designed to be a (4.0) zigzag SWCNT.

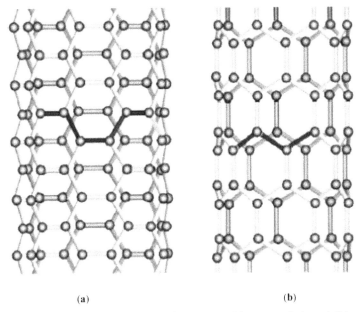

(a) **(b)**

FIGURE 16.11 Illustrations of the atomic structure (a) an armchair and (b) a ziz-zag nanotube.

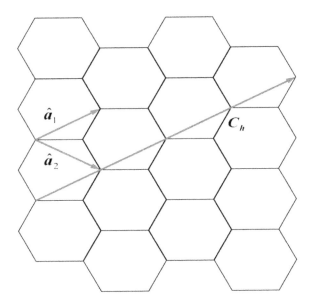

FIGURE 16.12 Basis vectors and chiral vector.

MWCNT can be considered as the structure of a bundle of concentric SW-CNTs with different diameters. The length and diameter of MWCNTs are different from those of SWCNTs, which means, their properties differ significantly. MWCNTs can be modeled as a collection of SWCNTs, provided the interlayer interactions are modeled by Van der Waals forces in the simulation. A SWCNT can be modeled as a hollow cylinder by rolling a graphite sheet as presented in Figure 16.13.

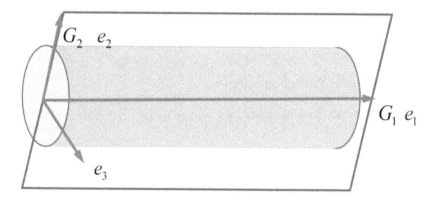

FIGURE 16.13   Illustration of a graphite sheet rolling to SWCNT.

If a planar graphite sheet is considered to be an undeformed configuration, and the SWCNT is defined as the current configuration, then the relationship between the SWCNT and the graphite sheet can be shown to be:

$$e_1 = G_1 \quad , \quad e_2 = R\sin\frac{G_2}{R} \quad , \quad e_3 = R\cos\frac{G_2}{R} - R \qquad (16.3)$$

The relationship between the integer's n, m and the radius of SWCNT is given by:

$$R = a\sqrt{m^2 + mn + n^2} \,/\, 2\pi \qquad (16.4)$$

where a = $\sqrt{3}a_0$, and a_0 is the length of a nonstretched C-C bond which is 0.142 nm [56].

As a graphite sheet can be "rolled" into a SWCNT, we can "unroll" the SWCNT to a plane graphite sheet. Since a SWCNT can be considered as a rectangular strip of hexagonal graphite monolayer rolling up to a cylindrical tube, the general idea is that it can be modeled as a cylindrical shell, a cylinder surface, or it can pull-back to be modeled as a plane sheet deforming into curved surface in three-dimensional space. A MWCNT can be modeled as a combination of a series of concentric SWCNTs with interlayer intera-atomic reactions. Provided the continuum shell theory captures the deformation at the macrolevel, the inner microstructure can be described by finding the appropriate form of the potential function which is related to the position of the atoms at the atomistic level. Therefore, the SWCNT can be considered as a generalized continuum with microstructure [35].

16.4.3 CNT COMPOSITES

CNT composite materials cause significant development in nanoscience and nanotechnology. Their remarkable properties offer the potential for fabricating composites with substantially enhanced physical properties including conductivity, strength, elasticity, and toughness. Effective utilization of CNT in composite applications is dependent on the homogeneous distribution of CNTs throughout the matrix. Polymer-based nanocomposites are being developed for electronics applications such as thin-film capacitors in integrated circuits and solid polymer electrolytes for batteries. Research is being conducted throughout the world targeting the application of carbon nanotubes as materials for use in transistors, fuel cells, big TV screens, ultrasensitive sensors, high-resolution Atomic Force Microscopy (AFM) probes, supercapacitor, transparent conducting film, drug carrier, catalysts, and composite material. Nowadays, there are more reports on the fluid transport through porous CNTs/polymer membrane.

16.4.4 STRUCTURAL DEVELOPMENT IN POLYMER/CNT FIBERS

The inherent properties of CNT assume that the structure is well preserved (large-aspect-ratio and without defects). The first step toward effective reinforcement of polymers using nanofillers is to achieve a uniform dispersion of the fillers within the hosting matrix, and this is also related to the as-synthesized nanocarbon structure. Secondly, effective interfacial interaction and stress transfer between CNT and polymer is essential for improved mechanical properties of the fiber composite. Finally, similar to polymer molecules, the excellent intrinsic mechanical properties of CNT can be fully exploited only if an ideal uniaxial orientation is achieved. Therefore, during the fabrication of polymer/CNT fibers, four key areas need to be addressed and understood in order to successfully control the microstructural development in these composites. These are: (i) CNT pristine structure, (ii) CNT dispersion, (iii) polymer—CNT interfacial interaction and (iv) orientation of the filler and matrix molecules (Figure 16.14).

Figure 16.14 Four major factors affecting the microstructural development in polymer/CNT composite fiber during processing [57].

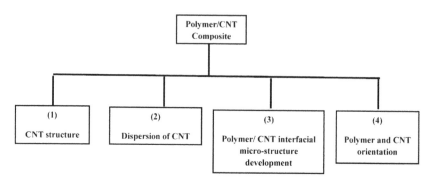

FIGURE 16.14   Four major factors affecting the microstructural development in polymer/ CNT composite fiber during processing.

Achieving homogenous dispersion of CNTs in the polymer matrix through strong interfacial interactions is crucial to the successful development of CNT/ polymer nanocomposite [58]. As a result, various chemical or physical modifications can be applied to CNTs to improve its dispersion and compatibility with polymer matrix. Among these approaches acid treatment is considered most convenient, in which hydroxyl and carboxyl groups generated would concentrate on the ends of the CNT and at defect sites, making them more reactive and thus better dispersed [59–60].

The incorporation of functionalized CNTs into composite membranes are mostly carried out on flat sheet membranes [61–62]. For considering the potential influences of CNTs on the physicochemical properties of dope solution [63] and change of membrane formation route originated from various additives [64], it is necessary to study the effects of CNTs on the morphology and performance.

16.4.5   GENERAL FABRICATION PROCEDURES FOR POLYMER/ CNT FIBERS

In general, when discussing polymer/CNT composites, two major classes come to mind. First, the CNT nanofillers are dispersed within a polymer at a specified concentration, and the entire mixture is fabricated into a composite. Secondly, as grown CNT are processed into fibers or films, and this macroscopic CNT material is then embedded into a polymer matrix [65]. The four major fiber-spinning methods (Figure 16.15) used for polymer/CNT composites from both the solution and melt include dry-spinning [66], wet-spinning [67], dry-jet wet spinning (gel-spinning), and electrospinning [68]. An ancient solid-state spinning approach has been used for fabricating 100 percent CNT fibers from both forests and aero gels. Irrespective of the processing technique, in order to develop high-quality fibers many parameters need to be well controlled.

All spinning procedures generally involve:

(i) Fiber formation, (ii) coagulation/gelation/solidification, and (iii) drawing/alignment.

For all of these processes, the even dispersion of the CNT within the polymer solution or melt is very important. However, in terms of achieving excellent axial mechanical properties, alignment and orientation of the polymer chains and the CNT in the composite is necessary. Fiber alignment is accomplished in postprocessing such as drawing/annealing and is key to increasing crystallinity, tensile strength, and stiffness [69].

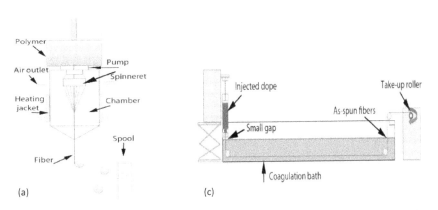

(a)   (c)

FIGURE 16.15   Schematics representing the various fiber processing methods (a) dry-spinning; (b) wet-spinning; (c) dry-jet wet or gel-spinning; and (d) postprocessing by hot-stage drawing.

16.5   COMPUTATIONAL METHODS

Computational approaches to obtain solubility and diffusion coefficients of small molecules in polymers have focused primarily upon equilibrium molecular dynamics (MD) and Monte Carlo (MC) methods. These have been thoroughly reviewed by several investigators [70–71].

Computational approach can play an important role in the development of the CNT-based composites by providing simulation results to help on the understanding, analysis and design of such nanocomposites. At the nanoscale, analytical models are difficult to establish or too complicated to solve, and tests are extremely difficult and expensive to conduct. Modeling and simulations of nanocomposites, on the other hand, can be achieved readily and cost effectively on even a desktop computer. Characterizing the mechanical properties of CNT-based composites is just one of the many important and urgent tasks that simulations can follow out [72].

Computer simulations on model systems have in recent years provided much valuable information on the thermodynamic, structural and transport properties of classical dense fluids. The success of these methods rests primarily on the fact that a model containing a relatively small number of particles is in general found to be sufficient to simulate the behavior of a macroscopic system. Two distinct techniques of computer simulation have been developed which are known as the method of molecular dynamics and the Monte Carlo method [73–75].

Instead of adopting a trial—and—error approach to membrane development, it is far more efficient to have a real understanding of the separation phenomena to guide membrane design [76–79]. Similarly, methods such as MC, MD and other computational techniques have improved the understanding of the relationships between membrane characteristics and separation properties. In addition to these inputs, it is also beneficial to have simple models and theories that give an overall insight into separation performance [80–83].

16.5.1   PERMEANCE AND SELECTIVITY OF SEPARATION MEMBRANES

A membrane separates one component from another on the basis of size, shape or chemical affinity. Two characteristics dictate membrane performance, permeability, that is the flux of the membrane, and selectivity or the membrane's preference to pass one species and not another [84].

A membrane can be defined as a selective barrier between two phases, the "selective" being inherent to a membrane or a membrane processes. The membrane separation technology is proving to be one of the most significant unit operations. The technology inherits certain advantages over other methods. These advantages include compactness and light weight, low labor intensity, modular design that allows for easy expansion or operation at partial capacity, low maintenance, low energy requirements, low cost, and environmentally friendly operations. A schematic representation of a simple separation membrane process is shown in Figure 16.16.

A feed stream of mixed components enters a membrane unit where it is separated into a retentate and permeate stream. The retentate stream is typically the purified product stream and the permeate stream contains the waste component.

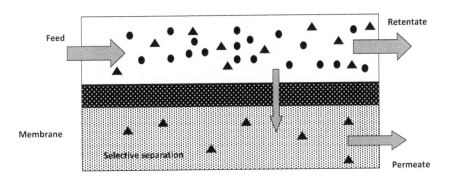

FIGURE 16.16   Schematic of membrane separation.

A quantitative measure of transport is the flux (or permeation rate), which is defined as the number of molecules that pass through a unit area per unit time [85]. It is believed that this molecular flux follows Fick's first law. The flux is proportional to the concentration gradient through the membrane. There is a movement from regions of high concentration to regions of low concentration, which may be expressed in the form:

$$J = -D\frac{dc}{dx}$$ (16.5)

By assuming a linear concentration gradient across the membrane, the flux can be approximated as:

$$J = -D\frac{C_2 - C_1}{L}$$ (16.6)

Where $C_1 = c(0)$ and $C_2 = c(L)$ are the downstream and upstream concentrations (corresponding to the pressures p_1 and p_2 via sorption isotherm $c(p)$, respectively, and L is the membrane thickness, as labeled in Figure 16.17.

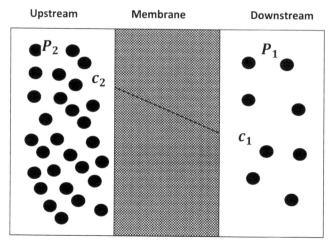

FIGURE 16.17 Separation membrane with a constant concentration gradient across membrane thickness L.

The membrane performance of various materials is commonly compared using the thickness, independent material property, the permeability, which is related to the flux as:

$$P = \frac{JL}{P_2 - P_1} = \left(\frac{C_2 - C_1}{P_2 - P_1} \right) D \tag{16.7}$$

In the case where the upstream pressure is much greater than the downstream pressure ($p_2 \gg p_1$ and $C_2 \gg C_1$) the permeability can be simplified so:

$$P = \frac{C_2}{P_2} D \tag{16.8}$$

The permeability is more commonly used to describe the performance of a membrane than flux. This is because the permeability of a homogenous stable membrane material is constant regardless of the pressure differential or membrane thickness and hence it is easier to compare membranes made from different materials.

By introducing a solubility coefficient, the ratio of concentration over pressure C_2/p_2, when sorption isotherm can be represented by the Henry's law, the permeability coefficient may be expressed simply as:

$$P = SD \tag{16.9}$$

This form is useful as it facilitates the understanding of this physical property by representing it in terms of two components:

Solubility which is an equilibrium component describing the concentration of gas molecules within the membrane, that is the driving force, and Diffusivity, which is a dynamic component describing the mobility of the gas molecules within the membrane.

The separation of a mixture of molecules A and B is characterized by the selectivity or ideal separation factor $\alpha_{A/B} = P(A)/P(B)$, the ratio of permeability of the molecule A over the permeability of the molecule B. According to Eq. (16.9), it is possible to make separations by diffusivity selectivity $D(A)/D(B)$ or solubility selectivity $S(A)/S(B)$ [85–86]. This formalism is known in membrane science as the solution—diffusion mechanism. Since the limiting stage of the mass transfer is overcoming of the diffusion energy barrier, this mechanism implies the activated diffusion. Because of this, the temperature dependences of the diffusion coefficients and permeability coefficients are described by the Arrhenius equations.

Gas molecules that encounter geometric constrictions experience an energy barrier such that sufficient kinetic energy of the diffusing molecule or the groups that form this barrier, in the membrane is required in order to overcome the barrier and make a successful diffusive jump. The common form of the Arrhenius dependence for the diffusion coefficient can be expressed as:

$$D_A = D_A^* \exp(-\Delta E_a / RT) \tag{16.10}$$

For the solubility coefficient the Van't Hoff equation holds:

$$S_A = S_A^* \exp(-\Delta H_a / RT) \tag{16.11}$$

Where $\Delta H_a < 0$ is the enthalpy of sorption. From Eq. (16.9), it can be written:

$$P_A = P_A^* \exp(-\Delta E_P / RT) \tag{16.12}$$

Where $\Delta E_p = \Delta E_a + \Delta H_a$ are known to diffuse within nonporous or porous membranes according to various transport mechanisms. Table 16.3 illustrates the mechanism of transport depending on the size of pores. For very narrow pores, size sieving mechanism is realized that can be considered as a case of activated diffusion. This mechanism of diffusion is most common in the case of extensively studied nonporous polymeric membranes. For wider pores, the surface diffusion (also an activated diffusion process) and the Knudsen diffusion are observed [87–89].

TABLE 16.3 Transport mechanisms

Mechanism	Schamatic	Process
Activated diffusion		Constriction energy barrier ΔE_a
Surface diffusion		Adsorption—site energy barrier ΔE_s
Knudsen diffusion		Direction and velocity $\bar{d}$ $\bar{u}$

Sorption does not necessarily follow Henry's law. For a glassy polymer an assumption is made that there are small cavities in the polymer and the sorption at the cavities follows Langmuir's law. Then, the concentration in the membrane is given as the sum of Henry's law adsorption and Langmuir's law adsorption

$$C = K_P P + \frac{C_h^* b_P}{1 + b_P} \tag{16.13}$$

It should be noted that the applicability of solution (sorption)-diffusion model has nothing to do with the presence or absence of the pore.

16.5.2 DIFFUSIVITY

The diffusivity through membranes can be calculated using the time-lag method [90]. A plot of the flow through the membrane versus time reveals an initial transient permeation followed by steady state permeation. Extending the linear section of the plot back to the intersection of the x-axis gives the value of the time-lag (θ) as shown in Figure 16.18.

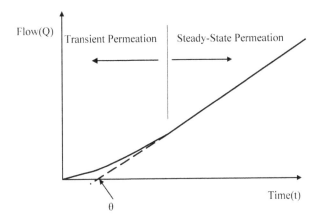

FIGURE 16.18 Calculation of the diffusion coefficient using the time-lag method, once the gradient is constant and steady state flow through the membrane has been reached, a extrapolation of the steady state flow line back to the x-axis where the flow is 0 reveals the value of the time lag (θ).

The time lag relates to the time it takes for the first molecules to travel through the membrane and is thus related to the diffusivity. The diffusion coefficient can be calculated from the time-lag and the membrane thickness as shown in Eq. (16.14) [91–92].

$$D = \frac{\Delta x}{6\theta} \qquad (16.14)$$

Surface diffusion is the diffusion mechanism which dominates in the pore size region between activation diffusion and Knudsen diffusion [93].

16.5.3 SURFACE DIFFUSION

A model that well described the surface diffusion on the pore walls was proposed many years ago. It was shown to be consistent with transport parameters in porous polymeric membranes. When the pore size decreased below a certain level, which depends on both membrane material and the permeability coefficient exceeds the value for free molecular flow (Knudsen diffusion), especially in the case of organic vapors. Note that surface diffusion usually occurs simultaneously with Knudsen diffusion but it is the dominant mechanism within a certain pore size. Since surface diffusion is also a form of activated diffusion, the energy barrier is the energy required for the molecule to jump from one adsorption site to another across the surface of the pore. By allowing the energy barrier to be proportionate to the enthalpy of adsorption, Gilliland *et al*, established an equation for the surface diffusion coefficient expressed here as :[94]

$$D_S = D_S^* \exp(\frac{-aq}{RT})$$ (16.15)

Where is a preexponential factor depending on the frequency of vibration of the adsorbed molecule normal to the surface and the distance from one adsorption site to the next. The quantity the heat of adsorption is $(q > 0)$ and a proportionality constant is $(0 < a < 1)$. The energy barrier separates the adjacent adsorption sites. An important observation is that more strongly adsorbed molecules are less mobile than weakly adsorbed molecules [95].

In the case of surface diffusion, the concentration is well described by Henry's law $c = Kp$, where K is $K = K_0 \exp(q/RT)$ [95–96]. Since solubility is the ratio of the equilibrium concentration over pressure, the solubility is equivalent to the Henry's law coefficient.

$$P_S = P_S^* \exp\left(\frac{(1-a)q}{RT}\right)$$ (16.16)

Which implies the solubility is a decreasing function of temperature. The product of diffusivity and solubility gives:

$$D_K = \frac{d}{3\tau}\bar{u}$$ (16.17)

Since $0 < a < 1$ the total permeability will decrease with increased temperature meaning that any increase in the diffusivity is counteracted by a decrease in surface concentration [95].

16.5.4 KNUDSEN DIFFUSION

Knudsen diffusion [95, 97–99] depending on pressure and mean free path which applies to pores between 10 Å and 500 Å in size [100]. In this region, the mean free path of molecules is much larger than the pore diameter. It is common to use Knudsen number $K_n = \lambda/d$ to characterize the regime of permeation through pores. When $K_n \ll 1$, viscous (Poiseuille) flow is realized. The condition for Knudsen diffusion is $K_n \gg 1$. An intermediate regime is realized when $K_n \approx 1$. The Knudsen diffusion coefficient can be expressed in the following form:

$$D_K = \frac{d}{3\tau}\bar{u}$$ (16.18)

This expression shows that the separation outcome should depend on the differences in molecular speed (or molecular mass). The average molecular speed is calculated using the Maxwell speed distribution as:

$$\bar{u} = \sqrt{\frac{8RT}{\pi m}} \tag{16.19}$$

And the diffusion coefficient can be presented as:

$$D_K = \frac{d}{3\tau}\bar{u} \tag{16.20}$$

For the flux in the Knudsen regime the following equation holds [101–102]:

$$J = n\pi d^2 \Delta p D_K / 4RTL \tag{16.21}$$

After substituting Eq. (16.20) into Eq. (16.21), one has the following expressions for the flux and permeability coefficient is:

$$J = \left(\frac{n\pi^{\frac{1}{2}} d^3 \Delta p}{6\tau L} \right) (\frac{2}{mRT})^{1/2} \tag{16.22}$$

$$P = \left(\frac{n\pi^{\frac{1}{2}} d^3}{6\tau} \right) (\frac{2}{mRT})^{1/2} \tag{16.23}$$

Two important conclusions can be made from analysis of Eqs. (16.22) and (16.23). First, selectivity of separation in Knudsen regime is characterized by the ratio $\alpha_{ij} = (M_j/M_i)^{1/2}$. It means that membranes where Knudsen diffusion predominates are poorly selective.

The most common approach to obtain diffusion coefficients is equilibrium molecular dynamics. The diffusion coefficient that is obtained is a self-diffusion coefficient. Transport-related diffusion coefficients are less frequently studied by simulation but several approaches using nonequilibrium MD (NEMD) simulation can be used.

16.5.5 MOLECULAR DYNAMICS SIMULATIONS

Conducting experiments for material characterization of the nanocomposites is a very time consuming, expensive and difficult. Many researchers are now concentrating on developing both analytical and computational simulations. Molecular dynamics (MD) simulations are widely being used in modeling and solving problems based on quantum mechanics. Using Molecular dynamics it is possible to study the reactions, load transfer between atoms and molecules. If the objective of the simulation is to study the overall behavior of CNT-based composites and structures, such as deformations, load and heat transfer mechanisms then the continuum mechanics approach can be applied safely to study the problem effectively [103].

MD tracks the temporal evolution of a microscopic model system by integrating the equations of motion for all microscopic degrees of freedom. Numerical integration algorithms for initial value problems are used for this purpose, and their strengths and weaknesses have been discussed in simulation texts [104–106].

MD is a computational technique in which a time evolution of a set of interacting atoms is followed by integrating their equations of motion. The forces between atoms are due to the interactions with the other atoms. A trajectory is calculated in a 6-N dimensional phase space (three position and three momentum components for each of the N atoms). Typical MD simulations of CNT composites are performed on molecular systems containing up to tens of thousands of atoms and for simulation times up to nanoseconds. The physical quantities of the system are represented by averages over configurations distributed according to the chosen statistical ensemble. A trajectory obtained with MD provides such a set of configurations. Therefore the computation of a physical quantity is obtained as an arithmetic average of the instantaneous values. Statistical mechanics is the link between the nanometer behavior and thermodynamics. Thus the atomic system is expected to behave differently for different pressures and temperatures [107].

The interactions of the particular atom types are described by the total potential energy of the system, U, as a function of the positions of the individual atoms at a particular instant in time\

$$U = U\left(X_i, \ldots, X_n\right) \tag{16.24}$$

where X_i represents the coordinates of atom i in a system of N atoms. The potential equation is invariant to the coordinate transformations, and is expressed in terms of the relative positions of the atoms with respect to each other, rather than from absolute coordinates [107].

MD is readily applicable to a wide range of models, with and without constraints. It has been extended from the original microcanonical ensemble formulation to a variety of statistical mechanical ensembles. It is flexible and valuable for extracting dynamical information. The Achilles' heel of MD is its high demand of computer time, as a result of which the longest times that can be simulated with MD fall short of the longest relaxation times of most real-life macromolecular systems by several orders of magnitude. This has two important consequences. (a) Equilibrating an atomistic model polymer system with MD alone is problematic; if one starts from an improbable configuration, the simulation will not have the time to depart significantly from that configuration and visit the regions of phase space that contribute most significantly to the properties. (b) Dynamical processes with characteristic times longer than approximately 10^{-7} s cannot be probed directly; the relevant correlation functions do not decay to zero within the simulation time and thus their long-time tails are inaccessible, unless some extrapolation is invoked based on their short-time behavior.

Recently, rigorous multiple time step algorithms have been invented, which can significantly augment the ratio of simulated time to CPU time. Such an algorithm is the reversible Reference System Propagator Algorithm (rRESPA) [108–109]. This algorithm invokes a Trotter factorization of the Liouville operator in the numerical integration of the equations of motion: fast-varying (e.g., bond stretching and bond angle bending) forces are updated with a short time step Δt, while slowly varying forces (e.g., nonbonded interactions, which are typically expensive to calculate, are updated with a longer time step Δt. Using $\delta t = 1\,fs$ and $\Delta t = 5\,ps$, one can simulate 300 ns of real time of a polyethylene melt on a modest workstation [110]. This is sufficient for the full relaxation of a system of C_{250} chains, but not of longer-chain systems.

A paper of Furukawa and Nitta is cited first to understand the NEMD simulation semiquantitatively, since, even though the paper deals with various pore shapes, complicated simulation procedure is described clearly.

MD simulation is more preferable to study the nonequilibrium transport properties. Recently some NEMD methods have also been developed, such as the grand canonical molecular dynamics (GCMD) method [111–112] and the dual control volume GCMD technique (DCV- GCMD) [113–114]. These methods provide a valuable clue to insight into the transport and separation of fluids through a porous medium. The GCMD method has recently been used to investigate pressure-driven and chemical potential-driven gas transport through porous inorganic membrane [115].

16.5.5.1   EQUILIBRIUM MD SIMULATION

A self-diffusion coefficient can be obtained from the mean-square displacement (MSD) of one molecule by means of the Einstein equation in the form [115]:

$$D_A^* = \frac{1}{6N_\alpha} \lim_{t \to \infty} \frac{d}{dt} \left(r_i(t) - r_i(0) \right)^2 \tag{16.25}$$

Where Na is the number of molecules, $r_1(t)$ and $r_1(0)$ are the initial and final (at time t) positions of the center of mass of one molecule i over the time interval t, and $(r_1(t) - r_1(0))^2$ is MSD averaged over the ensemble. The Einstein relationship assumes a random walk for the diffusing species. For slow diffusing species, anomalous diffusion is sometimes observed and is characterized by:

$$\left(r_i(t) - r_i(0)\right)^2 \propto t^n \tag{16.26}$$

Where $n < 1$ ($n = 1$ for the Einstein diffusion regime). At very short times ($t < 1$ ps), the MSD may be quadratic iv n time ($n = 2$) which is characteristic of "free flight" as may occur in a pore or solvent cage prior to collision with the pore or cage wall. The result of anomalous diffusion, which may or may not occur in intermediate time scales, is to create a smaller slope at short times, resulting in a larger value for the diffusion coefficient. At sufficiently long times (the hydrodynamic limit), a transition from anomalous to Einstein diffusion ($n = 1$) may be observed [71].

An alternative approach to MSD analysis makes use of the center-of-mass velocity autocorrelation function (VACF) or Green–Kubo relation, given as follows [116]

$$D = \frac{1}{3} \int (v_i(t).v_i(0)) dt \tag{16.27}$$

Concentration in the simulation cell is extremely low and its diffusion coefficient is an order of magnitude larger than that of the polymeric segments. Under these circumstances, the self diffusion and mutual diffusion coefficients of the penetrant are approximately equal, as related by the Darken equation in the following form:

$$D_{AB} = (D_A^* x_B + D_B^* x_A)\left(\frac{d \ln f_A}{d \ln c_A}\right) \tag{16.28}$$

In the limit of low concentration of diffusion $x_A \approx 0$, Eq. (16.28) reduces to:

$$D_A^* \equiv D_{AB} \tag{16.29}$$

16.5.5.2 NONEQUILIBRIUM MD SIMULATION

Experimental diffusion coefficients, as obtained from time-lag measurements, report a transport diffusion coefficient which cannot be obtained from equilibrium MD simulation. Comparisons made in the simulation literature are typically between time-lag diffusion coefficients (even calculated for glassy polymers without correction for dual-mode contributions and self-diffusion coefficients.

As discussed above, mutual diffusion coefficients can be obtained directly from equilibrium MD simulation but simulation of transport diffusion coefficients require the use of NEMD methods, that are less commonly available and more computationally expensive [117].

For these reasons, they have not been frequently used. One successful approach is to simulate a chemical potential gradient and combine MD with GCMC methods (GCMC–MD), as developed by Heffelfinger and coworkers [114] and MacElroy [118]. This approach has been used to simulate permeation of a variety of small molecules through nanoporous carbon membranes, carbon nanotubes, porous silica and self-assembled monolayers [119–121]. A diffusion coefficient then can be obtained from the relation:

$$D = \frac{KT}{F}(V) \tag{16.30}$$

16.5.6   GRAND CANONICAL MONTE CARLO SIMULATION

A standard Grand Canonical Monte Carlo (GCMC) simulation is employed in the equilibrium study, while MD simulation is more preferable to study the non-equilibrium transport properties [104].

Monte Carlo method is formally defined by the following quote as: Numerical methods that are known as Monte Carlo methods can be loosely described as statistical simulation methods, where statistical simulation is defined in quite general terms to be any method that utilizes sequences of random numbers to perform the simulation [122].

The name "Monte Carlo" was chosen because of the extensive use of random numbers in the calculations [104]. One of the better known applications of Monte Carlo simulations consists of the evaluation of integrals by generating suitable random numbers that will fall within the area of integration. A simple example of how a MC simulation method is applied to evaluate the value of π is illustrated in Figure 16.19. By considering a square that inscribes a circle of a diameter R, one can deduce that the area of the square is R^2, and the circle has an area of $\pi R^2/4$. Thus, the relative area of the circle and the square will be $\pi/4$. A large number of two independent random numbers (with x and y coordinates) of trial shots is generated within the square to determine whether each of them falls inside of the circle or not. After thousands or millions of trial shots, the computer program keeps counting the total number of trial shots inside the square and the number of shots landing inside the circle. Finally, the value of $\pi/4$ can be approximated based on the ratio of the number of shots that fall inside the circle to the total number of trial shots.

As stated earlier, the value of an integral can be calculated via MC methods by generating a large number of random points in the domain of that integral. Equation (16.31) shows a definite integral:

$$F = \int_a^b f(x)\, dx$$

(16.31)

Where $f(x)$ is a continuous and real-valued function in the interval $[a, b]$. The integral can be rewritten as [104]:

$$F = \int_a^b dx \left(\frac{f(x)}{\rho(x)} \right) \rho(x) \cong \frac{f(\xi_i)}{\rho(\xi_i)} \bigg|_\tau$$

(16.32)

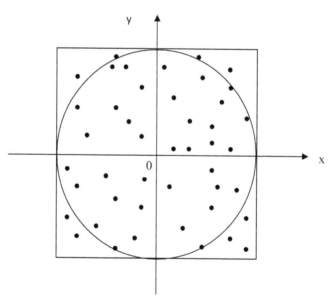

FIGURE 16.19 Illustration of the application of the Monte Carlo simulation method for the calculation of the value of π by generating a number of trial shots, in which the ratio of the number of shots inside the circle to the total number of trial shots will approximately approach the ratio of the area of the circle to the area of the square.

If the probability function is chosen to be a continuous uniform distribution, then:

$$\rho(x) = \frac{1}{(b-a)} \qquad a \le x \le b$$

(16.33)

Subsequently, the integral, F, can be approximated as:

$$F \approx \frac{(b-a)}{\tau} \sum_{i=1}^{\tau} f(\xi_i) \tag{16.34}$$

In a similar way to the MC integration methods, MC molecular simulation methods rely on the fact that a physical system can be defined to possess a definite energy distribution function, which can be used to calculate thermodynamic properties.

The applications of MC are diverse such as Nuclear reactor simulation, Quantum chromo dynamics, Radiation cancer therapy, Traffic flow, Stellar evolution, Econometrics, Dow Jones forecasting, Oil well exploration, VSLI design [122]

The MC procedure requires the generation of a series of configurations of the particles of the model in a way which ensures that the configurations are distributed in phase space according to some prescribed probability density.

The mean value of any configurational property determined from a sufficiently large number of configurations provides an estimate of the ensemble-average value of that quantity ; the nature of the ensemble average depends upon the chosen probability density.

These machine calculations provide what is essentially exact information on the consequences of a given intermolecular force law. Application has been made to hard spheres and hard disks, to particles interacting through a Lennard-Jones 12–6 potential function and other continuous potentials of interest in the study of simple fluids, and to systems of charged particles [123].

The MC technique is a stochastic simulation method designed to generate a long sequence, or "Markov chain" of configurations that asymptotically sample the probability density of an equilibrium ensemble of statistical mechanics [105, 116]. For example, a MC simulation in the canonical (NVT) ensemble, carried out under the macroscopic constraints of a prescribed number of molecules N, total volume V and temperature T, samples configurations r_P with probability proportional to $\exp[-\beta v(r_P)]$, with $\beta = 1/(k_B T)$, k_B being the Boltzmann constant and T the absolute temperature. Thermodynamic properties are computed as averages over all sampled configurations.

The efficiency of a MC algorithm depends on the elementary moves it employs to go from one configuration to the next in the sequence. An attempted move typically involves changing a small number of degrees of freedom; it is accepted or rejected according to selection criteria designed so that the sequence ultimately conforms to the probability distribution of interest. In addition to usual moves of molecule translation and rotation practiced for small-molecule fluids, special moves have been invented for polymers. The reptation (slithering snake) move for polymer chains involves deleting a terminal segment on one end of the chain and appending a terminal segment on the other end, with the newly created torsion angle being assigned a randomly chosen value [124].

In most MC algorithms the overall probability of transition from some state (configuration) m to some other state n, as dictated by both the attempt and the selection stages of the moves, equals the overall probability of transition from n to m; this is the principle of detailed balance or "microscopic reversibility." The probability of attempting a move from state m to state n may or may not be equal to that of attempting the inverse move from state n to state m. These probabilities of attempt are typically unequal in "bias" MC algorithms, which incorporate information about the system energetics in attempting moves. In bias MC, detailed balance is ensured by appropriate design of the selection criterion, which must remove the bias inherent in the attempt [105, 116].

16.5.7 MEMBRANE MODEL AND SIMULATION BOX

the MD simulations [125] can be applied for the permeation of pure and mixed gases across carbon membranes with three different pore shapes: the diamond pore (DP), zigzag path (ZP) and straight path (SP), each composed of micrographite crystalline. Three different pore shapes can be considered: DP, ZP, and SP.

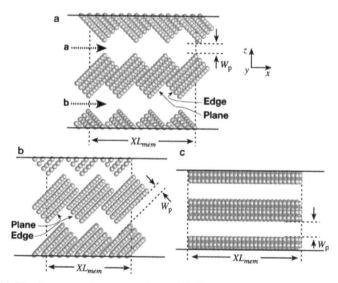

FIGURE 16.20   Three membrane pore shapes; (a) diamond path (DP), (b) zigzag path (ZP), (c) straight path (SP).

Figures 16.20(a), (b), and (c) shows the cross-sectional view of each pore shape. DP (A) has two different pore mouths; one a large (pore a) and the other a small mouth (pore b). ZP (B) has zigzag shaped pores whose sizes (diameters) are all the same at the pore entry. SP (C) has straight pores which can be called slit-shaped pores.

In a simulation system, we investigate the equilibrium selective adsorption and nonequilibrium transport and separation of gas mixture in the nanoporous carbon membrane are modeled as slits from the layer structure of graphite. A schematic representation of the system used in our simulations is shown in Figures 16.21(a) and (b), in which the origin of the coordinates is at the center of simulation box and transport takes place along the x-direction in the nonequilibrium simulations. In the equilibrium simulations, the box as shown in Figure 16.22(a) is employed, whose size is set as 85.20 nm × 4.92 nm × (1.675 + W) nm in x-, y-, and z-directions, respectively, where W is the pore width, that is, the separation distance between the centers of carbon atoms on the two layers forming a slit pore (Figure 16.21). L_{cc} is the separation distance between two centers of adjacent carbon atom; L_m is the pore length; W is the pore width, Δ is the separation distance between two carbon atom centers of two adjacent layers [126].

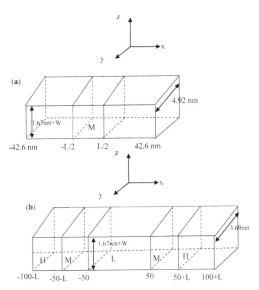

FIGURE 16.21 Schematic representation of the simulation boxes. The H-, L- and M-areas correspond to the high and low chemical potential control volumes, and membrane, respectively. Transport takes place along the x-direction in the nonequilibrium simulations. (a) Equilibrium adsorption simulations and (b) nonequilibrium transport simulations. L is the membrane thickness and W is the pore width.

The simulation box is divided into three regions where the chemical potential for each component is the same. The middle region (M-region) represents the membrane with slit pores in which the distances between the two adjacent carbon atoms (Lcc) and two adjacent graphite basal planes (Δ).

Period boundary conditions are employed in all three directions. In the non-equilibrium molecular dynamics simulations in order to use period boundary conditions in three directions, we have to divide the system into five regions as shown in Figure 16.22(b).

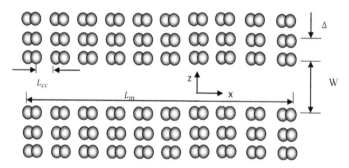

FIGURE 16.22   Schematic representation of slit pore.

Each symmetric box has three regions. Two are density control; H-region (high density) and L-region (low density) and one is free of control M-region which is placed between the H- and L- region. For each simulation, the density in the H-region, ρ_H, is maintained to be that of the feed gas and the density in the L-region is maintained at zero, corresponding to the vacuum.

The difference in the gas density between the H- and L- region is the driving force for the gas permeation through the M-region which represents the membrane.

The transition and rotational velocities are given to each inserted molecules randomly based on the Gaussian distribution around an average velocity corresponding to the specified temperature.

Molecules spontaneously move from H- to L-region via leap-frog algorithm and a nonequilibrium steady state is obtained at the M-region. During a simulation run, equilibrium with the bulk mass at the feed side at the specified pressure and temperature is maintained at the H-region by carrying out GCMC creations and destructions in terms of the usual acceptance criteria [28]. Molecules entered the L-region were moved out immediately to keep vacuum. The velocities of newly inserted molecules were set to certain values in terms of the specified temperature by use of random numbers on the Gaussian distribution.

16.6 CONCLUDING REMARKS

The concept of membrane processes is relatively simple but nevertheless often unknown. Membrane separation processes can be used for a wide range of applications. The separation mechanism in MF/UF/NF is mainly the size exclusion, which is indicated in the nominal ratings of the membranes. The other separation mechanism includes the electrostatic interactions between solutes and membranes, which depends on the surface and physiochemical properties of solutes and membranes. The available range of membrane materials includes polymeric, carbon, silica, zeolite and other ceramics, as well as composites. Each type of membrane can have a different porous structure. Nowadays, there are more reports on the fluid transport through porous CNTs/polymer membrane. Computational approach can play an important role in the development of the CNT-based composites by providing simulation results to help on the understanding, analysis and design of such nanocomposites. Computational approaches to obtain solubility and diffusion coefficients of small molecules in polymers have focused primarily upon molecular dynamics and Monte Carlo methods. Molecular dynamics simulations are widely being used in modeling and solving problems based on quantum mechanics. Using Molecular dynamics it is possible to study the reactions, load transfer between atoms and molecules. Monte Carlo molecular simulation methods rely on the fact that a physical system can be defined to possess a definite energy distribution function, which can be used to calculate thermodynamic properties. The Monte Carlo technique is a stochastic simulation method designed to generate a long sequence, or "Markov chain" of configurations that asymptotically sample the probability density of an equilibrium ensemble of statistical mechanics. So using from molecular dynamic or Monte Carlo techniques can be useful to simulate the membrane separation process depends on the purpose and the condition of process.

SYMBOL

Symbol	Definition
r_p	Pore radius
Γ	Surface tension
T	Thickness of the adsorbate film
V_L	Molecular volume of the condensate

a_1 and a_2	Unit vectors
G_1, G_2	The material coordinates of a point in the initial configuration
e_1, e_2 and e_3	The coordinates in the current configuration
R	The radius of the modeled SWCNT
J	Molecular flux
D	The diffusivity (diffusion coefficient)
$c(x)$	Concentration
X	Position across the membrane
P	Permeability
S	Solubility coefficient
K_P	Henry's law constant
B	Hole affinity constant
C_h^*	Saturation constant
Q	Flow
T	Time
Θ	Time-lag
Q	Heat of adsorption
A	Proportionality constant
aq	Energy barrier
K	Temperature dependent Henry's law coefficient
K_0	Proportionality constant
P_S^*	Constant
Λ	The mean free path of molecules
D	Pore diameter
K_n	Knudsen number
T	Pore tortuosity
$\bar{u}$	Average molecular speed
M	Molecular mass
N	Surface concentration of pores
Δp	Pressure drop across the membrane
L	Membrane thickness
D_A^*	Self-diffusion coefficient
Na	The number of molecules

f_A	Fugacity
c_A	The concentration of diffusant A
F	Applied external force
V	The center-of-mass velocity component
$\rho(x)$	An arbitrary probability distribution function
ξ_i	The random numbers generated for each trial
T	The number of trials
RO	Reverse osmosis
NF	Nanofiltration
UF	Ultrafiltration
MF	Microfiltration
MWCO	Molecular Weight Cut-Off
IUPAC	International Union of Pure and Applied Chemistry
SWNT	Single-Walled Carbon Nanotube
MWNT	Multiwalled Carbon Nanotube
CNT	Carbon nanotube
AFM	Atomic Force Microscopy
MD	Molecular Dynamics
MC	Monte Carlo
NEMD	Nonequilibrium MD
RESPA	Reference System Propagator Algorithm
GCMD	Grand Canonical Molecular Dynamics
DCV- GCMD	Dual-volume GCMD
MSD	Mean-Square Displacement
DP	Diamond Pore
ZP	Zigzag Path
SP	Straight Path

KEYWORDS

- **Computational methods**
- **Filtration**
- **Membrane**
- **Membrane types**

REFERENCES

1. Majeed, S., et al., Multi-walled carbon nanotubes (MWCNTs) mixed polyacrylonitrile (PAN) ultrafiltration membranes. *J. Membrane Sci.* **2012**, *403*, 101–109.
2. Macedonio, F.; and Drioli, E.; Pressure-driven membrane operations and membrane distillation technology integration for water purification. *Desalination.* **2008**, *223(1)*, 396–409.
3. Merdaw, A. A.; Sharif, A. O.; and Derwish, G. A. W.; Mass transfer in pressure-driven membrane separation processes, part ii. *Chem. Eng. J.* **2011**, *168(1)*, 229–240.
4. Van Der Bruggen, B.; et al., A review of pressure-driven membrane processes in wastewater treatment and drinking water production. *Environ. Prog.* **2003**, *22(1)*, 46–56.
5. Cui, Z. F.; and Muralidhara, H. S.; Membrane Technology: A Practical Guide to Membrane Technology and Applications in Food and Bioprocessing. Elsevier; **2010**, 288.
6. Shirazi, S.; Lin, C. J.; and Chen, D.; Inorganic fouling of pressure-driven membrane processes—a critical review. *Desalination.* **2010**, *250(1)*, 236–248.
7. Pendergast, M. M.; and Hoek, E. M. V.; A review of water treatment membrane nanotechnologies. *Energy Environ. Sci.* **2011**, *4(6)*, 1946–1971.
8. Hilal, N.; et al., A comprehensive review of nanofiltration membranes: Treatment, pretreatment, modelling, and atomic force microscopy. *Desalination.* **2004**, *170(3)*, 281–308.
9. Srivastava, A.; Srivastava, S.; and Kalaga, K.; Carbon nanotube membrane filters. In: Springer Handbook of Nanomaterials. Springer; **2013**, 1099–1116.
10. Colombo, L.; and Fasolino, A. L.; Computer-Based Modeling of Novel Carbon Systems and Their Properties: Beyond Nanotubes. Springer; **2010**, *3*, 258.
11. Polarz, S.; and Smarsly, B.; Nanoporous materials. *J. Nanosci. Nanotechnol.* **2002**, *2(6)*, 581–612.
12. Gray-Weale, A. A.; et al., Transition-state theory model for the diffusion coefficients of small penetrants in glassy polymers. *Macromolecules.* **1997**, *30(23)*, 7296–7306.
13. Rigby, D.; and Roe, R.; Molecular dynamics simulation of polymer liquid and glass. i. glass transition. *J. Chem. Phys.* **1987**, *87*, 7285.
14. Freeman, B. D.; Yampolskii, Y. P.; and Pinnau, I.; *Materials Science of Membranes for Gas and Vapor Separation.* Wiley.com; **2006**, 466.
15. Hofmann, D.; et al., Molecular modeling investigation of free volume distributions in stiff chain polymers with conventional and ultrahigh free volume: comparison between molecular modeling and positron lifetime studies. *Macromolecules.* **2003**, *36(22)*, 8528–8538.
16. Greenfield, M. L.; and Theodorou, D. N.; Geometric analysis of diffusion pathways in glassy and melt atactic polypropylene. *Macromolecules.* **1993**, *26(20)*, 5461–5472.
17. Baker, R. W.; Membrane Technology and Applications. John Wiley & Sons; **2012**, 592.
18. Strathmann, H.; Giorno, L.; and Drioli, E.; Introduction to Membrane Science and Technology. Wiley-VCH Verlag & Company; **2011**, 544.
19. Chen, J. P.; et al., Membrane Separation: Basics and Applications, in Membrane and Desalination Technologies, Wang, L. K.; et al., ed. Humana Press; **2008**, 271–332.
20. Mortazavi, S.; Application of Membrane Separation Technology to Mitigation of Mine Effluent and Acidic Drainage. Natural Resources Canada; 2008, 194.
21. Porter, M. C.; Handbook of Industrial Membrane Technology. Noyes Publications; **1990**, 604.
22. Naylor, T. V.; Polymer Membranes: Materials, Structures and Separation Performance. Rapra Technology Limited; **1996**, 136.
23. Freeman, B. D.; Introduction to Membrane Science and Technology. By Heinrich Strathmann. Angewandte Chemie International Edition, **2012**, *51(38)*, 9485–9485.
24. Kim, I.; Yoon, H.; and Lee, K. M.; Formation of integrally skinned asymmetric polyetherimide nanofiltration membranes by phase inversion process. *J. Appl. Polym. Sci.* **2002**, *84(6)*, 1300–1307.

25. Khulbe, K. C.; Feng, C. Y.; and Matsuura, T.; Synthetic Polymeric Membranes: Characterization by Atomic Force Microscopy. Springer; **2007**, 198.

26. Loeb, L. B.; The Kinetic Theory of Gases. Courier Dover Publications; **2004**, 678.

27. Koros, W. J.; and Fleming, G. K.; Membrane-based gas separation. *J. Membrane Sci.* **1993**, *83(1)*, 1–80.

28. Perry, J. D.; Nagai, K.; and Koros, W. J.; Polymer Membranes for Hydrogen Separations. MRS Bulletin; **2006**, *31(10)*, 745–749.

29. Yang, W.; et al., Carbon nanotubes for biological and biomedical applications. *Nanotechnology.* **2007**, *18(41)*, 412001.

30. Bianco, A.; et al., Biomedical applications of functionalised carbon nanotubes. *Chem. Commun.* **2005**, *5*, 571–577.

31. Salvetat, J.; et al., Mechanical properties of carbon nanotubes. *Appl. Phys. A.* **1999**, *69(3)*, 255–260.

32. Zhang, X.; et al., Ultrastrong, stiff, and lightweight carbon-nanotube fibers. *Adv. Mater.* **2007**, *19(23)*, 4198–4201.

33. Arroyo, M.; and Belytschko, T.; Finite crystal elasticity of carbon nanotubes based on the exponential cauchy-born rule. *Phys. Rev. B.* **2004**, *69(11)*, 115415.

34. Wang, J.; et al., Energy and mechanical properties of single-walled carbon nanotubes predicted using the higher order cauchy-born rule. *Phys. Rev. B.* **2006**, *73(11)*, 115428.

35. Zhang, Y.; Single-walled carbon nanotube modelling based on one-and two-dimensional Cosserat continua. University of Nottingham; **2011**.

36. Wang, S., Functionalization of Carbon Nanotubes: Characterization, Modeling and Composite Applications. Florida State University; 2006, 193.

37. Lau, K.-t., C. Gu, and D. Hui, A critical review on nanotube and nanotube/nanoclay related polymer composite materials. *Compos. Part B: Eng.* **2006**, *37(6)*, 425–436.

38. Choi, W., et al., Carbon nanotube-guided thermopower waves. *Mater. Today.* **2010**, *13(10)*, 22–33.

39. Iijima, S., *Helical microtubules of graphitic carbon.* nature, **1991**, *354(6348)*, 56–58.

40. Sholl, D. S.; and Johnson, J.; Making high-flux membranes with carbon nanotubes. *Science.* **2006**, *312(5776)*, 1003–1004.

41. Zang, J.; et al., Self-diffusion of water and simple alcohols in single-walled aluminosilicate nanotubes. *ACS Nano.* **2009**, *3(6)*, 1548–1556.

42. Talapatra, S.; Krungleviciute, V.; and Migone, A. D.; Higher coverage gas adsorption on the surface of carbon nanotubes: Evidence for a possible new phase in the second layer. *Phys. Rev. Lett.* **2002**, *89(24)*, 246106.

43. Pujari, S.; et al., Orientation dynamics in multiwalled carbon nanotube dispersions under shear flow. *J. Chem. Phys.* **2009**, *130*, 214903.

44. Singh, S.; and Kruse, P.; Carbon nanotube surface science. *Int. J. Nanotechnol.* **2008**, *5(9)*, 900–929

45. Baker, R. W.; Future directions of membrane gas separation technology. *Ind. Eng. Chem. Res.* **2002**, *41(6)*, 1393–1411.

46. Erucar, I.; and Keskin, S.; Screening metal–organic framework-based mixed-matrix membranes for CO2/CH4 separations. *Ind. Eng. Chem. Res.* **2011**, *50(22)*, 12606–12616.

47. Bethune, D. S., et al., Cobalt-catalysed growth of carbon nanotubes with single-atomic-layer walls. *Nature.* **1993**, *363*, 605–607.

48. Iijima, S.; and Ichihashi, T.; Single-shell carbon nanotubes of 1-nm diameter. *Nature.* **1993**, *363*, 603–605.

49. Treacy, M.; Ebbesen, T.; and Gibson, J.; Exceptionally high Young's modulus observed for individual carbon nanotubes. 1996.

50. Wong, E. W.; Sheehan, P. E.; and Lieber, C.; Nanobeam mechanics: elasticity, strength, and toughness of nanorods and nanotubes. *Science.* **1997**, *277(5334),* 1971–1975.

51. Thostenson, E. T.; Li, C.; and Chou, T. W.; Nanocomposites in context. *Compos. Sci. Technol.* **2005**, *65(3),* 491–516.

52. Barski, M.; Kędziora, P.; and Chwał, M.; Carbon nanotube/polymer nanocomposites: a brief modeling overview. *Key Eng. Mater.* **2013**, *542,* 29–42.

53. Dresselhaus, M. S.; Dresselhaus, G.; and Eklund, P. C.; Science of Fullerenes and Carbon nanotubes:Ttheir Properties and Applications. Academic Press; **1996**, 965.

54. Yakobson, B.; and Smalley, R. E.; Some unusual new molecules—long, hollow fibers with tantalizing electronic and mechanical properties—have joined diamonds and graphite in the carbon family. *Am. Sci.* **1997**, *85,* 324–337.

55. Guo, Y.; and Guo, W.; Mechanical and electrostatic properties of carbon nanotubes under tensile loading and electric field. *J. Phys. D: Appl. Phys.* **2003**, *36(7),* 805.

56. Berger, C.; et al., Electronic confinement and coherence in patterned epitaxial graphene. *Science.* **2006**, *312(5777),* 1191–1196.

57. Song, K.; et al., Structural polymer-based carbon nanotube composite fibers: Understanding the processing–structure–performance relationship. *Materials.* **2013**, *6(6),* 2543–2577.

58. Park, O. K.; et al., Effect of surface treatment with potassium persulfate on dispersion stability of multi-walled carbon nanotubes. *Mater. Lett.* **2010**, *64(6),* 718–721.

59. Banerjee, S.; Hemraj-Benny, T.; and Wong, S. S.; Covalent surface chemistry of single-walled carbon nanotubes. *Adv. Mater.* **2005**, *17(1),* 17–29.

60. Balasubramanian, K.; and Burghard, M.; Chemically functionalized carbon nanotubes. *Small.* **2005**, *1(2),* 180–192.

61. Xu, Z. L.; and Alsalhy Qusay, F.; Polyethersulfone (PES) hollow fiber ultrafiltration membranes prepared by PES/non-solvent/NMP solution. *J. Membrane Sci.* **2004**, *233(1–2),* 101–111.

62. Chung, T. S.; Qin, J. J.; and Gu, J.; Effect of shear rate within the spinneret on morphology, separation performance and mechanical properties of ultrafiltration polyethersulfone hollow fiber membranes. *Chem. Eng. Sci.* **2000**, *55(6),* 1077–1091.

63. Choi, J. H.; Jegal, J.; and Kim, W. N.; Modification of performances of various membranes using MWNTs as a modifier. *Macromole. Sym.* **2007**, *249–250(1),* 610–617.

64. Wang, Z.; and Ma, J.; The role of nonsolvent in-diffusion velocity in determining polymeric membrane morphology. *Desalination.* **2012**, *286*(0), 69–79.

65. Vilatela, J. J.; Khare, R.; and Windle, A. H.; The hierarchical structure and properties of multifunctional carbon nanotube fibre composites. *Carbon.* **2012**, *50(3),* 1227–1234.

66. Benavides, R. E.; Jana, S. C.; and Reneker, D. H.; Nanofibers from scalable gas jet process. *ACS Macro Lett.* **2012**, *1(8),* 1032–1036.

67. Gupta, V. B.; and Kothari, V. K.; Manufactured Fiber Technology. Springer; **1997**, 661.

68. Wang, T.; and Kumar, S.; Electrospinning of polyacrylonitrile nanofibers. *J. Appl. Polym. Sci.* **2006**, *102*(2): p. 1023-1029.

69. Song, K.; et al., Lubrication of poly (vinyl alcohol) chain orientation by carbon nano-chips in composite tapes. *J. Appl. Polym. Sci.* **2013**, *127(4),* 2977–2982.

70. Theodorou, D. N.; Molecular simulations of sorption and diffusion in amorphous polymers. *Plast. Eng.-New York.* **1996**, *32,* 67–142.

71. Müller☐Plathe, F.; Permeation of polymers—a computational approach. *Acta Polym.* **1994**, *45(4),* 259–293.

72. Liu, Y. J.; and Chen, X. L.; Evaluations of the effective material properties of carbon nanotube-based composites using a nanoscale representative volume element. *Mech. Mater.* **2003**, *35(1),* 69–81.

73. Gusev, A. A.; and Suter, U. W.; Dynamics of small molecules in dense polymers subject to thermal motion. *J. Chem. Phys.* **1993**, *99*, 2228.

74. Elliott, J. A.; Novel approaches to multiscale modelling in materials science. *Int. Mater. Rev.* **2011**, *56(4)*, 207–225.

75. Greenfield, M. L.; and Theodorou, D. N.; Molecular modeling of methane diffusion in glassy atactic polypropylene via multidimensional transition state theory. *Macromolecules.* **1998**, *31(20)*, 7068–7090.

76. Peng, F.; et al., Hybrid organic-inorganic membrane: solving the tradeoff between permeability and selectivity. *Chem. Mater.* **2005**, *17(26)*, 6790–6796.

77. Duke, M. C.; et al., Exposing the molecular sieving architecture of amorphous silica using positron annihilation spectroscopy. *Adv. Funct. Mater.* **2008**, *18(23)*, 3818–3826.

78. Hedstrom, J. A.; et al., Pore Morphologies in Disordered NanoporousTthin Films. Langmuir; **2004**, *20(5)*, 1535–1538.

79. Pujari, P. K.; et al., Study of pore structure in grafted polymer membranes using slow positron beam and small-angle X-ray scattering techniques. *Nuclear Instru. Methods Phys. Res. Sect. B: Beam Interact. Mater. Atoms.* **2007**, *254(2)*, 278–282.

80. Wang, X. Y.; et al., Cavity size distributions in high free volume glassy polymers by molecular simulation. *Polymer.* **2004**, *45(11)*, 3907–3912.

81. Skoulidas, A. I.; and Sholl, D. S.; Self-diffusion and transport diffusion of light gases in metal-organic framework materials assessed using molecular dynamics simulations. *J. Phys. Chem. B.* **2005**, *109(33)*, 15760–15768.

82. Wang, X. Y.; et al., A molecular simulation study of cavity size distributions and diffusion in para and meta isomers. *Polymer.* **2005**, *46(21)*, 9155–9161.

83. Zhou, J.; et al., Molecular dynamics simulation of diffusion of gases in pure and silica-filled poly (1-trimethylsilyl-1-propyne) [PTMSP]. *Polymer.* **2006**, *47(14)*, 5206–5212.

84. Scholes, C. A.; Kentish, S. E.; and Stevens, G. W.; Carbon Dioxide Separation Through Polymeric Membrane Systems for Flue Gas Applications. *Recent Patents Chem. Eng.* **2008**, *1(1)*, 52–66.

85. Wijmans, J. G.; and Baker, R. W.; The solution-diffusion model: a unified approach to membrane permeation. *Mater. Sci. Membranes Gas Vapor Separat.* **2006**, 159–190.

86. Wijmans, J. G.; and Baker, R. W.; The solution-diffusion model: A review. *J. Membrane Sci.* **1995**, *107(1)*, 1–21.

87. Way, J. D.; and Roberts, D. L.; Hollow fiber inorganic membranes for gas separations. *Separat. Sci. Technol.* **1992**, *27(1)*, 29–41.

88. Rao, M. B.; and Sircar, S.; Performance and pore characterization of nanoporous carbon membranes for gas separation. *J. Membrane Sci.* **1996**, *110(1)*, 109–118.

89. Merkel, T. C.; et al., Effect of nanoparticles on gas sorption and transport in poly (1-trimethylsilyl-1-propyne). *Macromolecules.* **2003**, *36(18)*, 6844–6855.

90. Mulder, M.; Basic Principles of Membrane Technology Second Edition. Kluwer Academic Publication; **1996**, 564.

91. Wang, K.; Suda, H.; and Haraya, K.; permeation time lag and the concentration dependence of the diffusion coefficient of CO_2 in a carbon molecular sieve membrane. *Ind. Eng. Chem. Res.* **2001**, *40(13)*, 2942–2946.

92. Webb, P. A.; and Orr, C.; Analytical Methods in Fine Particle Technology. Micromeritics Norcross, GA; **1997**, *55*, 301.

93. Pinnau, I.; et al., Long-term permeation properties of poly (1-trimethylsilyl-1-propyne) membranes in hydrocarbon—vapor environment. *J. Polym. Sci. Part B: Polym. Phys.* **1997**, *35(10)*, 1483–1490.

94. Jean, Y. C.; Characterizing free volumes and holes in polymers by positron annihilation spectroscopy. *Positron Spectroscopy of Solids.* **1993**, 1.

95. Hagiwara, K.; et al., Studies on the free volume and the volume expansion behavior of amorphous polymers. *Radiat. Phys. Chem.* **2000**, *58(5)*, 525–530.
96. Sugden, S.; Molecular volumes at absolute zero. Part II. Zero volumes and chemical composition. *J. Chem. Soc. (Resumed).* **1927**, 1786–1798.
97. Dlubek, G.; et al., Positron annihilation: A unique method for studying polymers. In: Macromolecular Symposia. Wiley Online Library; **2004**.
98. Golemme, G.; et al., NMR study of free volume in amorphous perfluorinated polymers: Comparsion with other methods. *Polymer.* **2003**, *44(17)*, 5039–5045.
99. Victor, J. G.; and Torkelson, J. M.; On measuring the distribution of local free volume in glassy polymers by photochromic and fluorescence techniques. *Macromolecules.* **1987**, *20(9)*, 2241–2250.
100. Royal, J. S.; and Torkelson, J. M.; Photochromic and fluorescent probe studies in glassy polymer matrices. *Macromolecules.* **1992**, *25(18)*, 4792–4796.
101. Yampolskii, Y. P.; et al., Study of high permeability polymers by means of the spin probe technique. *Polymer.* **1999**, *40(7)*, 1745–1752.
102. Kobayashi, Y.; et al., Evaluation of polymer free volume by positron annihilation and gas diffusivity measurements. *Polymer.* **1994**, *35(5)*, 925–928.
103. Huxtable, S. T.; et al., Interfacial heat flow in carbon nanotube suspensions. *Nature Mater.* **2003**, *2(11)*, 731–734.
104. Allen, M. P.; and Tildesley, D. J.; Computer Simulation of Liquids. Oxford University Press; **1989**.
105. Frenkel, D.; Smit, B.; and Ratner, M. A.; Understanding molecular simulation: From algorithms to applications. *Phys. Today.* **1997**, *50*, 66.
106. Rapaport, D. C.; The Art of Molecular Dynamics Simulation. Cambridge University Press; **2004**, 549.
107. Leach, A. R.; and Schomburg, D.; Molecular Modelling: Principles and Applications. Longman London; **1996**.
108. Martyna, G. J.; et al., Explicit reversible integrators for extended systems dynamics. *Mole. Phys.* **1996**, *87(5)*, 1117–1157.
109. Tuckerman, M.; Berne, B. J.; and Martyna, G. J.; Reversible multiple time scale molecular dynamics. *J. Chem. Phys.* **1992**, *97(3)*, 1990.
110. Harmandaris, V. A.; et al., Crossover from the rouse to the entangled polymer melt regime: signals from long, detailed atomistic molecular dynamics simulations, supported by rheological experiments. *Macromolecules.* **2003**, *36(4)*, 1376–1387.
111. Firouzi, M.; Tsotsis, T. T.; and Sahimi, M.; Nonequilibrium molecular dynamics simulations of transport and separation of supercritical fluid mixtures in nanoporous membranes. I. Results for a single carbon nanopore. *J. Chem. Phys.* **2003**, *119*, 6810.
112. Shroll, R. M.; and Smith, D. E.; Molecular dynamics simulations in the grand canonical ensemble: application to clay mineral swelling. *J. Chem. Phys.* **1999**, *111*, 9025.
113. Firouzi, M.; et al., Molecular dynamics simulations of transport and separation of carbon dioxide–alkane mixtures in carbon nanopores. *J. Chem. Phys.* **2004**, *120*, 8172.
114. Heffelfinger, G. S.; and van Swol, F.; Diffusion in Lennard□Jones fluids using dual control volume grand canonical molecular dynamics simulation (DCV□GCMD). *J. Chem. Phys.* **1994**, *100*, 7548.
115. Pant, P. K.; and Boyd, R. H.; Simulation of diffusion of small-molecule penetrants in polymers. *Macromolecules.* **1992**, *25(1)*, 494–495.
116. Allen, M. P.; and Tildesley, D. J.; Computer Simulation of Liquids. Oxford University Press; **1989**, 385.

117. Cummings, P. T.; and Evans, D. J.; Nonequilibrium molecular dynamics approaches to transport properties and non-newtonian fluid rheology. *Ind. Eng. Chem. Res.* **1992**, *31(5)*, 1237–1252.

118. MacElroy, J.; Nonequilibrium molecular dynamics simulation of diffusion and flow in thin microporous membranes. *J. Chem. Phys.* **1994**, *101*, 5274.

119. Furukawa, S.; and Nitta, T.; Non-equilibrium molecular dynamics simulation studies on gas permeation across carbon membranes with different pore shape composed of micro-graphite crystallites. *J. Membrane Sci.* **2000**, *178(1)*, 107–119.

120. Düren, T.; Keil, F. J.; and Seaton, N. A.; Composition dependent transport diffusion coefficients of CH_4/cf_4 mixtures in carbon nanotubes by non-equilibrium molecular dynamics simulations. *Chem. Eng. Sci.* **2002**, *57(8)*, 1343–1354.

121. Fried, J. R.; Molecular simulation of gas and vapour transport in highly permeable polymers. *Mater. Sci. Membranes Gas Vapour Separat.* **2006**, 95–136.

122. El Sheikh, A.; Ajeeli, A.; and Abu-Taieh, E.; Simulation and Modeling: Current Technologies and Applications. IGI Publishing; **2007**.

123. McDonald, I.; NpT-ensemble monte carlo calculations for binary liquid mixtures. *Mole. Phys.* **2002**, *100(1)*, 95–105.

124. Vacatello, M.; et al., A computer model of molecular arrangement in a n-paraffinic liquid. *J. Chem. Phys.* **1980**, *73(1)*, 548–552.

125. Furukawa, S.-I.; and Nitta, T.; Non-equilibrium molecular dynamics simulation studies on gas permeation across carbon membranes with different pore shape composed of micro-graphite crystallites. *J. Membrane Sci.* **2000**, *178(1)*, 107–119.

INDEX

Milton Keynes UK
Ingram Content Group UK Ltd.
UKHW031141141024
449569UK00024B/1152